Smart Cities and Homes
Key Enabling Technologies

Edited by

Mohammad S. Obaidat

Petros Nicopolitidis

ELSEVIER

AMSTERDAM • BOSTON • HEIDELBERG • LONDON
NEW YORK • OXFORD • PARIS • SAN DIEGO
SAN FRANCISCO • SINGAPORE • SYDNEY • TOKYO

Morgan Kaufmann is an imprint of Elsevier

Morgan Kaufmann is an imprint of Elsevier
50 Hampshire Street, 5th Floor, Cambridge, MA 02139, USA

British Library Cataloguing-in-Publication Data
A catalogue record for this book is available from the British Library

Library of Congress Cataloging-in-Publication Data
A catalog record for this book is available from the Library of Congress

ISBN: 978-0-12-803454-5

For information on all Morgan Kaufmann publications visit
our website at https://www.elsevier.com/

Working together
to grow libraries in
developing countries

www.elsevier.com • www.bookaid.org

Publisher: Todd Green
Acquisition Editor: Brian Romer
Editorial Project Manager: Amy Invernizzi
Production Project Manager: Punithavathy Govindaradjane
Designer: Maria Inês Cruz

To my family
Mohammad S. Obaidat

To the memory of my mother Stamatia
Petros Nicopolitidis

Contents

List of Contributors

A. Ahmad
Department of Electrical Engineering, COMSATS Institute of IT, Wah Campus, Wah, Pakistan

Ö.U. Akgül
Computer Engineering Department, Istanbul Technical University, Ayazaga, Istanbul, Turkey

O. Altintas
Toyota InfoTechnology Center, Tokyo, Japan

A. Anpalagan
Department of Electrical and Computer Engineering, Ryerson University, Toronto, ON, Canada

M.N. Avcil
Department of Computer Engineering, Marmara University, Istanbul, Turkey

A. Belghith
Department of Telecommunications, University of Sfax, Sfax, Tunisia

M. Berlingerio
IBM Research, Dublin, Ireland

A. Botea
IBM Research, Dublin, Ireland

E. Bouillet
IBM Research, Dublin, Ireland

S. Braghin
IBM Research, Dublin, Ireland

N. Buchina
Department of Computer Science, Eindhoven University of Technology, Eindhoven,
The Netherlands

F. Calabrese
IBM Research, Dublin, Ireland

B. Canberk
Computer Engineering Department, Istanbul Technical University, Ayazaga, Istanbul, Turkey

M. Cello
Electrical, Electronics and Telecommunication Engineering and Naval Architecture Department
(DITEN)—University of Genoa, Genoa, Italy

N. Cercone
Department of Electrical Engineering and Computer Science, Lassonde School of Engineering,
York University, Toronto, ON, Canada

B. Chen
IBM Research, Dublin, Ireland

P.J.L. Cuijpers
Department of Computer Science, Eindhoven University of Technology, Eindhoven,
The Netherlands

C. Degano
Research and Development Business Unit—Gruppo SIGLA S.r.l., Genoa, Italy

F. Dressler
Heinz Nixdorf Institute, Paderborn University, Paderborn, Germany

D. Gavalas
Department of Cultural Technology and Communication, University of the Aegean, Mytilene, Greece;
Computer Technology Institute & Press "Diophantus", Patras, Greece

Y. Gkoufas
IBM Research, Dublin, Ireland

T. Guelzim
Department of Computer Science and Software Engineering, Monmouth University, West Long
Branch, NJ, United States

F. Hagenauer
Heinz Nixdorf Institute, Paderborn University, Paderborn, Germany

S. Hu
Department of Electrical and Computer Science, Michigan Technological University, Houghton,
MI, USA

Y. Hu
Institute of Computing Technology, Chinese Academy of Science, Beijing, China

M. Iqbal
Department of Electrical Engineering, COMSATS Institute of IT, Wah Campus, Wah, Pakistan

Y. Jin
Department of Electrical Engineering and Computer Science, University of Central Florida,
Orlando, FL, USA

E. Kammoun
Department of Telecommunications, ENET'Com, University of Sfax, Sfax, Tunisia

B. Kantarci
Department of Electrical and Computer Engineering, Clarkson University, NY, United States

C. Konstantopoulos
Department of Informatics, University of Piraeus, Piraeus, Greece; Computer Technology
Institute & Press "Diophantus", Patras, Greece

C.A. Kyriakopoulos
Department of Informatics, Aristotle University, Thessaloniki, Greece

A. Lasek
Department of Electrical Engineering and Computer Science, Lassonde School of Engineering, York University, Toronto, ON, Canada

M. Laumanns
IBM Research, Zurich, Switzerland

X. Li
Institute of Computing Technology, Chinese Academy of Science, Beijing, China

Y. Liu
Department of Electrical and Computer Science, Michigan Technological University, Houghton, MI, USA

J.J. Lukkien
Department of Computer Science, Eindhoven University of Technology, Eindhoven, The Netherlands

M. Marchese
Electrical, Electronics and Telecommunication Engineering and Naval Architecture Department (DITEN)—University of Genoa, Genoa, Italy

J. Matthews
Department of Computer Science, Clarkson University, NY, United States

K.N. Muhammad
Department of Computer Engineering, Marmara University, Istanbul, Turkey

M. Naeem
Department of Electrical Engineering, COMSATS Institute of IT, Wah Campus, Wah, Pakistan; Department of Electrical and Computer Engineering, Ryerson University, Toronto, ON, Canada

R. Nair
IBM Research, Dublin, Ireland

P. Nicopolitidis
Department of Informatics, Aristotle University, Thessaloniki, Greece

T. Nonner
IBM Research, Zurich, Switzerland

M.S. Obaidat
Department of Computer and Information, Fordham University, Bronx, NY, United States

N. Omheni
Department of Telecommunications, ENET'COM, University of Sfax, Sfax, Tunisia

G. Pantziou
Department of Informatics, Technological Educational Institution of Athens, Athens, Greece; Computer Technology Institute & Press "Diophantus", Patras, Greece

G.I. Papadimitriou
Department of Informatics, Aristotle University, Thessaloniki, Greece

F. Podda
Research and Development Business Unit—Gruppo SIGLA S.r.l., Genoa, Italy

B. Sadoun
Department of Surveying & Geomatics Engineering, Al-Balqa' Applied University, Salt, Jordan

J. Saunders
Fuseforward Solutions Group, Vancouver, BC, Canada

Y. Shi
Department of Computer Science and Engineering, Notre Dame University, Notre Dame, IN, USA

C. Sommer
Heinz Nixdorf Institute, Paderborn University, Paderborn, Germany

M. Soyturk
Department of Computer Engineering, Marmara University, Istanbul, Turkey

M. Stolikj
Department of Computer Science, Eindhoven University of Technology, Eindhoven, The Netherlands

E. Varvarigos
Department of Computer Engineering and Informatics, Patras University, Patras, Greece

J. Wu
Department of Computer Science and Engineering, Notre Dame University, Notre Dame, IN, USA

F. Zarai
Department of Telecommunications, ENET'COM, University of Sfax, Sfax, Tunisia

About the Editors

Professor Mohammad S. Obaidat (Fellow of IEEE and Fellow of SCS) is an internationally well-known academic/researcher/scientist. He received his PhD and MS degrees in Computer Engineering with a minor in Computer Science from The Ohio State University, Columbus, Ohio, USA. Dr Obaidat is currently the Chair and Full Professor of Computer and Information Science at Fordham University, NY, USA. Among his previous positions are Chair of the Department of Computer Science and Director of the Graduate Program, Dean of the College of Engineering at Prince Sultan University, and Advisor to the President of Philadelphia University for Research, Development and Information Technology. He has received extensive research funding and has published *38 books* and over *600* refereed technical articles in scholarly international journals and proceedings of international conferences, and currently working on three more books. Professor Obaidat

has served as a consultant for several corporations and organizations worldwide. Mohammad is the Editor-in-Chief of the Wiley *International Journal of Communication Systems, the FTRA Journal of Convergence,* and *the KSIP Journal of Information Processing.* He is also an Editor of *IEEE Wireless Communications.* Between 1991 and 2006, he served as a Technical Editor and an Area Editor of *Simulation: Transactions of the Society for Modeling and Simulations (SCS)* International, *TSCS.* He also served on the Editorial Advisory Board of *Simulation.* He is now an editor of the *Wiley Security and Communication Networks Journal, Journal of Networks, International Journal of Information Technology, Communications and Convergence, IJITCC,* Inderscience. He served on the International Advisory Board of *the International Journal of Wireless Networks and Broadband Technologies,* IGI-global. Professor Obaidat is an associate editor/editorial board member of many other refereed scholarly journals including two IEEE Transactions, *Elsevier Computer Communications Journal, Kluwer Journal of Supercomputing, SCS Journal of Defense Modeling and Simulation, Elsevier Journal of Computers and EE, International Journal of Communication Networks and Distributed Systems,* The Academy *Journal of Communications, International Journal of BioSciences and Technology, International Journal of Information Technology,* and *ICST Transactions on Industrial Networks and Intelligent Systems.* He has guest edited numerous special issues of scholarly journals such as IEEE Transactions on Systems, Man and Cybernetics, SMC, *IEEE Wireless Communications, IEEE Systems Journal,* SIMULATION: Transactions of SCS, *Elsevier Computer Communications Journal, Journal of C & EE,* Wiley Security and Communication Networks, *Journal of Networks,* and *International Journal of Communication Systems,* among others. Obaidat has served as the steering committee chair, advisory Committee Chair and program chair of numerous international conferences.

He is the founder of two well-known international conferences: The International Conference on Computer, Information and Telecommunication Systems (CITS) and the International Symposium on Performance Evaluation of Computer and Telecommunication Systems (SPECTS). He is also the co-founder of the International Conference on Data Communication Networking, DCNET.

Between 1994 and 1997, Obaidat has served as distinguished speaker/visitor of *IEEE* Computer Society. Since 1995, he has been serving as an *ACM* distinguished lecturer. He is also an *SCS* distinguished lecturer. Between 1996 and 1999, Dr Obaidat served as an IEEE/ACM program evaluator of the Computing Sciences Accreditation Board/Commission, CSAB/CSAC. Obaidat is the founder and first Chairman of SCS Technical Chapter (Committee) on PECTS (Performance Evaluation of Computer and Telecommunication Systems). He has served as the Scientific Advisor for the World Bank/UN Digital Inclusion Workshop—The Role of Information and Communication Technology in Development. Between 1995 and 2002, he has served as a member of the board of directors of the Society for Computer Simulation International. Between 2002 and 2004, he has served as Vice President of Conferences of the Society for Modeling and Simulation International *SCS*. Between 2004 and 2006, Professor Obaidat has served as Vice President of Membership of the Society for Modeling and Simulation International *SCS*. Between 2006 and 2009, he has served as the Senior Vice President of *SCS*. Between 2009 and 2011, he served as the President of SCS. Four of his recent papers have received the best paper awards from IEEE AICCSA 2009, IEEE GLOBCOM 2009, IEEE DCNET 2011, and IEEE CITS 2013 International Conferences. Professor Obaidat has been awarded a Nokia Research Fellowship and the distinguished Fulbright Scholar Award. He received the SCS Outstanding Service Award for his excellent leadership, services, and technical contributions. Dr Obaidat received very recently the Society for Modeling and Simulation Intentional (SCS) prestigious *McLeod Founder's Award* in recognition of his outstanding technical and professional contributions to modeling and simulation. He received in Dec. 2010, the IEEE ComSoc—GLOBECOM 2010 Outstanding Leadership Award for his outstanding leadership of Communication Software Services and Multimedia Applications Symposium, CSSMA 2010. He received the Society for Modeling and Simulation International's (SCS) prestigious Presidential Service Award for his outstanding unique, long-term technical contributions and services to the profession and society. He was inducted to SCS Modeling and Simulation Hall of Fame—Lifetime Achievement Award for his technical contribution to modeling and simulation and for his outstanding visionary leadership and dedication to increasing the effectiveness and broadening the applications of modeling and simulation worldwide.

He has been invited to lecture and give keynote speeches worldwide. His research interests are wireless communications and networks, cybersecurity, telecommunications and networking systems, security of network, information and computer systems, security of e-based systems, performance evaluation of computer systems, algorithms and networks, green ICT, high performance and parallel computing/computers, applied neural networks and pattern recognition, adaptive learning, and speech processing. During the 2004/2005, he was on sabbatical leave as Fulbright Distinguished Professor and Advisor to the President of Philadelphia University in Jordan, Dr Adnan Badran. The latter became the Prime Minister of Jordan in April 2005 and served earlier as Deputy Director General of UNESCO. Professor Obaidat is a Fellow of the Society for Modeling and Simulation International *SCS*, and a Fellow of the Institute of Electrical and Electronics Engineers (*IEEE*). For more information, see: www.theobaidat.com

Dr Petros Nicopolitidis received the BS and PhD degrees in computer science from the Department of Informatics, Aristotle University of Thessaloniki, Greece, in 1998 and 2002, respectively. From 2004 to 2009, he was a lecturer in the same department, where he now serves as an Assistant Professor. He has published more than 90 papers in international refereed journals and conferences. He is coauthor of the book *Wireless Networks* (Wiley, 2003). His research interests are in the areas of wireless networks and mobile communications. Since 2007, he serves as an associate editor for the *International Journal of Communication Systems* published by Wiley. He is a senior member of IEEE.

Preface

OVERVIEW AND GOALS

We are witnessing a period of great worldwide urbanization and development. In populous countries such as China and India, which make up for about 40% of the World's population, people are moving to cities at alarming rates. In China alone, it is expected that over 250 million rural inhabitants will transfer to urban areas over the next decade or so. This will necessitate creating new infrastructure to house almost the corresponding of 75% of the present population of the United States in a matter of a few decades.

Cities in the 21st century will make up for about 85–90% of world population growth, 75–80% of affluence formation, and 60% of the overall power spending. It is essential to improve means and systems that enhance the community's quality of life including the built and natural environments, economic affluence, social stability and parity, educational prospects, and cultural, and entertainment opportunities, among others. The development and design of these systems must be robust, scalable, energy aware, environment friendly, and cost effective.

Smart cities can be created from scratch or can be developed gradually by improving the currently existing cities' infrastructures and key resources. Cities that utilize Information and Communication Technology (ICT) Systems have reported enhancements in power efficiency, water use, traffic congestion, environment protection, pollution reduction, senior citizens care, public safety and security, literacy rate, among others. However, there is a limit on what ICT technologies can do for making an already existing city a "smart city," especially when we deal with huge old cities.

Smart homes and cities have become an important research and development area in the 21st century due mainly to their significance to national and international health, economy, infrastructures and key resources, safety, transportation, environment, and security, among others.

There are many key enabling technologies that have played and are playing a great role in building and developing smart homes and cities. Among these are: the internet, wireless networks, and systems such as Wi-Fi, Bluetooth, and Zigbee; Smart Phones including LTE, 3G, 4G, and 5G cell systems; body area sensor networks, smart grids and renewable energy, optical fiber systems and high-speed networks, Internet of Things (IoTs), wireless sensor networks (WSNs), Vehicle Ad Hoc Networks (VANETs), global positioning systems (GPSs), geographical information systems (GISs), wireless navigation systems, world wide web (WWW), social networks, smart TV, radio frequency ID (RFID), sensor-enabled smart objects, actuators and sensors, cloud computing systems, intelligent transportation systems (ITSs), biometric systems, e-based systems including e-commerce, e-government, e-business and e-service systems, network infrastructures, data management systems, analytics, cyber security.

A smart city employs the prospects provided by pervasive and cooperative computing technologies to the benefits of human beings. In such a setting, digital environments are projected to perform a central role for handling the challenges of urbanization and demographic changes such as safety and security, energy distribution, mobility, comfort, health care especially for senior/elderly citizens, among others.

ICT systems have played a vital role in the emergence and development of smart cities and homes. The impressive advances in areas of information and wired and wireless communications technology have brought with them the prospect of embedding different hierarchies of smartness and intelligence in the modern home and cities.

Offering comfort and safe and healthy living with an intelligent form of collaboration with the residents has been the prime goal of smart and digital homes and cities. Contingent upon the settings, the communications may be multifaceted such as mobile agent based and context-aware services or they may be uncomplicated such as controlling the room temperature or its humidity level. Sophisticated situations include the delivery of position/location-aware information content of the resident of the digital home/city as well as his/her activities.

The availability of inexpensive low-power sensors, smart phones, tablets, RF IC chips, Wi-Fi systems, optical networks backbones and the embedded microprocessors/microcontrollers has made tremendous impact on digital homes and cities. The large quantity of sensors jointly manage and make the inferences from the collected data on the state of the home and city as well as the actions and behavior of the inhabitants.

As the worldwide life expectancy, especially in developed and newly industrialized countries is increasing, the percentage of senior/elderly citizens is increasing at an accelerated pace and most projections suggest that this increase worldwide will reach about 10 millions in the coming decade. Senior citizens usually live in care centers, hospitals, or their own homes with some relative supervision/care. Smart homes and cities can be used efficiently and economically in order to accommodate the needs of this population.

The increase of worldwide population, especially in populous countries and cities and the increase migration of citizens to cities have also brought with it challenges in transportation systems, health care, utility's supplies, learning and education, sensing city dynamics, computing with heterogeneous data sources, managing urban big data, and environmental protection issues including decreasing pollution and others.

The aims of this book are to introduce fundamental principles and concepts of key enabling technologies for smart cities and homes, disseminate recent research and development efforts in this fascinating area, investigate related trends and challenges, and present case studies and examples.

We will also investigate the advances, and future in research and development in smart homes and cities. We will review the fundamental techniques in the design, operation, and development of these systems. We will also emphasize the importance of ICT systems as enabling systems in this intriguing field.

This book attempts to provide a comprehensive guide on fundamental concepts, challenges, problems, trends models, and results in the areas of smart cities and home.

The chapters of the book have been written by worldwide experts in the field of smart cities and homes. Despite the fact that there are different authors of these chapters, we have made sure that the book materials are as coherent, articulate, and synchronized as possible in order to be easy to follow by all readers.

This book has been prepared keeping in mind that it needs to prove itself to be a valuable resource dealing with both the important core and the specialized issues in the areas. We have attempted to offer a wide coverage of related topics. We have attempted to make this book useful for both the academics and the practitioners alike.

We hope that the book will be a valuable reference for students, instructors, researchers, and industry practitioners, as well as federal agencies researchers and developers. Every chapter of the book is accompanied by a set of PowerPoint slides to be used by instructors.

ORGANIZATION AND FEATURES

The book consists of 17 chapters that deal with major aspects of key enabling technologies for smart cities and homes.

Chapter 1 is basically an introductory chapter that overviews key enabling technologies for smart cities and homes. This chapter discusses this new trend of connected objects in the context of smart homes and cities, their challenges, as well as a survey of related key enabling technologies.

Chapter 2 deals with wireless sensor networks (WSNs) applications to smart homes and cities. It gives representative examples of WSNs applications in smart homes and cities. Then, it discusses the access technology to be used for the applications and presents some protocols useful to provide better applications' performance in areas such as routing strategies, energy saving methods, and security protocols.

Chapter 3 addresses software defined things with emphasis on green network management for future smart city architectures.

Chapter 4 deals nomadic service discovery in smart cities. Smart cities blend the boundaries between infrastructure and consumer devices, and rely on their cooperation for making new applications possible. This poses a challenge since extremely resource-constrained devices, such as ubiquitous sensors and actuators, need to communicate with more powerful devices, such as smartphones and servers. Moreover, users move through this infrastructure, and their devices need to find and use services based on their current location. This chapter overviews key interoperability issues on the service discovery layer.

Chapter 5 contains a survey on enabling wireless local area networks (WLNs) technologies for smart cities. Wireless LANs enable better and healthier cities through the implementation of environment-based applications allowed by wireless media, robust environmental sensors, and low energy consumption. The chapter focuses on the IEEE 802.11 family of standards and discusses applications of WLANs to smart cities and homes, and the challenges posed.

Chapter 6 provides a comprehension discussion on the key enabling technologies for 4G and 5G systems. It describes briefly the LTE and the LTE-A features, handover management, and the emerging concept of "smart city" in the context of LTE systems.

Chapter 7 reviews cars a main ICT resource of smart cities. It sees a way out in the use of cars as a main information and communication technology (ICT) resource of smart cities, in a technology agnostic architecture termed Car4ICT. Cars are be envisioned to be available almost ubiquitously in future smart cities and flexibly used to mediate between users offering and consuming services. Car4ICT presents a flexible and extensible scheme for service discovery.

Chapter 8 addresses vehicular networks to vehicular clouds in smart cities. It presents a comprehensive survey of VANET applications in smart cities along with challenges, solutions, and existing implementations. Furthermore, it introduces the state of the art in vehicular clouds for smart cities following an introduction of various vehicular cloud architectures. Furthermore, open issues and future directions are presented to help stimulate future studies in this emerging research field.

Chapter 9 describes the impact of the integration of renewable energy and net metering technology on the smart home pricing cyberattack detection. Renewable energy and net metering change the grid energy demand, which is considered by the utility companies when designing the guideline price. If the cyberattack detection technique does not consider this, the detection performance would be degraded. This motivates to develop a new smart home pricing cyberattack detection framework, which handles this impact in the detection process.

Chapter 10 deals with smart home scheduling and cybersecurity. The chapter covers the issues related to vulnerability of the smart home infrastructure and assesses the detection technologies against pricing cyber attacks, which are based on machine-learning techniques and the advanced control theoretical partially observable Markov decision process. The salience of the algorithms is demonstrated through simulation results.

Chapter 11 reviews the optical network architecture for the next generation internet access. It presents a new network architecture that is suitable for providing internet access and intercommunication to the end users, overcoming interoperability issues and offering high available bandwidth to fulfill connectivity needs for the years to come. Moreover, methods that aggregate and predict traffic, which can coexist with the new architecture, are described in detail. These schemes offer better resource utilization, leading to higher network throughput or lower energy consumption according to current customer needs.

Chapter 12 addresses cloud computing systems as applied to smart cities and homes. The chapter discusses the basic concepts behind cloud computing and its application in the fields of smart homes and cities. In order to support this rapid change in information delivery and consumption, cloud computing has evolved from pure technical and narrow field applications to solve higher problem domains in the realms of smart homes and cities. Through standardized system architecture, communication, and information exchange, cloud technologies rely on instrumentation and interconnection to provide intelligent feedback and support new capacities such as digital convergence, energy management, or safety and security.

Chapter 13 offers a comprehensive review of algorithmic approaches for the design and management of vehicle-sharing systems. The focus is on one-way vehicle-sharing systems—wherein customers are allowed to pick-up a vehicle at any location and return it to any other station)—which best suits typical urban journey requirements. It reviews the related design and management of vehicle sharing systems and surveys algorithmic approaches.

Vehicle (bike or car) sharing represents an emerging transportation scheme, which may comprise an important link in the green mobility chain of smart city environments.

Chapter 14 provides a review of smart transportation systems (STSs) in critical conditions. The chapter analyzes the principal issues of the networking aspects for such applications and proposes a solution mainly based on software-defined networking (SDN). It evaluates the benefit of such paradigm in the mentioned context focusing on the incremental deployment of such solutions in the existing metropolitan networks and proposes a "QoS App" able to manage the quality of service on top of the SDN controller.

In the context of smart transportation systems (STSs) in smart cities, the use of applications that can help in case of critical conditions is a key point. Examples of critical conditions may be natural disaster events such as earthquakes, hurricanes, floods, and manmade ones such as terrorist attacks and toxic waste spills. Disaster events are often combined with the destruction of the local telecommunication infrastructure, if any, and this implies real problems to the rescue operations. These issues are crucial in having successful smart cities.

Chapter 15 deals with optimization classification and techniques of wireless sensor networks (WSNs) in smart grid. It presents the optimization classification of wireless sensor networks for smart grids with respect to different types of renewable energy sources, different modes of operations, types of optimization, and different geographical areas.

The emergence of smart city occurred primarily through the use of communication and networking technologies with sensors require better use of available resources. Smart grid is one of such applications that help monitor, control and make efficient use of power in a smart city. However, due to the interconnected nature of the smart grid with other technologies, it becomes a challenge to design a fully optimized wireless sensor network (WSN) for smart grid.

Chapter 16 presents the Docit, the first multimodal journey advising system that reasons about uncertainty in the network knowledge, creating journey plans optimized on the likelihood of arriving on time. This in turn could lead to an increased adoption of public transportation, helping remove cars from the road, and fight congestion in smart cities.

Multimodal travel is a ubiquitous part of living in a city. The operation of modern urban transportation networks can negatively be impacted by multiple factors, including poor traffic conditions caused by congestion and events on the road. In effect, transportation networks feature many types of uncertainty, such as variations in the arrival times of public transport vehicles.

Chapter 17 deals with smart restaurants and provides a survey on customer demand and sales forecasting. Smart cities use digital technology in order to improve prosperity, reduce costs, reduce consumption of the resources, and enhance quality and efficiency of urban services. The chapter is meant to shed some light on an aspect important to cities—management of smart restaurants.

TARGET AUDIENCE

The book primarily targets the student community. This includes the students of all levels—those getting introduced to these areas, those having an intermediate level of knowledge of the topics, and those who are already knowledgeable about many of the topics. In order to keep up with this goal, we have attempted to design the overall structure and content of the book to make it useful at all learning levels.

The secondary audience for this book is the research community, which includes researchers working in academia, industry, or government.

Finally, we have also taken into consideration the needs of those readers, typically from the industry and practitioners, who have quest for getting insight into the practical significance of the topics, expecting to discover how the spectrum of knowledge and ideas presented are relevant for smart cities and homes. In order to make the book useful in the classroom, a set of PowerPoint slides has been prepared for each chapter of the book.

Acknowledgments

We are immensely thankful to all the authors of the chapters of this book, who have worked hard to bring forward this unique resource on the key enabling technologies of smart cities and homes, to help students, instructors, researchers, and community practitioners. We would like to stress that, as the individual chapters of this book were written by different authors, the responsibility for the contents of each chapter lies with the respective authors.

We would like to thank Mr Brian Romer, the Elsevier executive acquisitions editor, who worked with us on the project from the beginning, for his suggestions and advice. We also would like to thank Elsevier publishing and marketing staff members, in particular Ms Amy Invernizzi, editorial project manager (Elsevier), for her great efforts. Finally, we would like to sincerely thank our respective families, for their continuous support and encouragement during the course of this project.

Mohammad S. Obaidat
Petros Nicopolitidis
January 8, 2016

INTRODUCTION AND OVERVIEW OF KEY ENABLING TECHNOLOGIES FOR SMART CITIES AND HOMES

T. Guelzim*, M.S. Obaidat†, B. Sadoun**

*Department of Computer Science and Software Engineering, Monmouth University, West Long Branch, NJ, United States; †Department of Computer and Information, Fordham University, Bronx, NY, United States; **Department of Surveying & Geomatics Engineering, Al-Balqa' Applied University, Salt, Jordan

1 INTRODUCTION

Recent statistics show that 50% of the world's population is now living in or around cities. These cities are responsible for three-fourths of the world's energy consumption and green house emission. By 2050 the world's population will grow by another 30% and the number of citizens living around cities will jump to 70% [1–25]. With this evolution in mind, it is no doubt that the current city services and their governance will fail to deliver adequate added value to citizens. City planners and urban architects have been looking into tackling this issue by setting a vision for the city of the future under the concept of smart city. The adoption of these new concepts requires the provisioning of a wide range of services in a dynamic and effective manner. The solution is to benefit from the current boom in networked information technologies [3]. For smart cities many of the current technology trends contribute to its expansion such as Internet of Things (IoT), public Wi-Fi, ubiquitous cellular coverage, 4G and 5G networks, and smartphones. IoT and smartphone are thought to reduce the digital divide for citizens and smart cities. Being ubiquitous and democratized, these two enabling technologies will allow services accessed by all. In the next 10 years, smart cities will hold many investments. It is estimated that roughly $100 billion will be required in order to develop supporting technologies for advancing smart cities. By 2020, most of these technologies will be deployed across the globe. In parallel to smart cities, a complementary trend is also evolving, which is smart home concept. Smart homes rely on information and communications technology (ICT) as well in order to take the home and living experience to the next level. Homes are becoming connected to the Internet and digital services through smart devices that offer services ranging from simple utility information such as weather forecast to intelligent algorithms to adapt and optimize energy consumption by the home.

As the worldwide life expectancy, especially in developed countries and newly industrialized counties, is increasing, the percentage of elderly citizens is increasing at an alarming rate and most projections suggest that this increase worldwide will reach about 10 million in the coming decade or so. Senior citizens usually live in care centers, hospitals, or their own homes with some relative supervision/care. Smart homes and cities can be used efficiently and economically in order to accommodate the needs of this population in a cost-effective manner [1].

Interest in Smart Homes and Cities has increased in recent years due to the following [1]:

1. impressive economic development in populous countries such as China, India, and Brazil (China and India make for 40% of the World Population);
2. increased use of ICT devices and technology by individuals and organizations worldwide;
3. greater interest in environment protection and in reducing CO_2 emission; Green Economy;
4. noticeable rise in the number of elderly/senior citizens over 65 years old in many countries, especially in Japan, Europe, and even China, who need smart homes and smart cities to make their life comfortable, and healthy at affordable cost;
5. rapid increase of the population of big cities.

2 TRENDS IN SMART CITIES AND HOMES
2.1 SMART CITIES

There are major characteristics that make a city "smart." They can be summarized as follows [4]:

- immersive city services through the use of real-time data sensing;
- knowledge engineering that enables the aggregation and parsing of all of the data;
- gaining access to data in a seamless manner that contains information from various interlinked domains.

As we further congregate in the cities, it has become primordial to make them not only green but also efficient [4] as to set the urban development for the next decade. Fig. 1.1 shows the pillars and trend topics around smart cities.

2.1.1 Smart mobility and smart traffic management

Smart mobility is an essential aspect of a smart city endeavor. Mobility and transportation shall be simplified for city residents and visitors. In addition, traveling out of the city shall be very simple as well. Any flow in or out of the city shall be seamlessly, but carefully, planned so as to provide comfort to all citizens [26]. From current research, this can be accomplished by focusing on the following elements:

- using shared but personalized transportation means
- smart routes and navigation
- contextualized travel information
- improving the travel experience

Smart traffic management is made possible given the advances in computing power to process, analyze, and provide an array of optimal decision in real time. Data is available through Global Positioning System (GPS)-enabled devices, image data from CCTV cameras on main roads and intersections, Geographic Information System (GIS)-enabled cars, weather and road conditions, historical data on road traffic, etc. Data analytics on the cloud is a viable solution as it allows the use of a solid and scalable infrastructure to do the computations. Areas of application in smart management application are as follows:

- reduce traffic congestion
- road and traffic safety
- population safety through smart analytics of road and suspicious car activities

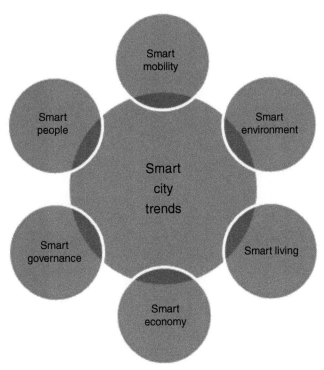

FIGURE 1.1 Smart City Trends

Through smart traffic management, in the United States alone, the elimination of traffic may result in the elimination of over 5 billion extra commuting hours for drivers and the saving of over $121 billion in total delay and fuel.

2.1.2 Smart environment

"The most profound technologies are those that disappear, they weave themselves into the fabric of everyday life until they are indistinguishable from it" [5]. Smart environment is a concept that emerged in the early 1990s where city residents are continuously interacting with objects and sensors seamlessly to better their lives.

An excerpt from Wikipedia [6]:

> The concept of smart environments evolves from the definition of Ubiquitous computing that, according to Mark Weiser, promotes the ideas of "a physical world that is richly and invisibly interwoven with sensors, actuators, displays, and computational elements, embedded seamlessly in the everyday objects of our lives, and connected through a continuous network."

There are many features to smart environment such as autonomy, adaptive behavior to environment, and interaction with humans in simple way. That being said, smart environment is not possible without

Table 1.1 Factors and Indicators for Smart Living in the Context of Smart City

	Indicator	Weighting (%)
Cultural facilities	3	14
Health conditions	4	14
Individual safety	3	14
Housing quality	3	14
Education facilities	3	14
Touristic attractiveness	2	14
Social cohesion	2	14

the rapid evolution of pervasive computing. The latter relies on four areas: device, networking, middleware, and applications [5]. Devices encompass intelligent appliances, cell phones, and setup boxes, among others. These devices collect and send data to cloud-based systems in order to process the data and provide intelligent insights to end users. Communication with these cloud systems passes though middleware, also qualified as pervasive, that interfaces these devices with the networking kernel. The middleware makes the heterogeneous infrastructure seamless to the users. On top of this infrastructure sit the applications that provide the actual added value services that then reach the end user through other IoT devices installed in the homes or via smartphones.

2.1.3 Smart living
A smart city is focused on providing and developing a desirable place to live, work, and spend time in. Quality of life is essential to the prosperity of the smart city. Table 1.1 describes the weighting of determining factors as defined by the European standards [7].

Health, housing, culture, safety, and education are essential factors to establishing smart city living. Without these basic factors, it is not possible to use ICT alone to enable a smart city project. One of the essential questions is how to ensure that the city remains livable and how to address people needs for longer.

2.1.4 Smart economy
The smart economy is new field that focuses on how a city is attractive as well as competitive with regard to factors such as innovation, art, culture, productivity, and most of all international appeal. This implies focusing on these economy vectors as described in Table 1.2 [8].

2.1.5 Smart governance
ICT plays a primary role in the governance of smart cities in order to create value for society [9]. Until the late 1990s, governance was viewed by international organization as form of political regime [10]. Nevertheless, this classical view is starting to be challenged and in some instances vanish as information systems are taking a big role in our daily lives and that of the cities infrastructures. It was suggested that governance practices be revisited by focusing on the following five pillars [11]:

- openness
- participation

Table 1.2 Smart Economy Vectors	
Smart Economy Vector	**Description**
Innovative revenue models	A smart economy is efficient and relies heavily on nonpolluting energy sources. It is efficient and recycles through waste and produces enough energy to sustain its needs. Most importantly, a smart economy allows sustainable growth to its citizens. This is ensured through innovation and entrepreneurship. The 21st-century economy is technology driven [7]. Start-ups can launch new services using little investment but without compromising the overall added value of their product
Sharing economy	Sharing economy can take many forms, most notably, using ICT to provide city residents with the information needed to distribute, share, and reuse the excess capacity in goods and services. The collaborative consumption model uses mobile apps and online platforms to support these innovative ideas. We have witnessed multiple examples of such innovative endeavors lately such as Uber that runs the largest taxi fleet without owning any taxi vehicle, social lending and crowd sourcing platforms, peer-to-peer accommodation (airbnb), and car sharing or pooling, among others. Encouraging the private sector and in particular private citizens to take over traditional or once regulated fields such as transportation allows introducing innovation and offering new breed of services that go along with the advances in technologies witnessed by the city residents

ICT, information and communications technology.

- accountability
- effectiveness
- coherence

In order to effectively implement these five concepts, it is primordial to rely on modern information systems for communicating with the city residents. The role of ICT in smart city governance is illustrated in Fig. 1.2.

Fig. 1.2 shows a framework that relies on ICT to create value for the society. It is codenamed "The smart city house" [9]. This model relies on a foundation that is composed of two parts: data and networking as the basis for the smart city endeavor on top of which sits three pillars whose role is to enable good governance, transform social organization, and inform or guide residents in their day-to-day choices. This foundation then enables smart and efficient services such as sustainable energy, fair employment, and better quality of life overall.

Smart governance further encloses better city planning, emergency management, budgeting, and forecasting based on real time data describing needs as well as changing priorities. In addition, it also relies on strategic orientation and better healthcare that reduces the impact of aging populations. At last, it ensures the aggregation and monitoring of energy production and consumption data in order to provide better management policies [9].

2.1.6 Smart people

As seen in previous sections, ICT is one of the main pillars for smart cities. Nevertheless, it is not the only one. For smart cities to thrive, the human factor has to be accounted for [12]. In this setup, city

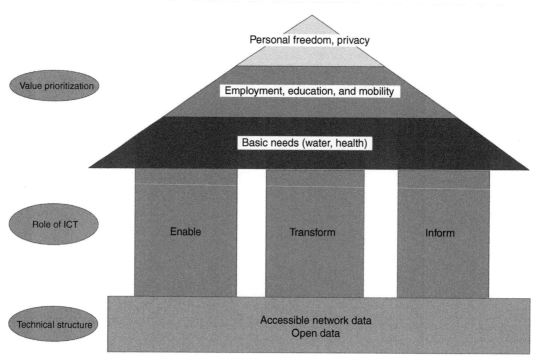

FIGURE 1.2 Role of ICT in Smart City Governance: Smart City House

residents have to possess additional technological skills that allow them to interact and benefit from their smart city as well as to improve it. According to a model by the European Union, Table 1.3 shows the factors and indicators that contribute to the concept of smart people [13].

The reader can quickly see that the qualification, learning capacity, and the diversity in the constitution of the society are important components, after which come the qualities such individual flexibility, creativity, and so on. This is due to the reliance on technology that requires people to possess

Table 1.3 Factors and Indicators for Smart People in the Context of a Smart City		
Factor	**Indicator**	**Weighting (%)**
Level of qualification	4	14
Affinity to life learning	3	14
Social and Ethnic plurality	2	14
Flexibility	1	14
Creativity	1	14
Cosmopolitanism and open mindedness	1	14
Participation in public life	1	14

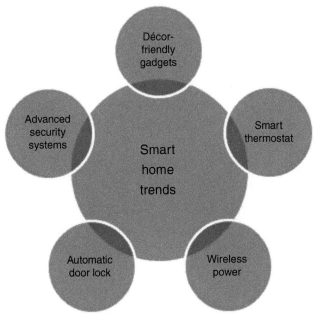

FIGURE 1.3 Smart Home Trends

primary skills first in order to use this technology but also to follow on the advances in day-to-day activities.

2.2 SMART HOMES

Making homes smart is a new trend [14]. A smart home will host a set of services that were once of a futuristic feet. Fig. 1.3 shows some major smart home trends.

2.2.1 Programmable and zone-based smart thermostat

In recent years, we have seen few start-ups that propose programmable thermostat such as Nest, recently acquired by Google. Nest offers a thermostat product that is fully automated and it uses smart machine learning algorithms to optimize energy consumption. For example:

- It will not heat or cool a home if there are no sensed occupants.
- It builds a schedule dynamically based on collected data from sensing current occupants.
- It is controllable via smartphone or via the web app for more advanced reporting.
- It is aesthetically pleasing and blends well with wall colors.

2.2.2 Wireless power

The smart home concept relies on a myriad of devices and objects (IoT). This adds an extra complexity of powering all of them by either wall plugs or Lithium batteries. Wireless power transfer (WPT) is used to transfer power over short distances by magnetic fields [15]. There are two techniques in WPT: nonradiative and radiative. Near-field nonradiative variant is already in use in consumer-oriented products such as electric toothbrush, smart cards, radio-frequency identification (RFID), and pacemakers. This variant is more suitable to enable smart homes because it tackles small devices in restrained areas. Radiative technique on the other hand is used in long-distance peer-to-peer (P2P) power transfer. It relies on either Laser beams or microwave technologies. These are, however, not suited for home use.

2.2.3 Automatic door locks
Automatic door locks are a logical extension to smart homes. Doors can be locked and unlocked using sensor technology to detect and authenticate a homeowner without the need for keys. This technology is already in use in modern cars to unlock cars on approach.

2.2.4 Advanced security system
Security is very important in today's technological home. Monitoring the home activity remotely using a smartphone, receiving alerts that the home was accessed when it should not, and sending a live feed to local enforcement authorities when a breach is detected and confirmed, all of these scenarios can become easier to implement using advanced security systems that rely on new technologies such as image and video processing and face detection.

3 CHALLENGES IN SMART CITIES AND HOMES
3.1 SECURITY
A common question that is commonly asked in conferences about smart cities is as follows:

> Who is responsible when a smart city ICT system crashes?
>
> **Dr. Simon Moores [27]**

   This question often goes unanswered because it is fundamentally linked to system security in overall that is complex to master. The attack surface is big especially if we take into account that most of these IoT objects in their initial version will require security patches that need firmware updates. Such a maintenance operation at a wide scale is vulnerable to classical configuration management difficulties. Provided that these devices are interconnected, we should see some major security breach consequence issues.

3.2 IOT CHALLENGES
IoT devices are the centerpiece of any future smart home and smart city project [24]. Although there are many benefits to IoT, there are many challenges to be solved [25]:

- With regards to data, there are many questions that are still challenging such as: Who owns the data? Who and how can we monetize the data? What standards/formats are to be used? With the absence of fully agreed-on protocols and standards today, it is unclear how all the companies that contribute to the provided service will share the responsibility over the data.
- Provided that there are many companies that contribute to the provided service, it is very important to define the boundaries of these companies in terms of customer relationship. If we take a thermostat example in a connected home, who will the customer perceive as the main relationship holder? The thermostat company or the utility company? There is no actual answer unless specific service-level agreements (SLAs) are drafted between the stakeholders of the service.

3.3 FRAGMENTATION OF STANDARDS

Moving to a smarter city requires city authorities to develop better governance principles and techniques in order to provide answers to the following questions [16]:

- How will the authorities set objectives for smart cities and measure their progress?
- How will information be shared between infrastructure services?
- What are the risks for moving to smart city services? And how can we mitigate them?
- How can a shared understanding and vision of a smart service between all stakeholders be created and delivered?

Answering these questions requires actions on multiple levels. These actions tackle regulating issues such as data use and common conceptual modeling and interoperability. There are few smart city standards that have emerged. Here is summary of major ones:

- PAS 180 aims at improving communication and understanding between different stakeholders in terms of vision, compatibility, and common organization issues.
- BS ISO 27000 is another standard that embraces the best practices in information security. This helps secure vast quantities of data that are transferred across many network boundaries.
- ISO/IEC 29100 deals with privacy risk management issues.

Having policy makers collaborate more closely on standards is of increased importance in order to have a comprehensive and coherent framework of work.

3.4 PROCESSING BIG DATA

As discussed in previous sections, smart homes and cities will rely on IoT devices, sensors, RFID chips, and smart electric meters, among others, in order to provide added value services to citizens and homeowners. These devices however generate large amount of data, big data, data sets, which are so big that traditional data processing techniques are not adequate to manage them.

3.5 SCALABILITY

With the rapid increase in the number of smart objects in either smart homes or cities, their heterogeneous interaction may cause scalability issues in large-scale deployments due to interoperability. This is a strong motive to speed up the convergence of standards that govern these smart objects. In addition, part of the scalability issues is the lack of test beds at scale that allow the validation of many proofs of concepts in real-world scenarios.

---

4 SURVEY OF MAJOR KEY ENABLING TECHNOLOGIES FOR SMART CITIES AND HOMES

4.1 INTERNET OF THINGS

The IoT is one of the major technologies that will shape the future of the digital world including Smart World, and Smart Cities and Homes. It is a mesh network of physical objects that either exchange data in p2p mode or communicate and relay information with the service provider [17]. There are many connected objects today such as electronic appliances (microwaves, cameras, refrigerators, etc.) that rely on RFID technology and state-of-the-art software and sensors [18] in their proper operation. IoT objects can be sensed and controlled across local area network (LAN) or wide area network (WAN) networks. This allows the creation of many products and opportunities to better integrate the physical infrastructure with the digital systems. Many experts expect > 10 billion IoT objects by year 2020 [19]. With such number of IoT devices in the world, it becomes primordial to work out a comprehensive IoT and digital business policy. Big industry players such as Microsoft, IBM, Cisco, Siemens, and Google already play an important role in helping draft and put in place such policies by offering cloud-based IoT services and devices.

4.2 SMART DUST

Smart dust objects are tiny Microelectromechanical Systems (MEMS). These are tiny robots that can detect anything from small vibrations to chemical composition. Smart dust is an extension to IoT objects [20]. They operate under a wireless network and are deployed in nonintrusive manner in countless areas. Their sizes allow them to establish communication within few millimeters. Many smart dust objects are hence required to cover large areas. Each one is composed of a semiconductor laser diode and MEMS beam reflector for communication and relay. These tiny objects are so light that they can move seamlessly from one area to another through air currents. Once they are deployed, it is very hard to detect their presence, but also it is harder to get rid of them. Here are few key applications of smart dust in the context of smart cities and homes:

- habitat monitoring
- indoor and outdoor environmental monitoring
- security and tracking of people and objects
- traffic monitoring and management
- human health and well-being monitoring

4.3 SMARTPHONES

Smart mobile devices are becoming the epicenter of people's lives [21]. Smartphones today contain a variety of chips and sensors such as GPS, gyroscope, microphone, camera, and accelerometer, among others, that are generating a lot of raw data [20]. Smartphones are essential building blocks in the smart home and city ICT infrastructure. They are used as a medium for relaying and accessing information as well as digital services [22]. Smartphones play a major role in smart cities and homes as briefly explained in the following:

- They are diminishing the gap between real and virtual world by making the captured data context aware.
- Mobile devices are integrated with services in the cloud, such as Google (Android) and iCloud (Apple) phones. This enables the use of off-load servers to increase the storing capacity of the phones.
- Advances in mobile access technologies such as 4G/Long-Term Evolution (LTE) networks allow connectivity everywhere and at very high speeds. This would enable mobility and simplify linking data across multiple domains.

In the context of a smart home, for instance [23], there are already some applications that run on smartphones to control many appliances in the home, such as TV sets, lights, window stores, garage doors, and security cameras. Moreover, users can operate their smartphones in order to interact with their city, receive live information, and connect with local authorities and public transportation systems.

4.4 CLOUD COMPUTING

Cloud computing has become a de facto platform to enable content delivery to consumers. Provided pervasive computing today, largely enabled by smartphones and IoT devices, massive amounts of data need to be processed in order to transform raw data into insightful information. Current computing paradigms are no longer suitable to such endeavor. Cloud computing has three main offers:

- *SaaS*: Software as a Service is what most people interact with today. It offers software in a subscription model for both end consumers and enterprises. SaaS simplifies the deployment model of software as it is no longer necessary to download and install the software or to worry about upgrade and maintenance issues. SaaS can be accessible via web or mobile apps, but also in services offered in high-tech cars such as Tesla or BMW.
- *PaaS*: Platform as a service is a software environment used mainly for development and execution runtime. It is defined as en environment for the development, integration, testing, and deployment of SaaS software. PaaS takes off the burden of managing software application servers and middleware, as well as other low-level technical aspects out of start-ups and entrepreneurs in order to concentrate their efforts in developing their products and services.
- *IaaS*: Infrastructure as a service is a fundamental layer in cloud computing. In this layer, vendors offer mutualized or dedicated hardware capacity to multiple users. Customers can request additional capacity dynamically through a simple management interface or based on preconfigured triggers (network traffic, CPU load, etc.). This is ideal for companies that require the optimization of their IT cost.

A combination of the aforementioned service cloud offering gives the emergences of other types commonly known as XaaS (everything as a service) such as storage as a service, communications as a service, network as a service, monitoring as a service, analytics as a service, data as a service, and so on.

4.5 BIG DATA AND OPEN DATA

A tentative estimation of the amount of digital information produced by mankind was 280 EB of data [19]. Fig. 1.4 shows how this data capacity is partitioned.

These are very large, complex, and changing data sets. If we take into consideration the digital rupture made by smartphones and tablets, the data it produces would rank in top 5% of data generators in today's information age.

In the context of smart homes and cities, often big data and open data are discussed hand in hand. While big data deals with large data sets and size, open data deals with accessible public data that is available to people, companies, and organizations and which can be freely used to develop added-value applications. Definition of Open data includes the following two basic features:

- The data must be publicly available for anyone to use.
- It must be licensed in a way that allows for its reuse.

The best example for a recent smart city is Songdo in South Korea. This smart, wired, and sustainable city was built at a staggering cost of $35 billion and it is entirely wired and connected to a mesh

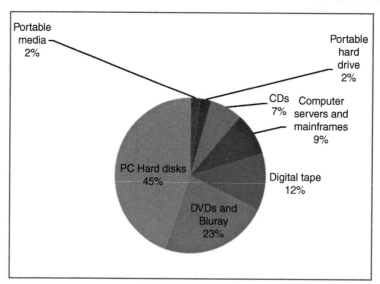

FIGURE 1.4 Global Information Storage Capacity [21]

network. The thousands of installed sensors generate large amounts of data continuously. To make sense of such large amounts of data, big data and data analytics are key technologies that transform raw data into insightful information.

4.6 SMART GRID

A smart grid is a modern electrical infrastructure that relies on ICT and digital networks in order to gather data about the grid, that is, produced and consumed electricity, and use that information to generate live consumption models to optimize the state of the grid in real time [24]. Following is a list of key smart grid features:

- *efficiency* of the grid by improving the *watt*age output and reducing the voltage on the lines whenever it is possible;
- *reliability* in smart grid attempts to improve fault detection and self-healing of the electricity network thus reducing the intervention of the maintenance technicians;
- *load balancing* in a situation where the total load of the grid varies during peak and low times (eg, the smart grid can warn smart client appliances to reduce their electricity consumption in peak times in order to use the saved capacity to serve other peak customers);
- *sustainability* in smart grid to allow the inclusion of renewable energy sources in the grid in a plug and play manner.

The field of smart grid is gaining a lot of attention and active research these days. There are major programs such as the IntelliGrid, which provides tools, methodology, and standards to provisioning for meters, and distribution of electricity. Other research includes Modern Grid Initiative, GridWise, GridWorks, Solar cities, and many others by various instances throughout the world.

5 EXAMPLES

5.1 SMART HOME DEVICES

This section describes some smart home devices that are already in the market today.

5.1.1 Nest thermostat

Nest is a start-up that specializes in the smart thermostat; they have been acquired by Google recently. Nest is a self-learning thermostat that will control the home heating system. What makes Nest smart is that it can learn your habits by sensing the presence in the house, and then adjust the temperature, the water heater cycles, etc. This results in an optimized energy bill for the homeowner.

5.1.2 Honeywell Lyric thermostat

The Honeywell Lyric thermostat is similar to the Nest thermostat, but instead of relying on motion detection, it relies on geofencing to detect if a person is available in the preset area. Lyric relies also on external weather and humidity levels to adjust the perfect temperature for the home.

5.1.3 Canary

Canary is a simple and intuitive out-of-the-box security system that allows monitoring of the home remotely as well as alerting for any intrusion. It is smart and capable of learning your habits, times when you are at home or not before alerting about an intrusion. Alert notifications are sent to the phone.

5.1.4 Goji

Goji is smart digital lock for the home. It is connected via Wi-Fi or Bluetooth and can be used to secure access to your home using face recognition technology. For instance, you may set picture of people that are authorized to access the home so they can authenticate on arrival. It can be used to record all access activities.

5.1.5 Aggregate smart home device controllers

Provided the multitude of smart devices that are installed in the home, it becomes a bit difficult to control them all separately from different interfaces. Some gadgets attempt to solve this issue and connect to all installed gadgets to provide a centric interface to controlling them. Some of these gadgets are Homey, Revolv, and smart things.

5.2 SMART CITY PROJECTS

This section describes some smart cities around the globe.

5.2.1 Amsterdam smart city

Amsterdam smart city is a public–private partnership initiated in 2009 to transform the Amsterdam Metropolitan into a smart city [28]. The Amsterdam smart city platform is composed of many initiatives; here are a few elements:

- *Dynamic traffic control*: This system relies on an advanced system to manage and monitor traffic; it is data driven and recently has been extended to span parking coordination. In Amsterdam, there are twice as many bikes as people.
- *Sharing economy*: Sharing is the new ownership! Examples include sharing car trips, taxi rides, and exchanging houses on trips.
- *Electric vehicles*: Amsterdam is an international frontrunner in this domain. This is partly because it has a solid infrastructure in addition to the construction of version 150 electricity points in the Harbor.
- *Energy efficiency*: The city developed a heating and cooling technology based on surrounding lake waters.
- *Smart grid*: "NieuwWest" is one of the most advanced grids in the world that has advanced features such as self-healing.

5.2.2 Smart city Barcelona
Barcelona is one of the most cosmopolitan cities in Europe and is ranked first smart city in the world as of 2015. It has become a leading tourist destination that proves the city's successful economic, social, and environmental policies. Barcelona has a vision to become self-sufficient, with productive neighborhood and producing zero emission. The city has an efficient bus network system and a bicycle sharing system. It has installed thousands of urban sensors and tourists can pay at most stores using Near Field Communication (NFC) technology (project Contactless). The city is transforming itself to an urban lab [29]. Multiple projects were launched to manage resources such as water with telemanaged irrigation. It has also a smart traffic light system that optimizes traffic in the city.

5.2.3 Smart city Birmingham
Digital Birmingham [30], United Kingdom, was launched in 2014 with a defined 49 actions plan grouped as follows:

- *Technology and place*: This involves the improvement of connectivity and sharing open data.
- *People*: This involves focusing on digital inclusion and improving resident technology skills. This is primordial in order to benefit from the smart city services.
- *Economy*: This involves digitalizing the social care, improving energy efficiency, as well as smart mobility.

5.2.4 Smart city Wien (Vienna)
Motivated by the rapid change in global climate and the shortage of natural resources, the city of Vienna decided to launch a complex endeavor to transform the city into a smart city [31]. This initiative is comprehensive and aims at:

- using ICT to drive innovation
- thriving in an ecofriendly environment for residents
- involving everyone through transparency

ICT and technology are used to expand the online services provided by the city in the aim of raising people's quality of life. Using data analytics in the field of transportation has allowed to foster the idea of car sharing and to optimize traffic. In this area, the SMILE project, for instance, allows promoting multimodality transportation systems, that is, using multiple transformation media to reach destination.

A resident can start by using a Taxi ride, then switch to a bus ride, and finish by a bike ride all seamlessly and in a coherent manner in order to optimize arrival time and/or travel cost. In order to allow third-party start-ups to propose and develop bleeding edge applications, the city has also made available open data for multiple services and scenarios. Thanks to this policy, many services are created in the field of geodata, traffic data, and ecology data.

6 CONCLUSIONS

Smart cities and homes concepts have been hot topics for over a decade. Changing lifestyles and the growing demand on useful and consistent services require a new approach that relies and benefits from ICT advances. Two major building block technologies that enable both smart cities and smart homes are cloud computing and IoT. Two major building block technologies that enable both smart cities and smart homes are cloud computing and IoT. The latter allows sensing the surrounding environment and exchanging data in order to provide digital services to end user or to infrastructure in order to help make optimized decisions about traffic flow, energy efficiency, home security, etc. Although there are many advantages to smart homes and cities, there still exist many challenges that make their implementation difficult. Some of these challenges are the security of data that transits across many components and interfaces, and the current fragmentation of standards that make it hard for different devices to be interoperable as well as deployed at scale. In addition, there is the challenge of making sense of the amount of data generated by all of the deployed IoT devices. Although these challenges exist, we have started to see a first batch of smart home devices that solve everyday's optimization issues such as energy efficiency, home security, and home automation. In the smart city front, major work is being done on smart grids and smart mobility. There are many successful pilots already in cities such as Amsterdam and Barcelona that have exceeded the initial objectives. Nevertheless, we still need to overcome the described challenges through standardized and interoperable protocols.

REFERENCES

[1] Obaidat MS. Key enabling ICT systems for smart homes and cities: the opportunities and challenges. Keynote speech. In: Proceedings of the 2014 IEEE international conference on network infrastructure and digital content (IC-NIDC 2015), Beijing, China; September 2015.

[2] Siemens. Wired for an urban world, <http://www.siemens.com/innovation/en/home/pictures-of-the-future/infrastructure-and-finance/smart-cities-trends.html>; June 2015 [online].

[3] Siemens. Why cities are getting smarter, <http://www.siemens.com/innovation/en/home/pictures-of-the-future/infrastructure-and-finance/smart-cities-facts-and-forecasts.html>; June 2015 [online].

[4] Singh S. Smart cities – a $1.5 trillion market opportunity, <http://www.forbes.com/sites/sarwantsingh/2014/06/19/smart-cities-a-1-5-trillion-market-opportunity/>; June 2014 [online].

[5] Weiser M. The computer for the 21st century. Sci Am 1991;265(3):94–104.

[6] Smart environment. In: Wikipedia, The Free Encyclopedia. Wikimedia Foundation, Inc, <https://en.wikipedia.org/wiki/Smart_environment>; July 2015 [online].

[7] TUWEIN Team, Technishe Universishe Wein. European smart cities, <http://www.smart-cities.eu/model_5.html>; July 2015 [online].

[8] Poloncarz MC. Initiatives for a smart economy, <http://www2.erie.gov/environment/sites/www2.erie.gov.environment/files/uploads/pdfs/SmartEconomy%20for%20Web3.pdf>; July 2013 [online].

[9] Ferro E, Caroleo B, Leo M, Osella M, Pautasso E. The role of ICT in smart cities governance. In: Conference for e-democracy and open government; 2013. p. 133.

[10] Kaufmann D, Kraay A, Zoido-Lobatón P. Governance matters. Finance Dev 2000;37(2):10–3.

[11] Armstrong KA. Rediscovering civil society: the European Union and the white paper on governance. Eur Law J 2002;8(1):102–32.

[12] Caragliu A, Del Bo C, Nijkamp P. Smart cities in Europe. J Urban Technol 2011;18(2):65–82.

[13] TUWEIN Team, Technishe Universität Wein. European smart cities, <http://www.smart-cities.eu/model_4.html>; July 2015 [online].

[14] Saha D, Mukherjee A. Pervasive computing: a paradigm for the 21st century. IEEE Comput 2003;36(3):25.

[15] Tseng R, von Novak B, Shevde S, Grajski KA. Introduction to the alliance for wireless power loosely-coupled wireless power transfer system specification version 1.0. In: Proceedings of the 2013 IEEE wireless power transfer conference, WPT'13; May 2013. p. 79, 83.

[16] Alfino S. The role of standards in smart cities, <http://www.bsigroup.com/LocalFiles/en-GB/smart-cities/resources/BSI-smart-cities-report-The-Role-of-Standards-in-Smart-Cities-UK-EN.pdf>; June 2013 [online].

[17] Atzori L, Iera A, Morabito G. The Internet of things: a survey. Comput Netw 2010;54(5):2787–805.

[18] Xiaohang W, Song Dong J, Chin CY, Hettiarachchi SR, Zhang D. Semantic space: an infrastructure for smart spaces. IEEE Pervasive Comput 2004;3:32–9.

[19] Hilbert M, López P. The world's technological capacity to store, communicates, and compute information. Science 2011;332(6025):60–5.

[20] Soldo D, Quarto A, Di Lecce V. M-DUST: an innovative low-cost smart PM sensor. In: Proceedings of the 2012 IEEE international, instrumentation and measurement technology conference, I2MTC'12; May 2012. p.1823, 1828.

[21] Balakrishna C. Enabling technologies for smart city services and applications. In: Proceedings of the 6th IEEE international conference on next generation mobile applications, services and technologies, NGMAST'12; 2012. p. 223–7.

[22] Al-Begain K, et al. IMS: a development and deployment perspective. New Jersey: John Wiley & Sons Ltd; 2009.

[23] Choudhury T, et al. The mobile sensing platform: an embedded system for activity recognition. IEEE Pervasive Comput 2008;7(2):32–41.

[24] Tsai C-W, Pelov A, Chiang M-C, Yang C-S, Hong T-P. A brief introduction to classification for smart grid. In: Proceedings of the 2013 IEEE international conference on systems, man, and cybernetics, SMC'13; October 13–16, 2013. p. 2905, 2909.

[25] Wikipedia, The Free Encyclopedia. Wikimedia Foundation, Inc, Internet of things, <https://en.wikipedia.org/wiki/Internet_of_Things>; July 2015 [online].

[26] RITA Office of Research, Development and Technology, U.S. Department of Transportation. 2012 urban mobility report released with new congestion measures, <https://www.rita.dot.gov/utc/utc/sites/rita.dot.gov.utc/files/utc_spotlights/pdf/spotlight_0313.pdf>; March 2013 [online].

[27] Moores S. Keynote speech: the challenges of making cities 'smart' in advanced democracies. IFSEC International 2015. London, UK; 2015.

[28] Amsmarterdamcity, <http://amsterdamsmartcity.com>; September 2015 [online].

[29] BCN smart city, <http://smartcity.bcn.cat/en>; September 2015 [online].

[30] Birmingham smart city, <http://www.birmingham.gov.uk/smartcity>; September 2015 [online].

[31] The city of Vienna, smart city Wien, <https://smartcity.wien.gv.at/site/en>; September 2015 [online].

WIRELESS SENSOR NETWORKS APPLICATIONS TO SMART HOMES AND CITIES

2

A. Belghith*, M.S. Obaidat†

*Department of Telecommunications, University of Sfax, Sfax, Tunisia; †Department of Computer and Information, Fordham University, Bronx, NY, United States

1 INTRODUCTION

Wireless Sensor Networks (WSNs) [1] connect our world more than we dream up. Noise and atmospheric pollutions, garbage level sensor, road traffic monitoring, and smart parking are some of the many WSN applications to smart cities. In smart homes, it is now difficult to avoid using home video message, alarm to mobile phone, door, window, and light control applications while providing comfortable and smart-economic life.

To ensure this admirable way of life, new challenges of WSNs arise. Which access technologies can be used in telecommunication systems for smart cities and homes? Wireless protocols have to be used to communicate with thousands of nodes while managing a trade-off between data transfer rate, speed, and power consumption.

The huge number of sensors requires data aggregation mechanisms in order to prevent information redundancy and high energy consumption, storage capacity, and communication bandwidth. Obviously, these aspects cannot be met given the limitations of WSNs.

Moreover, it is interesting to define efficient network discovery and intelligent path determination to obtain reliable routing protocols taking into account the characteristics of sensors in the smart cities and homes.

Reliability is also guaranteed by security aspects such as encryption, access control, and secure data aggregation. Encryption becomes vital as data exchanged are related to personal and confidential data. Access control avoids privacy information discloser, especially for remote home monitoring.

In this chapter, we first give some examples of WSN applications in smart homes and cities. Then, we discuss the access technology to be used for the applications. Finally, we present some protocols useful to provide better applications' performance such as the routing strategies, energy-saving methods, and security protocols.

2 WSN APPLICATIONS EXAMPLES

In this section, we present essential applications of WSNs for smart cities and homes. There are three key applications: energy saving, noise and atmospheric monitoring, and healthcare monitoring.

Smart Cities and Homes. http://dx.doi.org/10.1016/B978-0-12-803454-5.00002-X

2.1 ENERGY-SAVING APPLICATIONS

The huge energy consumption and the high fuel cost require an efficient use of the energy that becomes more and more rare. Fig. 2.1 shows the total consumption by End-Use sector in the United States [2].

Exact values of energy consumed in the electric power, residential, and transportation sectors in 1994 and 2014 are presented in Table 2.1. We note an increase of about 15% in the past 20 years and therefore it is essential to efficiently react to this growth.

The increase in energy consumption is not related only to the USA, but concerns all countries in the word such as in Finland where lighting consumes over 30% of the total electricity used in households for appliances [3] and in Egypt where the consumption of energy in a house can reach 1500 kWh/month [4].

Energy-efficient buildings have to be designed in order to significantly reduce the energy use, especially for heating and cooling. The decrease of the energy use can be performed when reducing the demand for energy by avoiding waste and implementing energy-saving measures. Waste energy avoidance can be performed by having good insulation means, air tightness, and ventilation. In Ref. [5], a procedure for measuring and reporting commercial building is proposed. This procedure consists of two steps. The first step is the measurement system design that identifies the performance metrics, the physical location of each measurement, the frequency of measurements, and the measurement

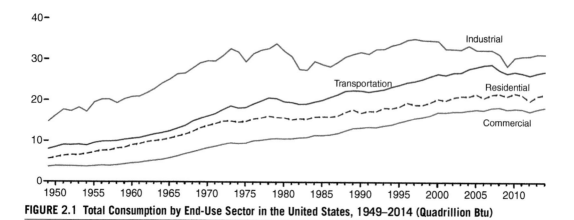

FIGURE 2.1 Total Consumption by End-Use Sector in the United States, 1949–2014 (Quadrillion Btu)

Table 2.1 Total Energy Consumed in 1994 and 2014 by Different Sectors in the United States

Total Energy Consumed by	In 2014 (Trillion Btu)	1994 (Trillion Btu)	Increase in Past 20 Years (%)
The electric power sector	38,520.472	32,398.714	15.9
The residential sector	21,530.939	18,111.572	15.88
The transportation sector	27,117.707	23,365.133	13.83

equipment. The second step is the data collection and analysis, which specifies how to monitor data, assemble data, and calculate monthly and annual metrics.

We note that the wasted energy reduction is not sufficient. It is also interesting that the buildings use sustainable and green sources of energy instead of finite fossil fuels. For example, a green building architecture is proposed in Ref. [6]. Monitoring the temperature, light intensity, and presence of persons provides information that is used by the control subsystem of the proposed architecture. The control subsystem can tune the energy to consume, inform the users about the status of the energy consumption periodically or when there are excessive energy consumption in real-time reports, and schedule flexible tasks. For example, when using smart meters in a building [7], the cost of energy varies depending on the period of day and therefore the control subsystem can schedule some house tasks when the price of energy becomes low. Finally, taking into account the atmospheric characteristics, we can program which type of energy can be used such as solar, wind, and geothermal energies.

Wireless sensors can also use green energy such as light, motion, and vibration to work. For example, GreenPark developed ultralow-power sensors that use environmental energy sources in order to reduce the energy consumption and extend the use of sensors in a building [8].

2.2 NOISE AND ATMOSPHERIC MONITORING

WSNs can be very helpful for the health of urban residents as they can monitor noise and atmospheric pollution. First experiences using WSNs for noise pollution monitoring are presented in Ref. [9]. These experiences are based on Tmote prototyping platform for collecting noise pollution data in both indoor and outdoor settings. Experimental results show the feasibility of using WSNs in noise monitoring. Therefore this kind of application can prevent expensive and complex tasks done especially by private entities.

The increase of noise pollution motivates researchers to continuously propose designs for noise pollution monitoring systems. For example, in Ref. [10], authors propose noise measurement instruments and techniques taking into account the transmitter, receiver, and atmosphere. Note that the noise follows a path depending on the atmosphere to reach the receiver. In Ref. [11], a design of an energy-harvesting noise-sensing WSN mote is proposed in order to mitigate and fight noise pollution. The proposed mote extension is able to detect noise levels in urban environments where there are multiple pulse loads. Experimental results show that the WSN mote provides an improvement of more than 300% in the analytically derived duty cycles.

Note that some commercial WSN devices for atmospheric pollution monitoring are presented in Ref. [12] such as Waspmote, Generic Ultraviolet Sensors Technologies and Observations (GUSTO), and CitiSense. Waspmote [13] can monitor several parameters to verify if the quality of air we breathe is healthy. These parameters consist of Nitrogen dioxide (NO_2), Carbon dioxide (CO_2), Methane (CH_4), and Hydrocarbons (Ethanol, Propane, Butane, etc.). GUSTO [14], based on the Differential Ultraviolet Absorption Spectroscopy (DUVASTM) technology [15], can measure and transmit urban pollutants such as NO_2, O_3, and benzene in real time.

Figs. 2.2 and 2.3 show Waspmote and GUSTO devices, respectively. We note that these devices do not have display. They send air pollution measurements to a collector for further analysis and investigation.

CitiSense [16], developed by a University of California-San Diego team, is a pollution monitoring system that can be integrated in smartphones. Therefore, it provides very smart useful service to

FIGURE 2.2 Waspmote Devices

FIGURE 2.3 GUSTO Device

FIGURE 2.4 Air Quality Monitoring by CitiSense

users in order to prevent staying in highly polluted places. Fig. 2.4 shows that CitiSense finds out that air quality in the current place is moderate. Note that real-time information about air pollution can be obtained using other sophisticated tools such as over the Google Map. However, information is not displayed for public users (only for users having authorization) [17]. Many areas deploy this technology such as in Qatar [18].

Like noise pollution monitoring, several system designs for air pollution were proposed. In Ref. [19], authors propose an air pollution system that monitors the air quality in real time while reducing the energy consumption of sensors using a power-saving strategy. The proposed power-saving strategy is described later. Sensors sense the pollution information and then compare the pollution level with defined standard reference values. If the pollution level sensed is high, then data are sent through transmitter and sensors wait 5 min before sensing the pollution information in the next time. Otherwise, when the sensed pollution level does not exceed the threshold values, sensors do not send data and wait 15 min before performing next measurements.

In Ref. [20], Kalaimani and Sakthivel propose a simple WSN-based air quality monitoring system (WSN-AQMS) for industrial areas. The proposed monitoring system controls and monitors the physical environment while reducing the energy consumption and the rate of data exchanged between sensors. It selects WSN components depending on the building's purpose, the number of nodes needed, and the options of evaluation. The architecture of the monitoring system uses Gas sensors, humidity sensors, and Global System for Mobile communications (GSM) modules for cellular communications.

2.3 HEALTHCARE MONITORING

In addition to the noise and atmospheric monitoring, which are useful for keeping health, WSNs can be also used for smart healthcare of residents while keeping their comfort and privacy. In Ref. [21], authors propose system architecture for residents' health monitoring. The proposed system integrates the existing medical technology in low-cost sensors in order to nurse elder and handicapped persons. The efficiency of this solution is based on the following proprieties: portability of the small devices (sensors), scalability of the number of devices used when decreasing the complexity of the functionalities, autoconfiguration capability especially when using Internet Protocol version 6 (IPv6) protocol [22], and the real-time response of the deployed sensors when measurements exceed specific thresholds.

Evidently, WSN monitoring is used in hospitals. Many projects are developed in this field [23] such as HealthGear (Microsoft project) [24], MobiHealth (European Commission project) [25], CodeBlue (Harvard University project) [26], and Wireless Sensor Network for Quality of Life (WSN4QoL) (Marie Curie project) [27].

HealthGear consists of a set of noninvasive physiological sensors that are connected via Bluetooth to cell phones. The sensors contain modules that measure many parameters such as the blood's oxygen level of users as well as the user respiration and motion. The measurements are then transmitted using Bluetooth in order to analyze them and represent the results in an interface (see Fig. 2.5).

MobiHealth profits from available technologies such as mobile medical sensors, public wireless network, and new services of *Universal Mobile Telecommunications System* (UMTS). It provides

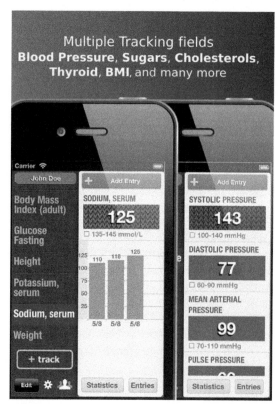

FIGURE 2.5 Interface of HealthGear

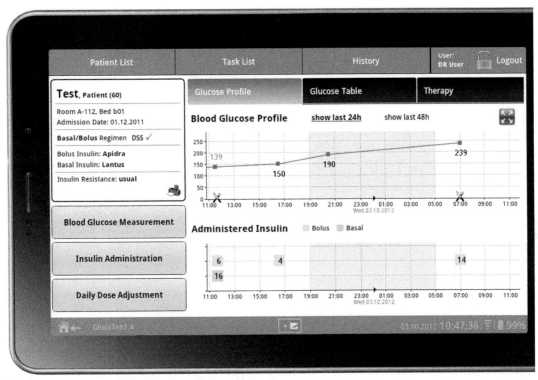

FIGURE 2.6 Interface of MobiHealth for Glucose Measurements

medical follow-up and medical research. Note that MobiHealth deals with many fields such as remote monitoring of glucose levels in Europe [28] and pregnancy telemonitoring in community centers and clinics in Zambia [29]. As an example, the interface of Diabetes management is presented in Fig. 2.6.

CodeBlue consists of wireless infrastructure that is deployed in emergency medical care environments. It integrates low-power wireless sensors, Personal Digital Assistants (PDAs), and Personal Computers (PCs). The wireless sensors integrate Global Positioning System (GPS) to track persons, an Electromyography (*EMG*) module that checks the health of the muscles for motion capture, and a mote-based pulse oximeter for transmitting periodic packets containing heart rate.

WSN4QoL aims to provide real-life implementations on the pervasive healthcare applications. The applications targeted by this project are for pacemaker, blood pressure, cochlear implant, and electroencephalogram (EEG).

3 ACCESS TECHNOLOGIES

In this section, we present three architectures for access technologies. These architectures are summarized in Ref. [30].

3.1 FIRST ACCESS TECHNOLOGIES ARCHITECTURE

The first architecture is presented in Fig. 2.7. It can be deployed at home, at work, and in hospitals when monitoring healthcare system. The wireless sensors perform measurement and send results to the personal server using Wireless Personal Area Network (WPAN) technologies such as IEEE 802.15.1 [31], IEEE 802.15.3 [32], and IEEE 802.15.4 [33]. The personal server can be a PDA and needs a second wireless interface, in addition to the WPAN interface, in order to connect to the home server. The connection with the home server is performed using Wireless Local Area Network (WLAN) technologies such as IEEE 802.11a/b/g/n.

The home server, called also the central server, has to be connected to the Internet in order to send data to the distant server. This connection has to be secured. Finally, the distant server gathers different information, analyzes it, and displays the final results. The final results can be obtained via web pages in order to facilitate the information access.

The well-known WPAN technology is Bluetooth, which is based on the IEEE 802.15.1 standard. This technology is designed for short range, low cost, and low energy consumption. It operates in the Industrial, Scientific, and Medical (ISM) frequency band of 2.4 GHz, but there are methods to reduce interference [1,34].

IEEE 802.15.3 represents the high rate of WPAN. It provides high throughput and therefore this technology is useful when the sensors need to send images and videos. Moreover, sensors using IEEE 802.15.3 can directly communicate between them and therefore this technology is useful when there is a need to create mesh networks [35]. However, this technology consumes more energy due to the overhead for communication links management [36].

IEEE 802.15.4 represents the low rate of WPAN. It provides low throughput, but enhances the reduction of the energy consumption. Moreover, this technology has sophisticated methods to prevent interference [37]. For example, the bad channels are not used for data transmission such as in Ref. [38] or all channels are used but with different probabilities depending on the channel quality such as in Refs. [1,39,40].

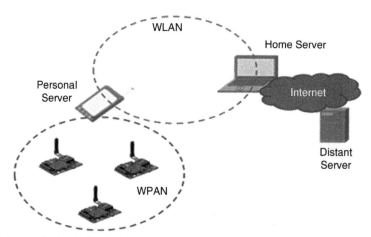

FIGURE 2.7 First Access Technologies Architecture: Use of WPAN and WLAN

Table 2.2 Main Characteristics of IEEE 802.11a/b/g/n

Standard IEEE	802.11a [41,95]	802.11b [43,102]	802.11g [44]	802.11n [45]
Frequency (GHz)	5	2.4	2.4	2.4 and 5
Modulation	OFDM	DSSS	OFDM and DSSS	MIMO–OFDM
Throughput (Mbit/s)	54	11	54	248
Range (m)	35	35	35	70
Publication date	October 1999	October 1999	June 2003	October 2009
Advantages	Good throughput	Low cost	Good throughput and better obstacle penetration	High throughput and better range
Drawbacks	Signals cannot penetrate obstacles	Medium throughput and sensitive to interferences	Wireless congestion	

DSSS, Direct Sequence Spread Spectrum; *MIMO*, Multiple Input Multiple Output; *OFDM*, orthogonal frequency division multiplexing.

Next, we briefly present the main characteristics of WLAN technologies (see Table 2.2). IEEE 802.11a [1,41] uses orthogonal frequency division multiplexing (OFDM) technique and works in the 5-GHz frequency band. The throughput can reach 54 Mbits/s. The main drawback of this standard is that the signals cannot penetrate obstacles such as walls and solid objects [42,95].

IEEE 802.11b uses the Direct Sequence Spread Spectrum (DSSS) technique and works in the 2.4-GHz frequency band. The throughput can reach 11 Mbits/s [46]. Note that the use of Carrier Sense Multiple Access/Collusion Avoidance (CSMA/CA) for physical channel access decreases the throughput [102]. Moreover, CSMA/CA is not fair [47]. The main advantage of IEEE802.11b is the low cost and so this technology is widely deployed. However, the throughput is not high and as the frequency band of 2.4 GHz is used, signals are very sensitive to interferences.

IEEE 802.11g uses both OFDM and DSSS techniques and works in the 2.4-GHz frequency band. The throughput can reach 54 Mbits/s. Like IEEE 802.11b, the use of CSMA/CA decreases the throughput. The main advantage of this technology is in its high throughput and the interoperability with the widely used standard IEEE 802.11b [48]. However, as the frequency band of 2.4 GHz is used by many types of equipment such as IEEE 802.11b wireless cards, phones, microwave ovens, and baby monitors, IEEE 802.11g suffers from wireless congestion [49].

IEEE 802.11n uses OFDM and Multiple Input Multiple Output (MIMO) techniques and works in the 2.4- and 5-GHz ISM frequency bands. MIMO enables the opportunity to spatially resolve multipath signals [50]. The throughput can reach 600 Mbits/s when using four antennas for transmission and four antennas for reception. Note that 4×4 is the maximum MIMO configuration allowed in IEEE 802.11n.

3.2 SECOND ACCESS TECHNOLOGIES ARCHITECTURE

The second architecture is presented in Fig. 2.8. This architecture utilizes only WPAN technologies and therefore it does not need the personal server. Therefore, this technology is the cheapest as it reduces the

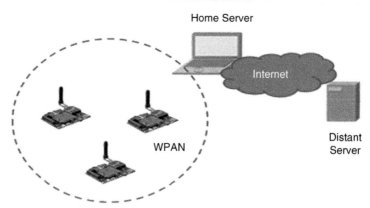

FIGURE 2.8 Second Access Technologies Architecture: Use of WPAN Only

number of devices needed. The wireless sensors send measurement results directly to the home server using WPAN technologies. However, the wireless sensors require more energy to access the home server as they have to increase Radio-Frequency (RF) output power. The increase of the RF power can also cause more collisions and hence more retransmissions. The increase of the retransmissions degrades the Quality of Service (QoS) and consumes more energy as the same data are transmitted several times.

3.3 THIRD ACCESS TECHNOLOGIES ARCHITECTURE

The third architecture is presented in Fig. 2.9. As in the first architecture, the wireless sensors send measurement results to the personal server. Then, the personal server gathers and forwards data to the

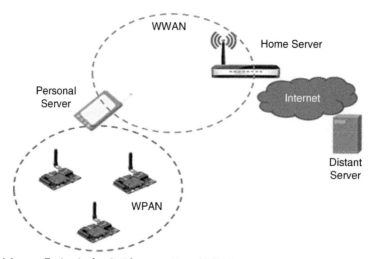

FIGURE 2.9 Third Access Technologies Architecture: Use of WPAN and WWAN

home server. Unlike in the first architecture, the connection between the personal server and the home server is performed using Wireless Wide Area Network (WWAN) technologies such as 2G, 2.5G, 3G, and 4G. Finally, the home server, connected to the Internet, sends data to the distant server.

Now, we briefly describe some WWAN technologies. The second generation (2G) of mobile networks started to be deployed in the beginning of the 1990s. The main 2G mobile network, and the most successful, by far, is the GSM [51,102]. The services were limited to voice and Short Message Service (SMS).

The so-called two-and-half generation (2.5G or 2G+) such as General Packet Radio Service (GPRS) [52] and Enhanced Data Rates for GSM Evolution (EDGE) [53,102] added packet data services and increased data rate. This generation is mainly used for Internet-style access and email. The theoretical maximum rate in the GPRS system is 115 Kbps while the EDGE system provides better theoretical maximum rate (up to 384 Kbps) [102].

Nevertheless, users need wireless high-speed Internet access. Moreover, users want to be able to access the Internet from a large area. The 3G system can support multimedia, data, video, and other services along with voice. The main 3G systems are the UMTS [54] and CDMA2000 [55,102]. The first deployment of CDMA2000 and UMTS took place in 2000–01.

Yet, the 3G systems are still in evolution. The first data rates were in the magnitude of 1 Mbps. Nowadays much higher data rates are expected in both uplink and downlink with the High Speed Downlink Packet Access (HSDPA) and the High Speed Uplink Packet Access (HSUPA) evolutions [56] (see, eg, Release 7 of UMTS). Apart from the displayed physical data rates, application-level data rates are smaller. For example, in 2007, the HSUPA could reach 1 Mbps for a File Transfer Protocol (FTP) application [57]. More recent versions of HSUPA have higher figures. The next step after 3G is (evidently) 4G or what is also known as Beyond 3G (B3G).

Long-Term Evolution (LTE) defined by third Generation Partnership Project (3GPP) Release 8 in 2008 is a very promising technology providing a high peak data rate of 163 Mbps in a channel bandwidth of 10 MHz and a low latency of 15 ms [58]. The enhancement of LTE, called LTE-Advanced (LTE-A), aims to reach a peak data rate of 1 Gbps in order to have a fourth-generation (4G) access technology. This technology continues to evolve through Release 13 that is planned to be completed in March 2016 although some features will be added [59]. This release includes advanced features such as supporting Advanced three Band Carrier Aggregation (three in Downlink/one in Uplink).

The different access technologies can be evaluated using experimental tests or simulation tools. Note that a web-based simulation tool, proposed in Ref. [60], provides a simulation environment with Network Simulation 2 (NS-2) [61] and includes WSN and Bluetooth modules that can be used to practically evaluate different access technologies for WSN networks.

4 ROUTING STRATEGIES

Routing strategies applied to WSN applications for smart cities and homes have to be simple but efficient in order to enhance the QoS supported by these applications. Moreover, it is essential to not propose complex strategies due to the limit capacities of sensors. An intelligent routing strategy is proposed in Ref. [62]. The proposed routing strategy is based on an efficient network discovery. The network discovery uses a multihop routing tree constructed by the Spanning Tree depending on the metric used. The construction of the Spanning Tree is performed using the Prim's approach [63]. After adding each sink node to the tree, tree edges are then iteratively selected in breadth-first search, based on the defined routing metric until all nodes are added. The proposed routing strategy uses messages for the discovery of the best parent, announcement of the potential parents (sending beacons periodically), association with the parent selected, and acknowledgment when the association is successful (ack is sent by the parent).

The best parent is determined based on the metric used. The choice of the metric depends on the application requirements. Metrics can be, for example, the number of hops, quality of signals, and residual energy. In Ref. [62], the defined metric combines between these three kinds of metrics using Fuzzy Logic (FL) decision approach. One example of the FL approach used in routing in WSN is presented in Ref. [64].

Experiments are run on the I3ASensorBed testbed [65]. This testbed, deployed in the Albacete Research Institute of Informatics, contains 47 nodes to emulate WSN applications in smart cities. Sensors can monitor temperature, humidity, CO_2, presence, door and window state (open or close), and energy consumption sensors. The testbed is accessible using a *WEB* interface to select nodes that perform applications, configure the nodes, and schedule application running. Note that there are other testbeds in the literature such as Wireless Sensor Network Testbeds (Wisebed) [66], Realnet [67], Twist [68], FIRE [69], and Neteye [70].

Experimental results show that the proposed routing strategy outperforms Hop Algorithm (HA) and Received Signal Strength Indicator Algorithm (RSSIA) in terms of energy consumption per delivered data packet (see Fig. 2.10) and packet delivery ratio (see Fig. 2.11). HA and RSSIA favor routes that have minimum number of hops to the receiver and the maximum RSSI of nodes, respectively.

The experimental results of energy consumed per delivered data packet can be explained by the fact that HA takes into account only the number of hops and therefore the number of delivered packet decreases as nodes do not have the same residual energy and the same quality of channel. On the one hand, the Modulation and Coding Scheme (MCS) used depends on the channel quality. An efficient MCS (so more data transmitted) is used when the channel quality is better. On the other hand, when

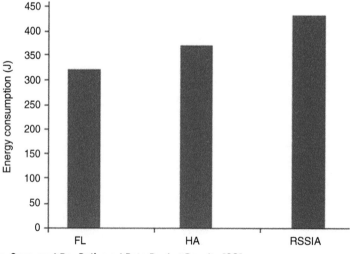

FIGURE 2.10 Energy Consumed Per Delivered Data Packet Results [62]

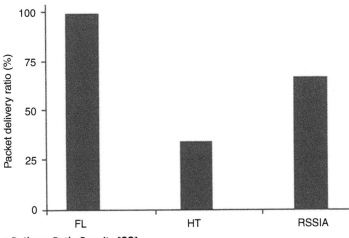

FIGURE 2.11 Packet Delivery Ratio Results [62]

data is forwarded by nodes having low residual energy, these nodes can shut down and so the amount of data transmitted decreases. For routing algorithm RSSIA, when the RSSI is the single parameter in selecting routes, data are forwarded by long routes (having high number of hops) and therefore require more total energy consumed.

The experimental results of the packet delivery ratio can be explained by the fact that HA favors routes having the minimum hops and therefore these hops have poor channel quality. When the channel quality is poor, the MCS used has to be robust and therefore lower number of data packets is transmitted. As RSSIA chooses long routes and consumes more energy, the packet delivery is lower than that provided by the FL strategy that combines several metrics in order to select routes having high quality.

The FL decision approach can also be performed in order to reduce the energy consumed. For example, in Ref. [71], authors propose a routing protocol using the FL approach depending on three parameters: the degree of closeness of node to the shortest path, the degree of closeness of node to sink, and the degree of energy balance. Simulation results show that the proposed routing protocol can reduce the energy consumed by 75% when compared to Greedy Perimeter Stateless Routing (GPSR) [72] and minimum transmission energy (MTE) [73]. These two routing protocols (GPSR and MTE) select routes depending only on the location of neighbors and therefore do not balance energy between nodes.

In Ref. [74], authors proposed an adaptive routing strategy taking into account many individual and environmental criteria in order to reduce the traffic congestion and the environmental impact. For example, an individual can choose a trip that contains several touristic areas or provides a safe route avoiding high-criminality areas.

To reduce the traffic congestion and the environmental impact, the route selection has to consider the system state depending on the characteristics of system areas [75,76]. The pattern used in Ref. [74] and located in the city of Milan (Italy) consists of four layers: traffic, pollution, crimes, and events. The traffic layer contains data about the total number of calls and texts generated over a period of 2 months. The pollution layer is based on data obtained by seven sensors. These sensors perform measurements each hour over the course of the past 2 months. The third layer contains 1276 crimes happened during the past year. Finally, the events layer contains 100,000 geolocated tweets generated over a 1-month period. These different layers aim to combine individual level (eg, crimes layer) and global level (eg, pollution layer). Note that Geographic Information System (GIS) can be used to enhance pattern definition.

As the different individuals do not have the same constraints of route selection, variable coefficients of static and dynamic constraints can be defined. Static constraints correspond to restrictions that do not change over time or change over large temporal scales while dynamic constraints correspond to rapid changes within the system itself such as the traffic flow, weather, or accidents.

Simulation results show that the proposed routing strategy decreases time to reach destination. In addition to the efficient route selection, sensors monitor the state of the city in real time and therefore automatically identify areas that are experiencing a temporary congestion and give authorities the possibility to rapidly take action when crimes or accidents occur.

5 POWER-SAVING METHODS

Energy management in smart homes and cities can be performed by different methods. In general, the visualization of power consumption reduces the energy consumption of cities and homes between 10% and 30% [77]. For example, optical reader can be used to read the power indicator of a utility revenue meter. In Ref. [78], Gagnon proposed a system that contains three components: a sensor that monitors the cyclical property of the indication and generates information depending on the energy consumption measurements, a transmitter that sends the information obtained, and a remote display that indicates the energy consumed when receiving information via the transmitter. The main drawback of this method is that the information collected is very limited as it depends only on the cyclical property of the indication.

Smart meters can replace optical reader in order to reduce the energy consumption [79]. The smart meters can access the information in real time and transmit data using wired or wireless networks. Evidently, for flexible use of the smart meter, it is recommended that smart meters enable wireless transmitter. However, the main drawback of smart meters is the low data refreshing rate (in general each 15 min) due to the communication constraints.

Traditional electric current and voltage probes can be installed inside the electrical panel of customers [80]. This method can collect high-quality information for the analysis of nonintrusive load monitoring. However, the installation of the measuring probes is difficult as it requires licensed professionals.

In Ref. [81], the authors propose the use of a new energy-saving method based on the magnetic sensor array technique. The measurement device for monitoring home energy use is presented in Fig. 2.12. It is composed of an array of magnetic sensors, an electric panel, and a conduit (containing conductor currents).

The main advantage of this method is that the installation is easy as there is no electrical contact to the conductors so a homeowner can easily install it. Moreover, data refreshing rate is high (each 1 s). Once the sensor is calibrated using the Power Line Communication (PLC) scheme [82], it collects the current information and then transmits the information collected to a receiver using 433-MHz low-power RF communication. Note that the PLC technology provides a self-contained system. The receiver is connected to a remote Internet server using Local Area Network (Ethernet) where information is analyzed in order to display results in an Internet Website.

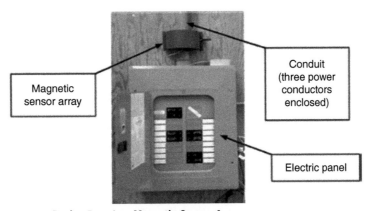

FIGURE 2.12 Measurement Device Based on Magnetic Sensor Array

Based also on the PLC scheme, a home energy management system (HEMS) is proposed in Ref. [83]. The proposed system does not require any additional electric construction as power lines are available in cities and houses. In addition to the energy saving in cities and home appliances, the proposed system takes into account the energy usage of the HEMS itself, in order to decrease the total system cost. The architecture of the proposed HEMS is presented in Fig. 2.13.

Sensors, called smart power trips nodes, perform measurements on the power consumption each 10 ms in order to obtain accurate measurements. Note that the accuracy of an electrical energy consumption monitoring system can be improved using a technique proposed in Ref. [84]. This technique is based on a network of Hall-effect wireless sensors attached to the wire in the electrical distribution panel. In order to mitigate the gain errors due to the distance between the sensor and

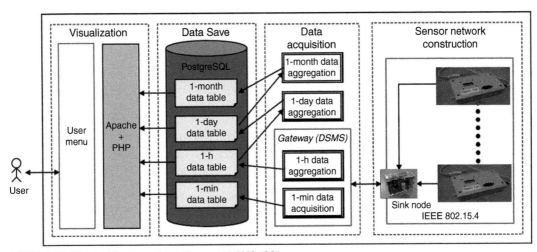

FIGURE 2.13 The Architecture of the Proposed HEMS [83]

the monitored wire, a single high-precision current transformer sensor has to be added at the main electrical input.

To reduce the energy consumption, the sensors send the average measurements obtained each 1 s. Moreover, they use a low-power wireless communication technology (IEEE 802.15.4) for transmitting results to the sink node. Note that the energy consumption can be enhanced when reducing the amount of the transmission of messages Clear to Send (CTS) [85]. When receiving data from the sensors, the sink node forwards it to a data acquisition component that includes a Data Stream Management System (DSMS) for data processing and data writing in the database. Finally, the results are displayed using web interfaces.

A similar architecture composed of three layers is proposed in Ref. [86] for the purpose of designing a zero-energy home using WSNs. The first layer is the sensors layer where it is essential to investigate the possible avenues of harvesting energy in homes and other buildings and so to efficiently localize the energy harvesting. The second layer is the communication and computation layer where the energy generation is analyzed in order to manage and control the energy system using Information and Communication Technologies (ICT). In order to reduce the energy consumption, sensors should have the ability to decide the actions to take in real time [87]. For example, a smart heating control system was adapted to the available sensors opportunistically in order to reduce the energy consumption while providing the comfort temperature [88]. Note that heating represents a great source of energy consumption. For example, in Switzerland, heating represents 65% of the total consumption in residential buildings [89]. The third layer is the storage layer using the small-scale energy storage devices such as batteries stationed in various locations in houses and cities and therefore the storage devices are distributed.

Finally, in order to evaluate the consumption of energy of the sensor itself, a sophisticated energy model has to be defined. This model has to take into account many energy sources such as:

- the radio transmission and reception energy representing the energy consumed for exchanged information;
- the control access energy representing the energy consumed when waiting the liberation of the physical channel and when exchanging control packets such as Request-To-Send (RTS) and CTS packets;
- the sensor sensing energy enabling the sensor to connect to the physical world for monitoring;
- the sensor logging energy representing the energy consumed for writing information in a database;
- the aggregation energy for processing in order to reduce the quantity of information exchanged and to eliminate redundancy;
- the transient energy representing the dissipated energy when the sensor changes from states (reception, transmission, idle, and sleep states [90]). Note that in the sleep state, the sensor consumes the lowest energy. In the reception and transmission states, the sensor is receiving and transmitting data, respectively. In the idle state, the sensor is awake but it is neither transmitting nor receiving its bursts.

Many energy models were proposed in the literature such as in Refs. [91–95]. The energy sources considerations of the energy models are presented in Table 2.3. Evidently, we recommend using the energy model proposed in Ref. [94] in order to use a realistic model.

Table 2.3 Energy Sources Considerations of Energy Models

Energy Sources	Model in Ref. [91]	Model in Ref. [92]	Model in Ref. [93]	Model in Refs. [94,102]
Radio transmission and reception energy	Yes	Yes	Yes	Yes
Control access energy	No	No	No	Yes
Sensor sensing energy	No	Yes	No	Yes
Sensor logging energy	No	No	No	Yes
Aggregation energy	Yes	Yes	Yes	Yes
Transient energy	No	No	Yes	Yes

6 SECURITY

Security in applications for smart homes and cities is almost similar to WSN application environment and therefore the security and privacy concerns are also similar. Security issues in WSNs are presented in Ref. [95] detailing specific and physical attacks as well as security protocols and requirements.

However, the communications in sensor networks applications in smart homes and cities is exposed to more serious threats especially when concerning privacy in homes and vital risks on health of persons. Moreover, in access technologies, many wireless networks can be used simultaneously and thus security in these technologies is a major challenge because of the different characteristics of security architectures used within each wireless network; see Refs. [95–97].

Among the attacks that can be harmful to WSN applications deployed in smart homes and cities, we cite the following:

- *Illegal access*: A user that does not have authorization may access the system when there is no consistent authentication.
- *Data alteration*: An attacker may modify the information transmitted in the network. Moreover, it can access the database and alter stocked information.
- *Fake data injection*: An attacker may insert erroneous information disrupting the system functioning and even causing damages.
- *Denial-of-Service (DoS) attacks*: An attacker may make the network or equipment unavailable by inserting a great amount of information and/or requests.
- *Replay attacks*: An attacker may replay the requests sent with false answers.

Attacks can be classified into passive or active. Passive attacks aim to obtain information without performing actions such as alteration and dropping and therefore it is too difficult to detect this kind of attacks. Active attacks are more harmful than passive attacks as they steal and alter information in addition to causing drops and even blocking of system functionality. For example, active attacks on hospital and healthcare WSN applications may conduce to life-threatening situations [98,103].

Note that any efficient and useful application has to consider the security. For example, a general architecture taking into account the security in the medical environment is proposed in Ref. [99]. Furthermore, the security is considered in all healthcare projects. In the CodeBlue project, in addition to the event delivery, filtering, aggregation, and handoff modules, a flexible naming and discovery scheme as well as authentication and encryption procedures is used (see Fig. 2.14 [26]).

In the WSN4QoL project, the middleware layer contains the security block (see Fig. 2.15 [27]). This block monitors the acknowledgment packets exchanged at the network layer in order to identify threats or equipment malfunctioning and instruct the MAC layer to encrypt data.

Therefore, in order to design robust applications, the following security requirements have to be considered:

- *Confidentiality*: As data is highly confidential especially in smart home applications, it is crucial to ensure the secrecy of data.
- *Integrity*: It is essential to guarantee that data received was not modified by attackers. This can be insured by hashing and digital signature. Error resilience as well as communication and data reliability are vital for medical applications [23], especially under chirurgical operations and emergency situations.
- *Availability*: Applications have to offer availability and therefore absolutely avoid system crash due to DoS or Distributed Denial-of-Service (DDoS) attacks. In fact, a lack of availability leads to several problems for saving lives in healthcare applications [100] or fatal accidents when attacking driving management applications in cities [95,101,103].

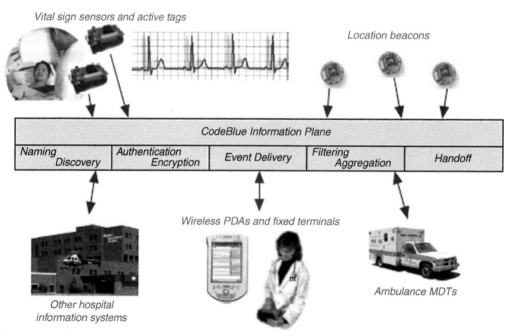

FIGURE 2.14 Main Modules in CodeBlue Project Architecture

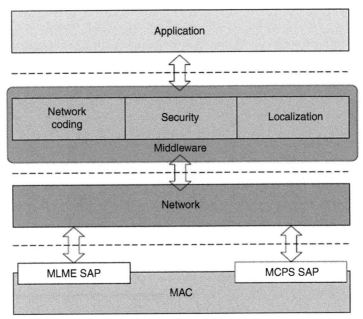

FIGURE 2.15 Protocol Stack Used in the WSN4QoL Project Architecture

- *Secure localization*: Things and persons have to be exactly located for efficient applications without unveiling the localizations for the safety and privacy of users.
- *Secure routing information*: Routes have to be safe and information has to cross only trusted equipment.
- *Intrusion protection and detection*: Any network is susceptible to intrusion and therefore prevention and detection techniques have to be deployed. The system has to be well protected and the source of attack has to be determined for further actions [102].

7 SUMMARY

In this chapter, we have presented different wireless sensor networks applications in smart homes and cities. We have focused on applications for energy saving, noise and atmospheric monitoring, and health-care monitoring. The energy saving is primordial as energy resources have become more and more scarce while the energy use significantly increases. The monitoring of noise, atmospheric, and healthcare aspects certainly makes lives more healthy and comfortable.

To perform these applications, we have presented three access technologies differing in networks and equipment used and therefore presenting various costs and efficiencies. Finally, in order to enhance applications' performance, we have discussed routing strategies, power-saving methods, and security requirements.

In order to enhance WSN applications for smart cities and homes, it is interesting to consider other challenges and techniques such as the use of IPv6 over Low-power Personal Area Network (6LoW-PAN), the determination of efficient services that not only connect sensors to the web but also build a service, and design of software architecture.

REFERENCES

[1] Obaidat MS, Misra S. Principles of wireless sensor networks. Cambridge, UK: Cambridge University Press; 2014.

[2] Energy consumption by sector. US Energy Information Administration (EIA), March 2015 monthly energy review. March 2015.

[3] Louis J-N, Caló A, Pongrácz E. Smart houses for energy efficiency and carbon dioxide emission reduction. In: The fourth international conference on smart grids, green communications and IT energy-aware technologies (ENERGY 2014), Chamonix, France; April 2014.

[4] Attia II, Ashour H. Energy saving through smart home. Online J Power Energy Eng 2011;2:223–7.

[5] Barley D, Deru M, Pless S, Torcellini P. Procedure for measuring and reporting commercial building energy performance. Technical report. National Renewable Energy Laboratory, NREL/TP-550-38601; October 2005.

[6] Corucci F, Anastasi G, Marcelloni F. A WSN-based testbed for energy efficiency in buildings. In: IEEE symposium on computers and communications (ISCC), Kerkyra, Greece; June 2011. p. 990–3.

[7] The Edison Foundation. Utility-scale smart meter deployments. Washington, DC: Institute for Electric Innovation (IEI); September 2014.

[8] Cees Links. After the smart phone: the smart home. White paper. GreenPeak; October 2014.

[9] Santini S, Ostermaier B, Vitaletti A. First experiences using wireless sensor networks for noise pollution monitoring. In: Workshop on real-world wireless sensor networks (REALWSN'08), Glasgow, Scotland; April 2008. p. 61–65.

[10] Bhusari P, Asutkar GM. Design of noise pollution monitoring system using wireless sensor network. Int J Software Web Sci 2013;1:55–8.

[11] Tan W, Jarvis S. On the design of an energy-harvesting noise-sensing WSN mote. EURASIP J Wireless Commun Networking 2014;167:1–18.

[12] Roseline RA, Devapriya M, Sumathi P. Pollution monitoring using sensors and wireless sensor networks: a survey. Int J Appl Innovation Eng Manage 2013;2(7):119–24.

[13] Odey AJ, Daoliang L. AquaMesh – design and implementation of smart wireless mesh sensor networks for aquaculture. Am J Netw Commun 2013;2:81–7.

[14] Richards M, et al. Grid-based analysis of air pollution data. Ecological modeling 2006;194:274–86.

[15] Hassard J, et al. Innovative multi-species sensor development for mobile/portable sensor networks. In: security & resilience for the public & private sectors, London, UK; 2010.

[16] Williams M, Villalonga P. ITI-SENSE: sensor technologies for air quality. In: Citizens' observatory coordination workshop, Brussels, Belgium; January 2013.

[17] Sirsikar S, Karemore P. Review paper on air pollution monitoring system. Int J Adv Res Comput Commun Eng 2015;4(1):218–20.

[18] Yaacoub E, Kadri A, Mushtaha M, Abu-Dayya A. Air quality monitoring and analysis in Qatar using a wireless sensor network deployment. In: The 9th international wireless communications and mobile computing conference (IWCMC), Sardinia, Italy; July 2013. p. 596–601.

[19] Mishra SA, Tijare DS, Asutkar GM. Design of energy aware air pollution monitoring system using WSN. Int J Adv Eng Technol 2011;1(2):107–16.

[20] Mansour S, Nasser N, Karim L, Ali A. Wireless sensor network-based air quality monitoring system. In: 2014 international conference on computing, networking and communications (ICNC), Honolulu, HI, February 3-6, 2014. p. 545–550.

[21] Virone G, et al. An advanced wireless sensor network for health monitoring. In: The transdisciplinary conference on distributed diagnosis and home healthcare, Arlington, VA; April 2006.

[22] Hyojeong S, Talipov E, Hojung C. IPv6 lightweight stateless address autoconfiguration for 6LoWPAN using color coordinators. In: IEEE international conference on pervasive computing and communications. PerCom 2009, Galveston, TX; March 2009. p. 1–9.

[23] Dhobley A, Ghodichor NA, Golait SS. An overview of wireless sensor networks for health monitoring in hospitals via mobile. Int J Adv Res Comput Commun Eng 2015;4(1):169–71.

[24] Oliver N, Flores-Mangas F. HealthGear: a real-time wearable system for monitoring and analyzing physiological signals. In: International workshop on wearable and implantable body sensor networks, Cambridge; April 2006. p. 64–7.

[25] Gautam KK, Gautam SK, Agrawal PC. Impact and utilization of wireless sensor network in rural area for health care. Int J Adv Res Comput Sci Software Eng 2012;2(6):93–8.

[26] Malan D, Thaddeus FJ, Welsh M, Moulton S. CodeBlue: an ad hoc sensor network infrastructure for emergency medical care. In: Workshop on applications of mobile embedded systems (WAMES 2004), Boston, MA; June 2004.

[27] Tennina S, et al. WSN4QoL: a WSN-oriented healthcare system architecture. Int J Distributed Sensor Netw 2014;2014:1–16.

[28] Kouroubali A, Chiarugi F. Developing advanced technology services for diabetes management: user preferences in Europe. In: Wireless mobile communication and healthcare, Mobihealth, Kos Island, Greece; October 2011. p. 69–74.

[29] MobiHealth in Zambia: focus on telemonitoring high risk pregnant women. Technical report. MobiHealth; August 2014.

[30] Milenković A, Otto C, Jovanov E. Wireless sensor networks for personal health monitoring: issues and an implementation. Comput Commun 2006;29:2521–33.

[31] Institute of Electrical and Electronics Engineers. IEEE Std 802.15.1-2005, wireless medium access control (MAC) and physical layer (PHY) specifications for wireless personal area networks (WPANs); June 2005.

[32] Institute of Electrical and Electronics Engineers. Part 15.3: wireless medium access control (MAC) and physical layer (PHY) specifications for high rate wireless personal area networks (WPAN); September 2003.

[33] Institute of Electrical and Electronics Engineers. IEEE Std 802.15.4-2006, wireless medium access control (MAC) and physical layer (PHY) specifications for low-rate wireless personal area networks (WPANs); September 8, 2006.

[34] Fredman A. Mechanisms of interference reduction for Bluetooth. pdf ebooks; February 2015.

[35] Khair MAI, Misic J, Misic VB. Piconet interconnection strategies in IEEE 802.15.3 networks. In: Zhang Y, Yang LT, Ma J, editors. Unlicensed mobile access technology: protocols, architectures, security, standards and applications. Boca Raton, FL: Auerbach Publications; 2009. p. 147–62. [chapter 8].

[36] Goratti L, Haapola J, Oppermann I. Energy consumption of the IEEE 802.15.3 MAC protocol in communication link set-up over UWB radio technology. Wireless Personal Commun 2007;40:371–86.

[37] Stabellini L, Parhizkar MM. Experimental comparison of frequency hopping techniques for 802.15.4-based sensor networks. In: The fourth international conference on mobile ubiquitous computing, systems, services and technologies (UBICOMM); 2010. p. 110–6.

[38] Popovski P, Yomo H, Prasad R. Strategies for adaptive frequency hopping in the unlicensed bands. IEEE Wireless Commun 2006;13(6):60–7.

[39] Hsu AC-C, et al. Enhanced adaptive frequency hopping for wireless personal area networks in a coexistence environment. In: Global telecommunications conference (GLOBECOM), Washington, DC, USA; November 2007. p. 668–72.

[40] Stabellini L, Shi L, Rifai AA, Espino J, Magoula V. A new probabilistic approach for adaptive frequency hopping. In: The 20th IEEE international symposium on personal indoor and mobile radio communications (PIMRC), Tokyo, Japan; September 2009. p. 2147–51.

[41] Institute of Electrical and Electronics Engineers. 802.11a-1999 high-speed physical layer in the 5 GHz band; October 1999.

[42] 802.11a. White paper. VOCAL Technologies; May 2012.

[43] Institute of Electrical and Electronics Engineers. 802.11b-1999 higher speed physical layer extension in the 2.4 GHz band; October 1999.

[44] Institute of Electrical and Electronics Engineers. IEEE 802.11g-2003: further higher data rate extension in the 2.4 GHz band; October 2003.

[45] Institute of Electrical and Electronics Engineers. IEEE 802.11n-2009 – amendment 5 enhancements for higher throughput; October 2009.

[46] Jun J, Peddabachagari P, Sichitiu M. Theoretical maximum throughput of IEEE 802.11 and its applications. In: Second IEEE international symposium on network computing and applications (NCA 2003), Cambridge, MA, USA; April 2003. p. 249–56.

[47] Wang X, Kar K. Throughput modelling and fairness issues in CSMA/CA based ad-hoc networks. In: The 24th annual joint conference of the IEEE Computer and Communications Societies (INFOCOM 2005), Miami, FL, USA; March 2005.

[48] Khanduri R, Rattan SS, Uniyal A. Understanding the features of IEEE 802.11g in high data rate wireless LANs. Int J Comput Appl 2013;64:1–5.

[49] Understanding and optimizing 802.11n. Technical report. Buffalo Technology; July 2011.

[50] Khanduri R, Rattan SS. Performance comparison analysis between IEEE 802.11a/b/g/n standards. Int J Comput Appl 2013;7:13–20.

[51] Lagrange X, Godlewski P, Tabbane S. Réseaux GSM. 5th ed. Paris: Hermes; September 2000.

[52] Seurre E, Savelli P, Pietri P-J. Gprs for mobile internet; December 2002.

[53] Halonen T, Romero J, Melero J. Gsm, Gprs and edge performance: evolution towards 3G/Umts. 2nd ed. West Sussex, UK: John Wiley & Sons; October 2003.

[54] Huber JF, Huber AJ. Umts and mobile computing; March 2002.

[55] Damnjanovic A, Vojcic B, Vanghi V. The CDMA2000 system for mobile communications: 3G wireless evolution; July 2004.

[56] Holma H, Toskala A. HSDPA/HSUPA for UMTS: high speed radio access for mobile communications; April 2006.

[57] Kafka H. Wireless broadband. Technical report. AT&T; November 2007.

[58] Realistic LTE experience: from peak rate to subscriber experience. White paper. Motorola; August 2009.

[59] 3GPP. Release 13 analytical view. RP-151569; September 2015.

[60] Saha BK, Misra S, Obaidat MS. A web-based integrated environment for simulation and analysis with NS-2. IEEE Wireless Commun 2013;20(4):109–15.

[61] The network simulator NS-2, <http://www.isi.edu/nsnam/ns/>, last visited in June 2015.

[62] Ortiz AM, et al. Intelligent routing strategies in wireless sensor networks for smart cities applications. In: IEEE international conference on networking, sensing and control, Evry, France; April 2013.

[63] Prim RC. Shortest connection networks and some generalizations. Bell Syst Tech J 2013;36:1389–401.

[64] Chiang S-Y, Wang J-L. Routing analysis using fuzzy logic systems in wireless sensor networks. Berlin, Heidelberg: Springer-Verlag; 2008. p. 966–73.

[65] Ortiz AM, Royo F, Galindo R, Olivares T. I3ASensorBed: a testbed for wireless sensor networks. Technical report. Albacete Research Institute of Informatics; December 2011.

[66] Fischer S. WISEBED – experimental facilities for wireless sensor. Technical report. University of Lübeck; December 2010.

[67] Albesa J, Casas R, Penella MT, Gasulla M. Realnet: an environmental wsn testbed. In: International conference on sensor technologies and applications, SENSORCOMM'07; 2007.

[68] Handziski V, Köpke A, Willig A, Wolisz A. Twist: a scalable and reconfigurable testbed for wireless indoor experiments with sensor networks. In: The 2nd international workshop on multihop ad hoc networks: from theory to reality; 2006.

[69] Karagiannis M, Chantzis K, Nikoletseas S, Rolim J. Passive target tracking: application with mobile devices using an indoors WSN future Internet testbed. In: The IEEE DCOSS workshop on building intelligence through IPv6 sensing systems, HOBSENSE; 2011.

[70] Ju X, Zhang H, Sakamuri D. NetEye: a user-centered wireless sensor network testbed for high-fidelity, robust experimentation. Int J Commun Syst 2012;25:1213–29.

[71] Jiang H, Sun Y, Sun R, Xu H. Fuzzy-logic-based energy optimized routing for wireless sensor networks. Int J Distributed Sensor Netw 2013;2013:1–8.

[72] Karp B, Kung HT. GPSR: greedy perimeter stateless routing for wireless networks. In: The 6th annual international conference on mobile computing and networking (MOBICOM '00); August 2000. p. 243–54.

[73] Heinzelman WR, Chandrakasan A, Balakrishnan H. Energy-efficient communication protocol for wireless microsensor networks. In: The 33rd annual Hawaii international conference on system sciences (HICSS'00); January 2000. p. 1–10.

[74] De Domenico M, Antonio L, Gonzales MC, Arena A. Personalized routing for multitudes in smart cities. EPJ Data Sci 2015;4:1–11.

[75] Wang P, et al. Understanding road usage patterns in urban areas. Scientific report. December 2012.

[76] Othmen S, Belghith A, Zarai F, Obaidat MS, Kamoun L. Power and delay-aware multi-path routing protocol for ad hoc network. In: Proceedings of 2014 IEEE international conference on information, computer and telecommunication systems, CITS 2014, Jeju Island, Korea; July 2014.

[77] Fischer C. Feedback on household electricity consumption: a tool for saving energy? Energy Efficiency 2008;1:79–104.

[78] Gagnon S. System and method for reading power meters. World Intellectual Property Organization (WIPO); October 2005.

[79] Lui T, Stirling W, Marcy H. Get smart. In: IEEE power energy magazine, vol. 8; June 2010. p. 66–78.

[80] Wallich P. Parsing power. IEEE Spectr 2013;50:23–4.

[81] Gao P, Lin S, Xu W. A novel current sensor for home energy use monitoring. IEEE Trans Smart Grid 2014;5(4):2021–8.

[82] Son Y-S, Pulkkinen T, Moon K-D, Chaekyu K. Home energy management system based on power line communication. IEEE Trans Consumer Electron 2010;56:1380–6.

[83] Hashizume A, Mizuno T, Mineno H. Energy monitoring system using sensor networks in residential houses. In: The 26th international conference on advanced information networking and applications workshops (WAINA), Fukuoka, Japan; March 2012. p. 595–600.

[84] Beaufort Samson G, Levasseur M-A, Gagnon F, Gagnon G. Auto-calibration of Hall effect sensors for home energy consumption monitoring. Electron Lett 2014;50:403–5.

[85] Panagiotakis A, Melidis P, Nicopolitidis P, Obaidat MS. Power-controlled reduction of exposed terminals in ad-hoc wireless LANs. In: Proceedings of the 2012 IEEE international conference on computer, information and telecommunication systems, CITS 2012, Amman, Jordan; May 2012. p. 122–6.

[86] Prasad RV, Rao VS, Niemegeers I, de Groot SH. Wireless sensor networks for a zero-energy home. MOBILIGHT 2011, LNICST 81; 2012. p. 338–46.

[87] Mattern F, Staake T, Weiss M. ICT for green – how computers can help us to conserve energy. In: The 1st international conference on energy-efficient computing and networking (e-Energy 2010), Passau, Germany; April 2010. p. 1–10.

[88] Kleiminger W, Beckel C, Santini S. Opportunistic sensing for efficient energy usage in private households. In: The smart energy strategies conference, Zurich, Switzerland; September 2011.

[89] Kemmler A, et al. Analysis of energy consumption by specific use. Technical report. Swiss Federal Office of Energy; September 2014.

[90] Belghith A, Belghith A, Molnar M. Enhancing PSM efficiencies in infrastructure 802.11 networks. Int J Comput Inf Sci 2007;5:13–23.

[91] Heinzelman WR, Chandrakasan A, Balakrishnan H. An application-specific protocol architecture for wireless microsensor networks. IEEE Trans Wireless Commun 2002;1:660–70.

[92] Zhu J, Papavassiliou S. On the energy-efficient organization and the lifetime of multi-hop sensor networks. IEEE Commun Lett 2003;7:537–9.

[93] Mille MJ, Vaidya NH. A mac protocol to reduce sensor network energy consumption using a wakeup radio. IEEE Trans Mobile Comput 2005;4:228–42.

[94] Halgamuge MN, Zukerman M, Ramamohanarao K. An estimation of sensor energy consumption. Electromagn Res B 2009;12:259–95.

[95] Othmen S, Zarai F, Obaidat MS, Belghith A. Re-authentication protocol from WLAN to LTE. In: IEEE global communication conference, GLOBECOM 2013, Atlanta, GA, USA; December 9–13, 2013.

[96] Ellbouabidi I, Ben Ameur S, Smaoui S, Zarai F, Obaidat MS, Kamoun L. Secure macro mobility protocol for new generation access network. In: The 2014 IEEE international on wireless communications and mobile computing conference, IWCMC 2014, Nicosia, Cyprus; 2014. p. 518–23.

[97] Stankovic JA, et al. Wireless sensor networks for in-home healthcare: potential and challenges. In: high confidence medical device software and systems (HCMDSS) workshop. Department of Computer Science; 2005.

[98] Jara AJ, Zamora MA, Skarmeta AFG. An architecture based on internet of things to support mobility and security in medical environments. In: The 7th IEEE consumer communications and networking conference (CCNC'10); January 2010. p. 1–5.

[99] Huang YM, et al. Pervasive, secure access to a hierarchical sensor based healthcare monitoring architecture in wireless heterogeneous networks. IEEE J Selected Commun 2009;27:400–11.

[100] Tavares J, Velez FJ, Ferro JM. Application of wireless sensor networks to automobiles. Meas Sci Rev 2008;8:65–70.

[101] Fatema N, Brad S R. Security requirements, counterattacks and projects in healthcare applications using WSNs – a review. Int J Comput Netw Commun 2014;2:1–9.

[102] Nicopolitidis P, Obaidat MS, Papdomitriou G, Amprtsis A. Wireless networks. Hoboken, NJ: Wiley; 2013.

[103] Obaidat MS, Boudriga N. Security of e-systems and computer networks. Cambridge, UK: Cambridge University Press; 2007.

SOFTWARE DEFINED THINGS: A GREEN NETWORK MANAGEMENT FOR FUTURE SMART CITY ARCHITECTURES

3

Ö.U. Akgül, B. Canberk

Computer Engineering Department, Istanbul Technical University, Ayazaga, Istanbul, Turkey

1 INTRODUCTION

The existing city applications mostly depend on reactive human-centralized controls. However, the infeasibility of this framework for the future's denser and more complicated cities is obvious [1]. In order to overcome this infeasibility, the real-time monitoring data of wireless sensor nodes (WSNs) and the user feedback are added to control cycle of the cyber physical devices (CPD) within the cities. Maintaining connections between CPDs, WSNs, and users brings out the idea of smart cities. Even though there is not a complete definition of the smart cities, the most common aspect of it is the connections between different subsystems of the city, for example, surveillance system and the traffic control system. This completely connected structure increases the computational and physical resource demand [2]. This new approach to the city management integrates the machine-centralized state detection–failure prevention model and user control. More specifically, the automatized systems continuously monitor the behaviors of their systems and detect the state changes and the anomalies in their systems. The control framework tries to predict the next failure in the system and tries to prevent it. Such a control framework presents a continuous observation on the parameters and machine-based parameter detection and processing that handles a far larger amount of parameters. Unlike digital city management, smart cities present more accurate and more efficient failure detection and recovery specifications. The participation of the user aspects to this autonomous structure improves the user experience and enables exceptional action selection. From this point of view, the smart cities can be investigated under two major regions, that is, data plane and the control plane of the city.

First of all, the design and implementation of the control plane is the most challenging part of the smart city deployment. The future's smart cities will cover large variety of devices, for example, WSNs, CPDs, Internet of Thing (IoT) devices, and mobile users. This huge variety of devices will use different technologies to connect to the Internet, as seen in Fig. 3.1. The heterogeneity of smart city devices (SCDs) will bring the scalability and interoperability problems [4]. The network management structure has to maintain the efficiency and the success of the communication between different devices and systems. The heterogeneity of the device types and their protocols should not affect the integrity of the city scope network [5]. Moreover, the intermediary steps of the data gathering, processing, and performing actions should be invisible to the users. More specifically, the machine-centralized

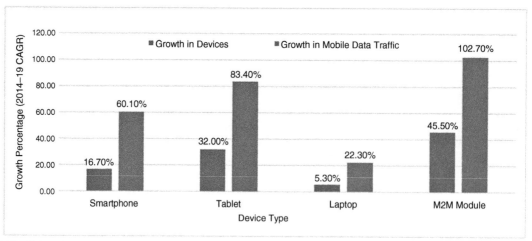

FIGURE 3.1 Comparison of Global Device Unit Growth and Global Mobile Data Traffic Growth [3]

control framework demands an abstraction of the users from the system parameters. For instance, when the control framework tries to optimize the temperature at a home, the user should not be concerned about the sensor data, for example, the thermometer value or the state of the air conditioner. From a user perspective, the main objective is receiving the ultimate result without concerning the intermediary steps. The communication between cyber physical systems would remove the necessity of a user control in the midlevel operations. For example, the communication between air conditioner and the number of thermometers around the room would remove the user-based iterative control process in order to keep the heat higher than a specific value. Another important challenge in control plane is the management of gathered data. The vast amount of SCDs continuously collect user and environment-specific and usually private data. Processing such a huge amount of data and storing the results are the main challenges [6].

Another aspect of the data collection process is maintaining the data security. The collected private data will face many possible attacks. Due to these malicious attempts, the service providers have to enable a secure environment [7]. Additionally, interoperability is another challenge in the smart cities [8]. The smart city infrastructure contains a huge number of switches and network equipment that have different objectives and use different operating systems (OSs). The control framework should be able to perform efficient operations using these devices. Finally and most importantly, the network control framework should be able to adjust the power consumption and present an energy-efficient network topology without a performance degradation [9]. The energy efficiency downfall is usually due to the controller's lack of capability in following the dynamic needs of the network. In order to maintain the Quality of Service (QoS) requirements, the network providers usually overprovide the network resources. This mismatch between resource demands and supply causes an unnecessarily high power dissipation. The network controller needs dynamic traffic observations and adaptive resource allocations [10].

Second, every kind of monitoring and performing devices including the mobile devices and CPDs forms the data plane of the smart network. Recently, the sensor devices and controllers are used in basically every layer of the city management from surveillance systems to farming techniques. In addition to the city-related sensor devices, consumer electronics are also increasing in the cities. The decrease in the electronic device manufacturing costs increased the consumer rate that uses at least one smart electronic device that can communicate through the Internet. By definition SCD covers basically every electronic device that carries a radio-frequency (RF) receiver, for example, intrasensor and intersensor networks, smartphones, smart buildings, or wearable electronics. As a generic approach, SCD types are divided into two main groups, that is, reactive and active devices. Reactive devices are dumber devices that only receive some information and act something. In contrast with reactive devices, active SCDs are smarter devices that are capable of pursuing a complete communication between other SCD or human operators and perform required actions. The generic objective of the reactive SCDs (R-SCDs) is increasing their energy efficiency without a sacrifice of performance [11]. The active SCDs have unique and application-based performance metrics and the power management usually has secondary importance [12].

2 SMART CITY DATA PLANE CHALLENGES

Smart city network proposes a massive capacity and throughput improvement and a better QoS in terms of latency, battery consumption, device cost, and reliability. In order to manage the vast amount of devices, the network densification and the autonomous network management frameworks are considered. The idea of connected SCDs would be effective in many aspects. The control and preventative applications would eliminate possible dangers and failures before they actually happen. Abstraction of the intermediate steps would increase the user satisfaction. Moreover, connection between different kinds of devices enables the determination of a more realistic approach toward the physical processes. The correlations between different data could be processed more accurately and effective models could be applied. However, the application of SCD network contains many challenges.

2.1 COMPATIBILITY BETWEEN SMART CITY DEVICES

In order to maintain connection with the generalization, the communication in a SCD network topology can be modeled with a few subroutines, that is, the communication between SCD and the network element (NE; switch), the communication between different NEs, and finally the communication between the NE and the controller. The controller is the device that tries to use the collected raw data from the SCD network. The connection between the NE and the controller would not be different from the recent communication framework. However, the communication between the NE and the SCD and the results of this communication cause the main challenge. Future's smart cities contain huge number of smart device networks, which will produce enormous amount of raw data. Since most of the SCDs are application based designed and dedicated to certain objectives, they all contain different structures and capabilities [13]. In the communication level, a certain layer structure is expected for these SCDs that demands a certain electronic capabilities. However, such a demanding framework will increase the expenditures on both design and implementation steps. For example, since the future's smart cities will have billions of sensor nodes, making these devices more complex than they need to be will increase both Capital Expenditures (CAPEX) and operational expenditures (OPEX) that will eventually make this sensor-based observation idea unrealistic. However, using application-defined simple and cheaper SCD would cause device to switch compatibility issues. The existing infrastructure demands a certain layered structure for these devices [14].

Additionally, the SCD networks contain basically all kinds of consumer electronics. Most of these devices carry plug and play specification [2]. So the NEs have to detect the existence of a new SCD in their network and have to determine a new set of rules for this new device. Such a determination process would also be impossible due to the hardware limitations. The NEs would need to be more complex that will increase the CAPEX and also more complexity would make them open to failure. Moreover, the communication infrastructure will also be demanding. In addition to the heterogeneity in SCD network, the communication infrastructure also contains NEs from different manufacturers that carry different embedded rules and applications. The system designers have to consider all that heterogeneity of network plane. As the number of SCD increases and they communicate more frequently, the massive network load will trigger the deployments of more NEs and make the heterogeneity problem more crucial. In the conventional communication infrastructure the controllers have to recognize a vast amount of switches. Apart from the heterogeneity between devices, the standardization heterogeneity between different countries and even cities will be problematic.

2.2 SIMPLICITY

The heterogeneity of SCDs and dynamic nature of them causes additional complexity issues. Since many SCDs are application-based and dedicated devices, they demand specific applications and special processes for each case. These demands cause complexity in management as the recent switches carry their control plane (their OS) inside them with a hard link to their data plane. This hard link prevents the implementation-based applications to be applied to the switches. This challenge brings out the necessity of simplicity in management. In order to overcome the management problem, software-based application control is necessary. The engineers should be able to integrate specific functions to NEs in order to handle this application-based SCD network. The design of the active networks is a result of this research. By theory, the software engineers would transmit specific packets to NEs that would add special application functions to these devices. However, it is not feasible for a large network as the engineers have to control the whole network manually. This manual control will also bring static characteristics to the network, as the network will lose its ability to adapt itself for anomalies autonomously. Moreover, the security is also a problem as wrong or bad applications would be injected to NEs. Additionally, the bandwidth and storage requirements complicate the hardware design of switches. Most of the SCD applications demand simple yet important preprocessing stages that would actually determine the packet's destination.

From the network management point, the network administrator has to consider system issues such as efficiency, fairness, and load balancing that results in high complexity. The key network performance metrics, that is, latency, reliability, speed, mobility, or spectrum usage, have to be considered in the network configurations. Since SCD would demand very different QoS requirements, the necessary algorithms to detect and respond to these demands will became more complex. However, as the complexity of the applied algorithms increases, the speed of the optimal configuration will decrease. This trade-off between configuration speed and complexity of the applied algorithms also appears in the power management. Higher complexity usually means more calculations and this leads to inefficient battery management for NEs. Design and implementation process of SCD network management devices is also challenging. By definition, SCD covers basically every electronic device that can connect to the network. This broad device coverage complicates the communication structure and the necessary protocols. Moreover, SCDs can demand specific applications or routines. Programmers should be able to implement their application libraries and programs to the network controllers. In the existing topology management framework, since each device carries its own OS and the necessary applications, application-specific functions and programs have to be integrated at the design process of the controllers or switches. This persistent structure is also infeasible for SCD networks as it basically demands the change of whole infrastructure at each device update.

2.3 MOBILITY AND GEOGRAPHIC CONTROL

In the future, smart cities will be depending on SCDs to perform various tasks, that is, perception, observation, execution, etc. Due to this dependency, the smart devices will be distributed in a large geographic region. These distributed devices continuously produce a vast amount of data that should be transmitted between these devices simultaneously. This massive packet load brings out the necessity of a high communication capacity. As these devices travel through a large region, they use many switches and links. In conventional network topology, each switch makes a decision according to a local knowledge. However, a possible link failure pushes the switch to make a new route to transfer the packet.

Their broad task definition will make the smart devices to be distributed in a large geographic region. As the geographic coverage area of an SCD network grows, the number of NEs between SCD and the controller will increase. In the conventional network topologies, each NE makes its own routing tables. When it receives a packet, based on the embedded rules and the specifications, the NE decides where to route the packet. However, each NE could have a local knowledge without a complete understanding of the network topology. This lack of generality makes it weak to the possible failures. For instance, a possible link failure will push the NE to make a new route to transfer the packet. This fluctuation in packet's route would have enormous effects on the latency and jitter. In real-time applications of SCD network, such QoS problems would lead to not only bad user experience but also wrong decisions. Especially for the medical applications of SCD, such misjudgments could end in fatal situations. Such problems could only be solved with the usage of a dynamic traffic engineering. In addition to the necessity of traffic management, the network management should have a strong mobility management. The smart devices can join the network in very broad areas and can move rapidly. For instance, in a commercial SCD network application, for example, fleet management, the company may demand the instant locations and the fuel consumptions of their truck fleet. In their trade route, the smart devices in truck will pass many switches and the communication has to be kept continuous. In order to perform such a strong mobility control a centralized control mechanism and a complete view of global network topology are necessary.

3 SOFTWARE DEFINED NETWORK-BASED SMART CITY NETWORK MANAGEMENT

The main challenges of the future applications of SCD networks in smart cities can be listed as follows:

- lack of central control that has a complete information of the topology;
- integration and application simplicity;
- excessive amounts of system needs;
- compatibility between SCDs and NEs.

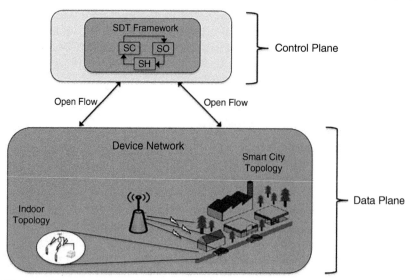

FIGURE 3.2 SDN-Based Network Control; Software Defined Things

In order to overcome these challenges software-centered control structures need to be developed. In this development process the Software Defined Network (SDN) is a promising framework. SDN proposes a logically centralized framework that separates the control and the data planes. The connections between these two separate planes are performed by the Open Flow protocol (Fig. 3.2).

3.1 CENTRALIZED CONTROL

One of the most important specifications of the SDN is the separation of the data and the control plane. The data plane contains dummy switches, whereas the control plane covers all the OSs. The control plane presents a centralized control over the data plane that contains lots of NEs. This logically centralized control framework has complete information of the network topology, that is, the distribution of NEs in 2D space and their workloads. Therefore, SDN has the capability to distribute the workload fairly and optimize the network throughput [15]. Moreover, SDN's control plane is also in connection with the user-defined applications that enables the applications to remotely connect the network devices. Due to this connection, it allows specific applications to optimize the network resources according to their needs. Applications can directly access the control plane to gather information about the network state or make specific reservations for network resources such as bandwidth or changing priorities of specific IP addresses.

Additionally, the logically centralized controller enables centralizing the policy control for all devices on the network. In the recent network infrastructure, to deploy a new network policy or a new protocol, all network equipment has to be updated separately. This discrete upgrade process is not applicable since it is not feasible to upgrade each device separately and it also increases the OPEX. Thanks to its logically centralized control plane, SDN allows easy deployment of new protocols and network services as a result of high operation abstraction. Moreover, since the controller can determine the network state perfectly, it allows dynamic application management. As previously stated, one of the most important reasons of the SCD deployment in smart cities is the necessity of continuous and complete monitoring of the SCD. However, in order to enable successful power management, the monitoring of noncritical parameters may be scaled down in off-peak times in the day, for example, night times.

3.2 SIMPLICITY AND INERRABILITY

One of the most crucial problems of the classical networks is their persistent structure. The classical NEs use their embedded functions and specifications to determine an action to the received packet. However, this persistency of the embedded software structure constraints the possible applications especially in SCD networks. SCD networks are designed in an application-specific manner. This design paradigm requires different actions for different cases. Even some SCD packets may need preprocessing stages or packets received from a specific SCD may need to send to different controllers to manage the situations. For instance, a heat sensor may send the data to a predicting controller until the measured data stays below a certain value, $T_{critical}$, whereas it sends the packets to an action controller after the heat exceeds this critical value. These kinds of implications are application-specific architectures and implementing them into the SCD would be infeasible and expensive. However, the SDN model enables this kind of application-specific actions.

The application plane of the SDN proposes a specific action space for each different implementation. Any software engineer can write an application for the SDN controller that can define a new set of actions for specific cases. The SDN controller is challenging in terms of security. First of all, the collected data needs to be encrypted and also the admission process has to be revised in order to improve privacy. Moreover, the application integration ability is challenging in terms of security. The attackers can manipulate the controller by submitting their malicious applications. Nevertheless, this kind of attacks is prevented by defining the usable primitives and the templates to write an application. By this way, the system provider could passively control the extent of the user applications. The SDN proposes an open and simple programmable network topology. Implementing new functions or new specifications to the network controller is as simple as downloading a new application to the network controller [16]. This simplicity in the distribution of the new applications and libraries presents an ease of use for the users. The generic network topologies demand licensed programs and unique, application-specific libraries to meet the application requirements. This method increases the capital and the OPEX. However, the simplicity that SDN presents enables the programmers to implement generic programs and libraries and openly share these applications with the other users.

Finally and most importantly, SDN presents the concept of "reconfiguration on the fly" specification [17]. The newly applied actions will be available while the network is actually being used. In SDN, dummy switches carry routing tables that is similar to the classical NE. However, in SDN, the newly received and unknown packets are buffered and an action request is sent to the control plane using Open Flow. Using the applications and the libraries, the controller chooses a new set of action and sends it to the dummy switch. As this mechanism is repeated for each unknown packet, the new applications and new action sets will be able while the system is running. By this "reconfiguration on the fly" capability, the optimization of the applications performance will be simpler due to the autonomous network provisioning and related configuration challenges. Moreover, SDN supports scalable and effective programming with diverse device services and raw data streams [13].

3.3 VIRTUALIZATION

Abstraction is a critical challenge in both network applications and SCD applications. The Abstraction within the smart city has two major parts. First, the abstraction of the services is important to improve user experience. One of the most important aspects of the smart cities is improving the quality of the user life. In order to manage this aspect the involvement of the users to the network management is necessary by presenting specific data and necessary feedbacks. But being a part of the control cycle would not imply involving in each state. The system would need to operate lots of functions and optimization processes in order to maintain an efficient and sustainable network topology. Human involvement in the intermediate steps would make the control framework open for failure and slow down the optimization process. Instead of demanding user control on each state, the network controller should accept the user-specific data as a network objective and the user feedback as a generic network state.

Second, the network resources and the network applications should be virtualized from each other. In order to maintain security and increase the user experience, the user-demanded applications that are implemented to the network controller should not notice the system resources they used and the constraints of the system. Most of the SDN applications are using cloud-based solutions, which presents a virtual infinite calculation capacity and service rate. These virtual resources increase the network service efficiency while easing the design and implementation of new service applications and decrease the necessary time to develop these applications. Moreover, the virtualization of discrete and usually geographically distant resources provides more control to the SDN over the system resources while decreasing the CAPEX and OPEX.

3.4 COMPATIBILITY

The increasing number of mobile devices, the participation of CPDs to the wireless network, and the increasing demand for performance improvement lead to the deployment of SDN-based smart city networks. It proposes a massive capacity and throughput improvement and a better QoS in terms of latency, battery consumption, device cost, and reliability. In order to manage the vast amount of devices, the network densification and the autonomous network management frameworks are considered. Despite the promises of the SDN-based smart city networks, many challenges are also coming to light.

In the future's smart city applications, vast amount of SCD networks will be applied. Such an enormous number of SCD will continuously produce raw monitoring and control data that will be transferred by delay-bounded QoS requirements through a large number of NEs. The compatibility between SCD and NEs and also between NEs is crucial. SCD concept covers a large variety of devices. Due to this reason, it contains many different OSs and protocols. Enabling a flat structure in terms of communication protocols and OS is not a feasible solution because of CAPEX and OPEX. The interoperability and the harmony in communication should be provided from the communication infrastructure. Moreover, especially interdevice communication should be autonomously managed and adjusted in terms of necessary communication protocol. Nevertheless, the interconnection between different network equipment is also challenging. Like SCDs, network equipment concept covers a large variety of switches. Different producers usually present different communication sequences and primitives and the network providers should be able to handle all these different protocols. In terms of network infrastructure the problem has two stages.

First of all, the compatibility between existing infrastructures needs to be managed. The geographic growth of the cities increased the distance between two connected devices. In order to reach destination

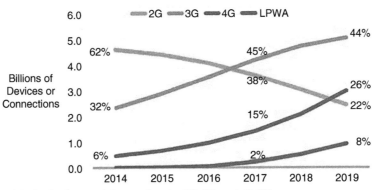

FIGURE 3.3 Global Mobile Devices and Connections by 2G, 3G, and 4G [3]

device, the packet has to pass many switches. The compatibility between different switches becomes a major aspect in order to achieve a fully integrated network. Second, the compatibility between existing infrastructure and the newly deployed infrastructure is also crucial. The number of devices in the network is increasing exponentially (Fig. 3.3). In order to sustain the connectivity, new network equipment is needed to be incorporated into the network infrastructure. However, the changing and increasing needs of users and service providers trigger the changes in the network equipment design. Due to this difference, the newly applied infrastructure should be able to cooperate with the existing infrastructure while performing specific tasks and actions.

SDN presents a simple yet efficient methodology to overcome the compatibility issues. In the generic network management frameworks, the network equipment contains both the data plane and the control plane. It has its own objectives, action sets, and algorithms. The interoperability becomes a problem due to this self-operable structure. SDN divides the control plane and the data plane. The data plane contains SCDs and dumb switches, whereas the SDN controller has the control plane. All network equipment will be dumb devices that do not carry specific action sets or objectives. The controller has these objectives and using the Open Flow protocol, it adjusts the actions of the switches. This division of control and data plane degrades the interoperability issues as the dumb switches will demand application-based assignments from the controller. Additionally, the SCDs would not need to carry any specific structure since the received data can be preprocessed by the SDN controller. As the SDN is a software-based control framework, the network architecture would be dynamically adjusted according to the needs of the network and the user expectations.

3.5 CHALLENGES OF SDN IN SMART CITY APPLICATIONS

Separation of control and the data planes can fulfill many expectations from the next-generation smart city networks such as simplicity in integration, development, and compatibility within a heterogeneous network. However, there are lots of open issues. First of all, as in all complex networks, the resource management is a challenging task. In future's smart cities, the network topology will be dynamically changing. Because of this dynamic nature, the devices will be competing for network resources. Even though the SDNs virtualize the network resources, an adaptive and fair management algorithm will be necessary. Determination of user expectations is an essential part of the network design. Different users would produce different types of data that would demand unique processes (eg, monitoring, prediction) and face various types of problems (eg, latency). The network controller has to carry the specifications of self-analyzing and dynamic configuration of the network resources while handling with a mix of network objectives. Increasing the user satisfaction from the network services by using a simple, robust, and low-cost network controller is a major challenge in future's smart cities.

In addition to the wireless resources in the network, the power consumption is another challenging issue to be handled. The power consumption within the network should be investigated under two major topics, that is, energy efficiency of the backhaul system and the energy efficiency of the network topology. According to the latest researches, the power consumption of the network is mainly because of the network infrastructure. A centralized control algorithm that can optimize the network topology to maximize the energy efficiency is relatively basic for SDN. However, most of the devices carry plug and play specification that changes the network topology rapidly. Due to that, the network management framework has to support autodiscovery capability as well as adaptive power control mechanism. The dynamic mode selection would perform adaptive energy consumption specification; however, the offloading techniques to the underlay network have to be investigated. The energy efficiency of the overall network topology is another challenge in the power management framework. The future's smart cities will be counting on SCDs in monitoring and analyzing. However, SCDs are battery-driven objects and the durability of these devices will be the main concern of the network management. Therefore, the network management will need to control the network topology to increase the lifetime of these devices. On another perspective, the decision process of where to locate these devices is another problem. A brute force solution is to use a vast amount of WSN and CPD devices to cover the whole geographic region. However, this solution will result in massive capital and OPEX. A lesser coverage would lead to a more cost-efficient application; nevertheless, it may cause low performance.

4 SOFTWARE DEFINED THINGS FRAMEWORK

The application of SDN in SCD management is an effective solution to the infrastructure-based problems. In a smart city, many services and applications are going to be observed with SCD and will be sent to the SCD controllers. Even though the SCD applications would demand a complete communication with the controller, R-SCDs will not require any specific response from the controller. Instead, they will demand only a power optimization framework to increase their lifetime. In a classical network, this optimization can be solved in the controller. The controller can provide a complete on/off schedule for the devices and optimize the power consumption. However, with huge number of R-SCDs an enormous traffic load would be observed in the network, which would end in bad resource management.

In this chapter, to provide a SDN-based SCD management framework, we provide a self-adaptive R-SCD network management framework. By performing the optimization process at the control plane layer, we are preventing a huge traffic load. Additionally, by dynamically changing the network structure, we are decreasing the response time.

4.1 REACTIVE SMART CITY DEVICE MANAGEMENT

R-SCDs are dummier electronics that cannot usually perform actions or can execute some simple tasks. Like sensor structures, R-SCDs usually make certain observations such as heat or pressure and with the detection of a specific situation, they transmit a constant data packet, usually an acknowledge packet. The concept of situation, or more generally event, has to be defined for this kind of devices. In this chapter, two possible event types are considered, that is, continuous events and discrete events. Continuous events can be deterministically modeled. In these events, the detection of a previous event assures the possible detection of a future event. For instance, motion is a continuous event. The change of location is continuous in time domain. More specifically, in a 1D space, to detect an obstacle at x location at time t, it has to be in $x + 1$ or $x-1$ location at $t-1$. In contrast with the continuous events, discrete events are chaotic events. They cannot be modeled deterministically. This event type is usually applicable to modeling consumer electronics. For example, the coffee machine tries to detect a discrete event, existence of hot coffee. However, even if one coffee machine detects hot coffee at time t, another coffee machine adjacent to it would not detect hot coffee at time $t + 1$.

R-SCDs try to detect events, continuous or discrete, and in case of detection, they send acknowledge message. The operation cycle of these devices changes the main objective of these devices. The main objective of these devices is to increase the number of events for a long time interval. To increase the number of observed continuous events the SCDs have to be placed at remote locations, whereas they need to cover the complete observation environment concurrently.

Even though for most of the applications the SCDs cannot work identical job, it is possible to model them in a layered structure. On the same layer devices performs identical tasks while devices from different layers observe different events. Conventional approach to solve this optimization problem is activating all SCDs simultaneously that decreases the battery life of the IoT devices [11]. In Fig. 3.4, the changes of IoT network lifetime with the communication ratio are measured. In a more specific sense, since the reactive IoT devices communicate in times of event detection, Fig. 3.4 emphasizes the network lifetime's dependency on the number of observed events. In a high event detection environment, by keeping all IoT devices in active mode, the lifetime of the network is decreased by 40%. Even though such a decrease may be acceptable for some intracity applications, most of the R-IoT devices are used in

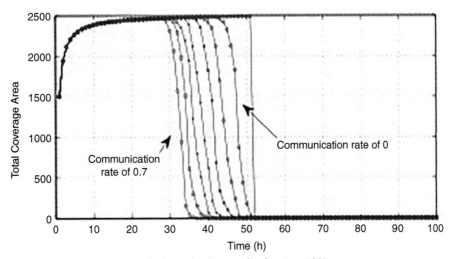

FIGURE 3.4 Variation of SCD Network Lifetime With Communication Rate [11]

suburban regions. Changing batteries frequently would increase the OPEX, and as the R-SCD network cannot be working until their batteries are changed, it will also cause decrease in QoS.

In order to increase the durability of network the battery management of the SCD network has to be managed. A simple and yet efficient approach is scheduling the R-SCD on/off times. On behalf of maintaining the connectivity with the generality, in this chapter three device states are considered, that is, on, off, and dead. When a device is in on state, it observes the environment and in case of an event detection, it transmits an acknowledge packet. In off state, device is in sleep mode and does not observe the environment. It is a battery-saving mode. Since it does not observe the environment, it stops transmitting and only keeps on listening. A R-SCD in off mode expects a wake call from the controller. Finally a device changes its state to the dead mode if its battery is exhausted. An energy-based scheduling algorithm is necessary for R-SCD networks; however, observation of the complete network is a challenge. Since SCDs have plug/play capacity, the topology of the network dynamically changes. In addition to the dynamically changing network topology, the network controller is also a problem. Optimization process requires a large amount of calculation and maintaining such a hardware system will increase the CAPEX. Finally the network should be able to adapt itself when a new device is added to the network. The existing framework demands a reset to configure the network topology that is acceptable for most cases since such a dynamic configuration process produces huge amount of calculation load. Managing the R-SCD network with a discrete management structure is expensive and for most of the situations infeasible. However, a centralized management with larger resources will efficiently manage the R-SCD network. SDNs enable smart solution in R-SCD network management.

As previously stated, SDN proposes a centralized control. This centralized OS controls dummy switches that expect an action when they encounter a new device. This action expectation model enables a control mechanism in the dynamically changing network structure. When a new device connects to the network, since the switch will not be able to find this device in the flow table, it will demand an action from the SDN controller. SDN controller will be able to decide the kind of the device and add it to the correct device list. Since R-SCDs cover a large group of electronic devices, SDN controller tries to cluster devices with the same capabilities and duties in order to optimize them. After adding device to a suitable list, it optimizes the network topology for this device type. The optimization process is performed by transmitting the control packages to the state changing devices and the necessary flow entries to the switches.

4.1.1 The number of necessary devices

Former studies proved that sleep scheduling improves the overall energy efficiency of the network. However, these studies did not consider the event observation rate in the network. The total coverage area increases with the number of working R-SCDs. And since the only way to increase the number of observed events is covering more regions, the direct proportion between working R-SCD count and the number of observed events is a valid argument. Nevertheless, the necessity of an energy management framework is an important challenge and the sleep scheduling is a promising method. Based on these arguments, a dynamic scheduling algorithm is applied in this chapter. Since applications are directly in touch with the SDN controller, each application could select a guaranteed detection rate that will determine the number of necessary working devices. The probability of observing an event can be formalized as in Eq. (3.1), where $P(C|x)$ denotes the expected observation rate, N is the device count, R is the radius of coverage for each R-SCD, and A is the total coverage area:

$$P(C \mid x) = \frac{N \times \pi \times R^2}{A} \tag{3.1}$$

In actual implementations, the NR^2 term has to be normalized since the total coverage area will not be equal to the theoretical coverage area due to collusions. A normalization parameter, φ, is defined as in Eq. (3.2):

$$\varphi = \frac{\sum_{n=1}^{N} \pi \times R_n^2}{N \times \pi \times R^2}, \quad \forall n \in N \tag{3.2}$$

Integrating Eqs. (3.1 and 3.2), the number of necessary working devices can be determined as in Eq. (3.3), where $P = P(C \mid x)$:

$$N = \frac{P \times A}{\varphi \times \pi \times R^2} \tag{3.3}$$

$$\sigma = \sqrt{P - P^2} \tag{3.4}$$

The variance of P that is more likely a mathematical result more than an optimization problem can be calculated as in Eq. (3.4). This parameter represents the missed events that will decrease the QoS.

4.1.2 Software defined reactive devices

The proposed framework tries to optimize the network topology such that the network can cover as much region as possible with the minimum energy. When a new R-SCD is connected to the network, it sends a "hello" package to the switch. When the switch receives that package, the MAC address of this R-SCD is compared with the existing MAC addresses in the flow table. Since this device is newly connected to the network, its address will not be available in the flow table. So the switch will inform the controller and wait for an action. The proposed framework will be working in the controller. When this action request is received, the controller will first determine this device's task. Task determination is the main clustering mechanism as the on/off mode scheduling would be effective only among devices with equivalent tasks. For simplicity, in this study we will consider only a single device task.

After the clustering process, the SDN controller will add this device to a correct list and start an optimization process among the devices in the list. In the previous sections, two different event types were introduced. In order to increase the observation rate of the discrete events the network management has to activate as much device as possible, since the determination of possible events is not possible. However, the continuous events are deterministic. This deterministic nature emphasizes that a possible event observed at location $(x_1; y_1)$ will probably be observed at an adjacent location like $(x_2; x_3)$. Despite the fact that such a determination would be beneficial especially for vector determination processes, it would not be efficient for many applications. Thanks to SDN structure, it is possible to let the application to select the P parameter that directly affects the number of active devices. For applications for which the determination of an existing event is more important than determining a motion vector for this event, selection of remote R-SCDs is a better solution. The device selection processes are performed with two calculation parameters, that is, Conflict Parameter and Spatial Correlation Coefficient.

4.1.3 Conflict parameter calculation

Dense deployment of R-SCDs causes conflicts in their coverage area. More specifically, two or more R-SCDs may cover the same region simultaneously. However, depending on the conflicting ratio, one of these conflicting devices would be enough to cover the area. The continuous event detection rate increases by remote devices and the discrete event detection rate affected only by the number of working devices. Since the main objective is to cover the largest area using a constant number of devices (N), selection of nonconflicting devices would be efficient. The conflict parameter (ξ_i) is used to measure how much unique area the ith device is covering. The conflict parameter for device i can be calculated as given in Eq. (3.5):

$$\xi = \sum_j \frac{D_{ij} \times (D_{ij} < 2R) + 2R \times (D_{ij} > 2R)}{2R}, \quad \forall i, j \in N \tag{3.5}$$

In Eq. (3.5), the numerator $[D_{ij} \times (D_{ij} < 2R) + 2R \times (D_{ij} > 2R)]$ calculates the distance between two R-SCDs. If their coverage areas are not conflicting each other's, then this term will be equal to $2R$. By normalizing this term with the optimal case that is equal to $2R$ a conflict ratio is calculated. For i^{th} device, this ratio is calculated for every jth device. After summing these conflict ratios, the conflict parameter is calculated. The optimization process tries to select highest parameter valued devices.

4.1.4 Spatial correlation coefficient calculation

The conflict parameter is effective in the configuration of the network topology. However, most of the continuous events follow a sequential path that results in the detection of the same event by lots of devices. For example, in a R-SCD network that is designed to detect a motion in a region, when a R-SCD detects a motion, an adjacent device will detect the same motion. Both of them are going to acknowledge the server about the motion, which will be a waste of power since the object of the R-SCD network is the detection of motion in an environment, not to determine the motion vector of the event. Due to this mismanagement, the working R-SCDs would not be effective and they would probably consume all their power with a minimum detection rate. With the objective of overcoming this mismanagement, a Spatial Correlation Coefficient (I_j) is designed.

This spatial parameter presents the distribution of the active devices in the environment. I_i can be calculated as presented in Eq. (3.6):

$$I_i = \frac{\phi_k \times \sum \phi \times D_{ki}}{\sum \phi^2} \tag{3.6}$$

where subscript k denotes the state of the kth device and D_{ki} denotes the Euclidean distance between k^{th} and i^{th} devices.

4.2 SELF-ORGANIZATION ALGORITHM

The proposed application uses a self-organizing structure to adjust the R-SCD network to the dynamic nature of the devices. The self-configuration framework is presented in Algorithm 3.1. At this part the controller activates all R-SCDs to optimize the network. After the optimization, according to their locations the controller calculates their I_K parameters. Since the considered structure contains only the same devices, a constant coverage area is considered. Based on this consideration the conflict parameters are calculated. After the self-configuration, the self-optimization is applied. In the self-optimization part (Algorithm 3.2), the main objective is to optimize the network in terms of power consumption. With this objective, controller first lists the devices according to their conflict parameters. The scheduling algorithm of the controller has an objective to select the R-SCD devices with the highest unique coverage area.

Require: (x, y)
Ensure: ξ_i, I_k
1: **for** $n \leftarrow 1$ to N **do**
2:      $\phi \leftarrow 1$
3: **end for**
4: Calculate ξ
5: Calculate I_K

ALGORITHM 3.1 Device Control Algorithm (Self-Configuration)

Require: ξ_i, I_k
Ensure: ϕ
1: List devices
2: Calculate N (t)
3: **while** $N_{actv} < N(t)$ or not All Devices **do**
4:    **if** $\xi_i = N$ **then**
5:       $\phi \leftarrow 1$
6:       $N_{actv} + 1 \leftarrow N_{actv}$
7:    **else**
8:       **if** $\xi_i = \xi_j + 1$ **then**
9:          **if** $I_i > I_j$ **then**
10:             $\phi_i \leftarrow 1, \phi_i + 1 \leftarrow 0$
11:          **else**
12:             $\phi_i + 1 \leftarrow 1, \phi_i \leftarrow 0$
13:          **end if**
14:       **else**
15:          **if** $\xi_i > xi_j + 1$ **then**
16:             $\phi_i \leftarrow 1, \phi_i + 1 \leftarrow 0$
17:          **else**
18:             $\phi_i + 1 \leftarrow 1, \phi_i \leftarrow 0$
19:          **end if**
20:       **end if**
21:    $N_{actv} + 1 \leftarrow N_{actv}$
22:    **end if**
23: **end while**
24: **if** $N_{actv} < N(t)$ **then**
25:    **for** $N(t) - N_{actv} > 0$ **do**
26:       **if** $\phi_i = 0$ **then**
27:          $\phi_i \leftarrow 1$
28:          $N_{actv} + 1 \leftarrow N_{actv}$
29:       **end if**
30:    **end for**
31: **end if**
32: Update $\psi(t)$

ALGORITHM 3.2 Device Control Algorithm (Self-Optimization)

Require *Passive signal from ith device*
Ensure ξ_i, l_k
1: Remove i from list
2: $N - 1 \leftarrow N$
3: **for** $n \leftarrow 1$ to N **do**
4:    $\phi \leftarrow 1$
5: **end for**
6: Calculate ξ
7: Calculate l_K

ALGORITHM 3.3 Device Control Algorithm (Self-Healing)

At the beginning of the algorithm, the controller tries to select the nonconflicting devices and activates all these devices. In practical applications because of the dense deployment, the number of nonconflicting R-SCD would be 0. After activating all the nonconflicting devices, the controller tries to select from the set of conflicting devices. In this process the controller compares the two adjacent R-SCD's I_K parameters. As a higher I_K parameter represents a higher change to observe a chaotic event, the controller activates the device with a higher I_K. This selection process continues until $N_{act} = N(t)$ condition is satisfied or the end of the list. If the controller reaches the end of the list without satisfying the condition, then it starts to sequentially activate the devices from the list until $N_{act} = N(t)$ condition is satisfied. When the new R-SCD network topology is determined, the normalization coefficient (t) is calculated.

The changes in the R-SCD network topology are handled by the self-healing mechanism of the framework. The changes in the network topology can happen only due to battery exhaustion. The self-healing mechanism (Algorithm 3.3) is a subfunction that recalculates the ξ_i and I_K parameters according to the new topology.

---

5 CONCLUSIONS AND FUTURE RESEARCH

The smart cities brought many good opportunities such as low latency over time-critical applications, better resource management, or more trustable control framework. However, the huge data load from the sensor and the actuator devices and economic inadequacies caused the necessity of handling the network devices and applications in different layers. Data layer contains CPDs and sensor nodes, whereas control layer contains the necessary applications and OS. However, this virtually centralized yet actually distributed topology brought essential problems.

In this chapter, we presented an energy-aware coverage optimization framework that makes smart coverage decisions to handle the energy optimization. In the first part of the chapter we briefly discussed the main drawbacks and the benefits of SDN-based network management in smart cities. Due to its suitability to the needs, we covered an SDN-based approach and worked on an energy-based optimization framework to solve the coverage problem. We first explored the concept of necessary coverage area to guarantee a certain amount of event detection and then using this parameter, we calculated the conflict ratio and Local Indicator of Spatial Association (LISA) coefficient to cover the optimization problem. Based on the MATLAB simulations we observed huge improvements in energy efficiency without an accountable loss of coverage.

The future studies of this study can be investigated in two main folds, that is, improving self-awareness and improving the covered arguments. First of all, this chapter shows a generic understanding over the self-awareness concept in future's smart cities. In this structure, the city should not only analyze the variable data but also anticipate future failures or needs and be able to make deterministically modelable logical (smart) decisions to handle this decision. The framework in this chapter is a self-aware energy model that considers only existing network state and does not anticipate any possible actions in the near future. Apart from an improvement in self-awareness, second, a more generic analysis and a more generic handler will be investigated. As previously explained, many assumptions are made throughout the chapter. We are hoping to improve our work by simply generalizing our study and making the necessary additions to overcome the needs to these assumptions.

REFERENCES

[1] Saint A. The rise and rise of the smart city [smart cities urban Britain]. Eng Technol 2014;9(9):72–6.

[2] Jin J, Gubbi J, Marusic S, Palaniswami M. An information framework for creating a smart city through internet of things. IEEE Internet Things J 2014;1(2):112–21.

[3] Visual networking index report: global mobile data traffic forecast update. 20132018, Technical report. Cisco; 2014.

[4] Cano J, Hernandez R, Ros S. Distributed framework for electronic democracy in smart cities. Computer 2014;47(10):65–71.

[5] Zanella A, Bui N, Castellani A, Vangelista L, Zorzi M. Internet of things for smart cities. IEEE Internet Things J 2014;1(1):22–32.

[6] Wenge R, Zhang X, Dave C, Chao L, Hao S. Smart city architecture: a technology guide for implementation and design challenges. Chin Commun 2014;11(3):56–69.

[7] Martinez-Balleste A, Perez-Martinez P, Solanas A. The pursuit of citizens' privacy: a privacy-aware smart city is possible. IEEE Commun Mag 2013;51(6):136–41.

[8] Vilajosana I, Llosa J, Martinez B, Domingo-Prieto M, Angles A, Vilajosana X. Bootstrap-ping smart cities through a self-sustainable model based on big data flows. IEEE Commun Mag 2013;51(6):128–34.

[9] Zhu Z, Lu P, Rodrigues J, Wen Y. Energy-efficient wideband cable access networks in future smart cities. IEEE Commun Mag 2013;51(6):94–100.

[10] Djahel S, Doolan R, Muntean GM, Murphy J. A communications-oriented perspective on traffic management systems for smart cities: challenges and innovative approaches. IEEE Commun Surv Tutorials 2015;17(1):125–51.

[11] Akgul OU, Canberk B. Self-organized things (sot): an energy efficient next generation network management. Comput Commun 2016;74:52–62.

[12] Liu P, Peng Z. China's smart city pilots: a progress report. Computer 2014;47(10):72–81.

[13] Li F, Vogler M, Sehic S, Qanbari S, Nastic S, Truong H-L, Dustdar S. Web-scale service delivery for smart cities. IEEE Internet Comput 2013;17(4):78–83.

[14] Vlacheas P, Giaffreda R, Stavroulaki V, Kelaidonis D, Foteinos V, Poulios G, Demestichas P, Somov A, Biswas A, Moessner K. Enabling smart cities through a cognitive management framework for the internet of things. IEEE Commun Mag 2013;51(6):102–11.

[15] Erel M, Arslan Z, Ozcevik Y, Canberk B. Grade of service (gos) based adaptive flow management for software defined heterogeneous networks (sdhetn). Comput Netw 2015;76(0):317–30.

[16] Nunes B, Mendonca M, Nguyen X-N, Obraczka K, Turletti T. A survey of software-defined networking: past, present, and future of programmable networks. IEEE Commun Surv Tutorials 2014;16(3):1617–34.

[17] Xing T, Xiong Z, Huang D, Medhi D. Sdnips: enabling software-defined networking based intrusion prevention system in clouds. In: 10th international conference on network and service management (CNSM); November 2014. p. 308–11.

NOMADIC SERVICE DISCOVERY IN SMART CITIES

4

M. Stolikj, J.J. Lukkien, P.J.L. Cuijpers, N. Buchina

Department of Computer Science, Eindhoven University of Technology, Eindhoven, The Netherlands

1 INTRODUCTION

The concept of "Smart Cities" refers to cities that enhance their processes by digital technologies, thus reducing cost and increasing well-being [1]. In general, the term "smart" refers to improvements through digital technology and can be applied to any process or service (eg, smart energy or smart transport). The main technological drivers for the emergence of Smart Cities are the vast embedding of electronics in physical objects, their imminent *connectivity*, and the availability of *data-based services.*

Connectivity includes the general ability for devices—embedded or not—to communicate. We are used to the connectivity of phones, game computers, and other handhelds to the Internet, although this trend started less than 10 years ago. The Internet is now being extended to include the increasing number of devices that are embedded in the physical world, in what is commonly referred to as the "Internet of Things" (IoT). The connectivity thus is the basis of data-based services that are services backed up by large-scale data collection. For example, predictive (smart) energy management is based on data collected over a long period of time.

In this chapter we concentrate on connectivity, and on finding and using digital services within a smart city context. A smart city has a digital infrastructure for use by its own internal processes but also available to mobile users (individuals, vehicles, buses) and more static users (shops, public facilities). Digital services are available through this infrastructure and can be as varied as the following examples:

1. information services such as the website of the shop I am looking at now;
2. parking spot location and status, within the vicinity, plus the ability to reserve one;
3. access to local lighting actuators;
4. access to local air quality measurements.

This list can obviously be extended with (location-dependent) services of interest to mobile users. If we classify both service providers (SPs) and users as either static or mobile, then the static/static case is much alike the current Internet while the other three cases require new approaches to service discovery (SD) and naming in order to facilitate an efficient matching of SP and user.

In Fig. 4.1 the process and goal of SD is highlighted: service seekers issue queries matched by SPs in a distributed context where neither one knows the existence of the other. Service discovery protocols (SDPs) achieve this goal by defining (1) a common language for describing services and selection criteria; (2) a common protocol for exchanging service descriptions between service clients (SCs) and SPs; and (3) rules for matching service descriptions with the selection criteria.

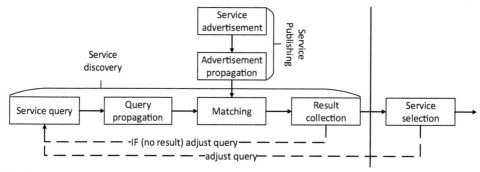

FIGURE 4.1 The Service Discovery Process

The service seeker issues a query that is matched by the service advertisement of the service provider. In a distributed context, these tasks can be performed by different entities.

One solution to this problem is to direct queries to a search engine such as Google. This requires services to be indexed timely, in a centralized way that is challenging in particular for the mobile case. The advantage is that this works regardless of the location of the service seeker (called *client* from now on), but it has as a downside a lack of scalability and also privacy concerns, since the information becomes globally available. We adopt as a requirement that the mobile user connects to a local access point (eg, via Wi-Fi, or via an overlay connection) and that discovery is confined to the local network scope. This is what we call *nomadic SD*.

In this chapter we aim to identify a suitable protocol for SD in Smart Cities as a context for the IoT. This protocol must therefore work for very constrained devices and networks. Typical examples of constrains are small memory capacity (both RAM and ROM, as shown in Table 4.1), lack of constant power supply (ie, battery powered or energy harvesting), slow processor speed, considerable packet loss, and large latencies for access. An important requirement is that it follows the Internet Engineering Task Force (IETF) standardization procedures, in order to guarantee wide acceptance. From the plethora of protocols, we focus on the Multicast Domain Name System (mDNS) [2] with DNS-Based Service Discovery (DNS-SD) [3], a standards-based protocol, with light footprint, good scalability, and wide usage. We motivate this choice with an examination of requirements for Smart Cities and a review of recent work in SD.

Table 4.1 Characteristics of Constrained Devices

Constraint Boundaries				
Nonvolatile Memory (eg, ROM) (KiB)	**Volatile Memory (eg, RAM)**	**CPU Frequency (MHz)**	**Platform (Bit)**	**Examples**
≤1	0 B	0	8	Passive RFID tags
≤64	≤2 KiB	≤16	8	ATmega644RFR2
≤128	≤10 KiB	≤16	16	RC230X, TI MSP430
≤256	≤128 KiB	≤96	32	MC13224, STM32W

DNS-SD is an extension of DNS that uses DNS resource records (RRs) to describe services in a domain name such as fashion. The service description can further specify the service protocol and additional context. Similarly, mDNS is an extension of the DNS protocol for local area networks. The main differences between mDNS and the original DNS protocols are as follows: (1) naming information is stored locally, on each node in the network; (2) resolution queries are sent multicast instead of unicast; and (3) each node directly responds to queries.

The mDNS/DNS-SD protocol in its current form contains several issues, which hinder its performance in Low-Power and Lossy Networks (LLNs). First, mDNS/DNS-SD assumes SPs to be constantly online, which is impossible for battery-operated devices. Second, the services are described using very descriptive, long names. As a result, service descriptions are relatively large, and require packet fragmentation when used over LLNs with small payloads, as IEEE 802.15.4. Finally, the protocol is limited in the query language, and it facilitates only coarse selection criteria in terms of service type. In networks with many similar devices, as in the IoT, this increases the traffic load.

In this chapter, we propose solutions for the stated three issues with mDNS/DNS-SD. In Section 2 we cover related work on SD, identify selection criteria, and give an overview of different approaches. In Section 3 we describe the mDNS/DNS-SD protocol and identify its key issues when used in large networks. We present protocol extensions for sleeping nodes in Section 4, and a context tag model that supports more specific queries in Section 5. We present a new compression method in Section 6 and give concluding remarks in Section 7.

2 RELATED WORK

2.1 BACKGROUND AND TERMINOLOGY

The main parties involved in SD are the SC and the SP. The SC provides the selection criteria for SD, while the SP hosts a specific service, called a service instance. SD protocols often use an intermediate party for storing and accessing service descriptions, commonly referred to as a Service Repository (SR), directory, or broker.

Several studies have focused on identifying the key properties of SD protocols for LLNs [4,5]. Based on these studies, we select the following features to characterize SD protocols for the IoT: architecture, message scope during SD and/or service advertisement/registration, service description language, overhead traffic, and interoperability with existing protocols. As additional features we also consider awareness of mobile SPs, caching support, detailed service descriptions, and energy and resource demands.

The communication patterns among the three parties during SD define the architecture of SD protocols. We distinguish three types of architectures: centralized, distributed, and hierarchical. In a centralized architecture, all service descriptions are stored in one or more SRs. SPs register their services in the SR, and SCs look for services in SRs. In a distributed architecture, service descriptions are stored locally on each SP. Therefore, SCs need to inquire all SPs whether they satisfy the selection criteria. Hierarchical architectures are a combination of both previous architectures. They organize SPs in subgroups or clusters, and elect cluster heads as SRs for the given group. Therefore, SPs register services with cluster heads, and SCs contact only cluster heads during SD.

Each of the three architectures has benefits and drawbacks. Centralized SD protocols take most of the workload concerning SD from SPs, as they only need to initially register services, and SCs need to inspect only a single SR. However, centralized SD protocols require a more powerful SR, which has to be preconfigured in the entire network, and represents a single point of failure. Distributed SD protocols require more complex actions from all nodes participating in the SD protocol, which can influence their lifetime. Furthermore, messages exchanged are usually broadcast based, and may flood the network. Finally, hierarchical architectures depend on clustering algorithms for creating the initial clusters and selecting appropriate cluster heads. While communication is limited within clusters, additional overhead is introduced while clusters are being established and cluster heads are elected.

2.2 RELATED WORK

Previous research in service description and SDPs for LLNs can be roughly categorized into two groups: general SD protocols and their application in LLNs, and custom SD protocols for LLNs. Naturally, SD protocols for LLNs differ considerably compared to similar protocols for high-capacity networks. Therefore, we will first cover related work on both general-purpose SD protocols, before describing custom SD protocols for LLNs. The main characteristics of all covered protocols are summarized in Tables 4.2 and 4.3. Then, we present related work on SD architectures and frameworks for IoT-specific deployments. Finally, we will discuss the applicability of existing protocols in the IoT domain.

2.2.1 General-purpose SD protocols

mDNS/DNS-SD is a widely used standard for SD in local area networks. It uses a distributed architecture, with services described as domain names. mDNS/DNS-SD has been implemented in all major operating systems for high-capacity devices, as well as in resource-constrained devices [17,18]. Further optimizations of the protocol have been proposed [19] to reduce message sizes in order to support low-power networks better. We describe the protocol in more detail in Section 3.

Java Intelligent Network Interface (JINI) [7] is a centralized SD protocol for the Java programming environment. JINI uses Java interfaces for describing services and Remote Method Invocation (RMI) for accessing them. Service descriptions are stored in lookup servers, which are discovered using multicast messages. Due to the memory requirements of the Java virtual machine, JINI is rarely used in LLN.

Simple Service Discovery Protocol (SSDP) [6] is part of the Universal Plug and Play (UPnP) specification, a commonly used software stack for consumer electronics devices. It is a hierarchical protocol, where participating entities register and resolve services using control points. The messaging protocol is based on the Hypertext Transfer Protocol (HTTP) standard, with multicast messages for advertising and discovering new services, and unicast responses. Services are described as Uniform Resource Locators (URLs) for control and eventing. The overhead brought by the transport protocol used in SSDP is unsuitable for low-capacity networks. Therefore, most efforts for interconnecting consumer electronics devices using SSDP with low-capacity networks have relied on some form of gateways [20,21] to translate messages between the two networks.

Service Location Protocol (SLP) [22] is another standardized SDP for local area networks. SLP has been designed to scale from small, decentralized networks to large corporate networks, by supporting fully distributed or centralized operation. SLP defines three types of entities: User Agents (UAs) that request services, Service Agents (SAs) that provide services, and Directory Agents (DAs) that cache service advertisements. DAs are optional, but if they exist in a network, both SAs and UAs are obliged to use them. The proposed Internet Protocol version 6 (IPv6) over LoWPAN (6LoWPAN) adaptation of

Table 4.2 Comparison of Service Discovery Protocols (part 1)

Protocol	Architecture	Discovery Mechanism	Registration Mechanism	Overhead	Description	Interoperability
mDNS/ DNS-SD [2,3]	Distributed	Multicast	Multicast	DNS packets	DNS-SD	Yes
UPnP [6]	Hierarchical	Multicast or unicast	Multicast	Heavy transport protocol	URL	Yes, gateways
Jini [7]	Centralized	Multicast or unicast	Multicast or unicast	Heavy protocol	Java	Yes
SLP [8]	Distributed or centralized	Multicast or unicast	Multicast or unicast	Translation agents	String, XML	Yes
SDP [9]	Distributed	Unicast	Unicast	SDP server discovery (unknown)	UUID	No
DEAPSpace [10]	Distributed	Broadcast	Proactive	Periodic advertisements	None	No
Konark [11]	Distributed	Multicast	Multicast	Periodic advertisements	XML	No
SANDMAN [12]	Hierarchical	Unicast	Unicast	Cluster maintenance, sleep announcements	None	No
Sleeper [13]	Distributed	Broadcast	Broadcast	Proxy election, periodic advertisements	Taxonomy	No
OSAS [14]	Centralized	Unicast	Unicast	Periodic advertisements (heartbeat)	None	No
Anwar et al. [15]	Hierarchical	Unicast	Unicast	Cluster maintenance	ENUM/DNS	DNS
TRENDY [16]	Centralized or hierarchical	Multicast	Multicast	Periodic updates, group leader election	CoAP, URI	No

DNS, Domain Name System; *ENUM*, Electronic Number Mapping; *JINI*, Java Intelligent Network Interface; *mDNS/DNS-SD*, Multicast Domain Name System with DNS-Based Service Discovery; *OSAS*, Open Service Architecture for Sensors; *SDP*, Service Discovery Protocol; *SLP*, Service Location Protocol; *UPnP*, Universal Plug and Play; *URI*, Uniform Resource Identifier; *URL*, Uniform Resource Locator; *XML*, Extensible Markup Language; *UUID*, Universally Unique Identifier.

the protocol called Simple Service Location Protocol (SSLP) [23] introduces a new Translation Agent (TA) that translates messages between a low-capacity 6LoWPAN network and a high-capacity network running SSLP. DAs are used as caching entities, which limit the scope of advertisements and queries. In both protocols, discovery and advertisement messages are multicast if DAs are not used, and responses

Table 4.3 Comparison of Service Discovery Protocols (part 2)

Protocol	Localization	Mobility	Proxy Support	Energy-Aware	Low Resource
mDNS/DNS-SD [2,3]	Domain name or service description	Time-to-live expiry, service invalidation	Yes	No	Yes
UPnP [6]	Domain name	Time-out expiry, periodic advertisements	No	No	Yes
Jini [7]	No	Time-out expiry	Yes	No	No
SLP [8]	Yes	Time-out expiry	Yes	No	Yes
SDP [9]	Yes (one hop)	No	No	Yes	Yes
DEAPSpace [10]	Yes (one hop)	Periodic advertisements	Yes	Yes	Yes
Konark [11]	Yes, in description	Time-out expiry, periodic advertisements	No	Yes	Unknown
SANDMAN [12]	Yes, in clusters	Moving clusters	Cluster heads	Yes	Unknown
Sleeper [13]	Yes, in description	Time-out expiry, service invalidation	Yes	Yes	No
OSAS [14]	No	Time-out expiry	Yes	Yes	Yes
Anwar et al. [15]	Yes, in identifier	Time-out expiry	Yes	No	Unknown
TRENDY [16]	Yes, description	Periodic advertisements	Yes	Yes	Yes

JINI, Java Intelligent Network Interface; *mDNS/DNS-SD*, Multicast Domain Name System with DNS-Based Service Discovery; *OSAS*, Open Service Architecture for Sensors; *SDP*, Service Discovery Protocol; *SLP*, Service Location Protocol; *UPnP*, Universal Plug and Play.

are always unicast. Unfortunately, the 6LoWPAN adaptation of the SSLP protocol seems abandoned. Furthermore, SLP has been extended with proxy agents [24], for connecting high- and low-power networks, and context support such as proximity services [25].

2.2.2 SD protocols for LLNs

Many SDPs have been designed with resource-constrained devices in mind. These protocols can operate in a single-hop network [9,10], clusters of nodes [12,26], centralized environments [14], or 6Low-PAN networks [15,16,27]. We now give an overview of some of the most popular SD protocols for resource-constrained devices.

SDP [9] is part of the Bluetooth standard for locating available server applications, and learning about their characteristics in one-hop ad hoc networks. Services are resolved through SDP servers, which run on every Bluetooth device offering services. The protocol itself does not specify how SDP servers are selected by clients, or how they detect when they become unavailable.

DEAPSpace [10] is a proactive distributed SD protocol for use in single-hop networks. It uses broadcast messages to advertise all known services to neighboring devices at regular intervals. Since each device advertises services offered by other neighboring devices, low-powered nodes may choose to have low intervals. The algorithm does not scale in large networks, due to (1) broadcast storms in multihop environments with large number of nodes and (2) large size of advertisements when many SPs coexist.

Konark [11] is a distributed SD and service delivery protocol for ad hoc networks. It uses multicast messages for both periodic service advertisements and SD queries. Services are described in Extensible Markup Language (XML), similar to the Web Services Description Language (WSDL), and delivered in the form of URLs. Konark has been designed and tested for ad hoc networks of high-capacity devices, and the underlying protocols used, such as HTTP, are not suitable for low-capacity networks.

SANDMAN [12] is an energy-aware hierarchical SDP. It assumes that an arbitrary set of mobile devices within a network have similar mobility patterns, and can therefore be grouped in a cluster. Then, from the cluster, only one device answers to SD queries, while the others can sleep. SANDMAN does not include a cluster management protocol, and it is therefore difficult to estimate its general applicability.

Sleeper [13] is a distributed SDP, using proxies for delegating answers to SD queries. This way, low-capacity nodes can endure long sleeping times while other nodes can still discover their services. Sleeper supports various description options for services, including geographic location, metadata, and ontology information. Service descriptions are enriched with popularity metrics, which are used by proxy nodes to select which advertisements will be cached. Services can be discovered proactive or reactive, that is, using periodic service advertisements or explicit SD queries. Sleeper does not cover the election process for proxy nodes. The protocol is evaluated on high-capacity devices using IEEE 802.11b, and it is therefore unknown how it performs in low-capacity networks.

Open Service Architecture for Sensors (OSAS) uses a centralized approach for SD [14]. It relies on a Resource Manager (RM) for holding service advertisements, which are populated from low-capacity nodes at regular intervals. Since there is one RM, unicast messages are used at all times except when discovering new nodes. A centralized solution is useful for small networks, such as body sensor networks or smart office spaces, but does not scale to large numbers. Introducing multiple RMs would make this protocol similar to the other clustering protocols.

The Electronic Number Mapping (ENUM) SD protocol [15] has been proposed as a possible solution for SD in 6LoWPAN networks. It uses compressed entries for describing service end points, which are then resolved using standard DNS queries. The architecture is hierarchical and assumed to be preconfigured, with sensor nodes connected to master nodes, which are globally addressable and interconnected with gateway nodes. The protocol uses unicast messages between sensor nodes and master nodes, but does not specify how master nodes are detected.

Fast and Energy Efficient Service Provisioning (FESP) [27] is a protocol for managing already discovered, frequently used services in 6LoWPAN. It assumes that initial discovery has already been carried out by another protocol. Afterward, it improves SD latency by (1) sending SD queries only to immediate, two-hop neighbors; and (2) as last resort, sending queries to a local gateway. While such metrics can be seen as optimizations to other SDPs, relying on every node to cache all popular service descriptions is inappropriate in low-capacity devices.

TRENDY [16] is a centralized context-aware SDP for 6LoWPANs. Similar to SLP, it has one central DA, but other SAs and UAs are organized in clusters, with Group Leaders (GLs) on top. Unlike cluster heads, GLs are not elected by surrounding nodes, nor store service advertisements. They are designated by the DAs, and merely monitor the status of SAs. The entire protocol is built on top of Constrained Application Protocol (CoAP) [28] and services are described as Uniform Resource Identifiers (URIs) with variable parameters, including location, type, etc. Caching and proxy support are taken from the CoAP base. Certain information such as the DA address has to be either hardcoded or discovered through another protocol.

2.2.3 SD frameworks and architectures for the IoT

Several works have focused on the issues of SD within the IoT domain. These works focus on some of the specifics carried by the IoT domain, such as context awareness or scalability, and reuse existing SD protocols within the proposed solutions. We now give an overview of some of these works.

In Ref. [29], a framework for context-aware SD for the IoT is presented. The framework is developed from an ontology-based context model that captures the different entities in the IoT, their relations, the data they produce, and the uncertainty and temporal characteristics. The reasoning about the model is realized through Dynamic Bayesian networks, which enables the framework to derive high-level context from different sources. The entire framework is conceptual, and is not linked to a particular protocol or implementation.

In Ref. [30], a semantic enhanced service proxy framework for the IoT cloud is proposed. The framework uses the concept of sensor-as-a-service, where smart objects expose their functionalities as services in the IoT cloud through the framework as an intermediate proxy. The framework uses event-based communication with the smart objects. The information from the smart objects is enriched with semantic knowledge, gathered from several ontologies, and virtualized as enriched services to outside clients.

In Ref. [31], a distributed peer-to-peer discovery service for IoT is proposed. The solution leverages a distributed hast table (DHT) over a peer-to-peer overlay, for describing and discovering for services. The services can be described using multiple attributes that reflect different context, and the queries can be range-based or multiattribute. The evaluation of the solution is performed in a synthetic environment, with an actual implementation planned for the future.

In Ref. [32], a combined architecture for SD in the IoT is presented. The architecture separates between local SD, confined in isolated LLNs, and global SD, between geographically distributed entities. Within LLNs, SD is done using the mDNS/DNS-SD protocol. The information about available services within LLNs is condensed at the edge of the network, in the so-called IoT gateway. This gateway becomes part of a global peer-to-peer overlay network, which enables discovery at a large scale.

Digcovery [33] is an architecture for global SD in the IoT. The proposed architecture uses DNS-SD queries for SD. At the central point in the architecture lies a middleware layer, which collects service descriptions from different types of sources, and uses ontologies for mapping them into compatible DNS-SD descriptions. The middleware exposes services using a representational state transfer (REST) application programming interface (API) based on CoAP.

The work presented in this chapter complements the last two architectures, as it improves mDNS/DNS-SD operation within LLNs.

2.2.4 Summary: solutions for the IoT

Obviously, the SD research field is very fragmented, with various protocols under development or in use. Such heterogeneity is unsuitable for the IoT, as an intermediate TA would be required between any set of devices that use a different SD protocol. Therefore, we believe that accepting a single solution for an SD protocol would be beneficial for easier integration in the IoT. This view has been backed by the community, with calls for "commitment to openness from companies" [34], "curated openness – standardization of a few core functions" [35], and acceptance and use of common standards [36,37].

Unfortunately, none of the existing protocols are ideal. They lack in features, are too specific for certain tasks, or simply have not been accepted in the community. In Ref. [38], based on the availability, expressiveness, and resource requirements, three of the described protocols have been identified as suitable candidates: mDNS/DNS-SD, SSLP, and TRENDY, with TRENDY selected as the best one. However, TRENDY acts as an extension of the CoAP protocol, which imposes requirements at the application layer. Even though CoAP is used heavily in the IoT at the moment, other application protocols in use, such as Message Queuing Telemetry Transport (MQTT), Extensible Messaging and Presence Protocol (XMPP), and Efficient XML Interchange (EXI), cannot be discarded. When a different application protocol is in use, implementing a CoAP-based SDP would be a large burden.

In the same study, mDNS/DNS-SD was discarded due to lack of context awareness and heavy traffic. However, recent activities in the IETF standardization body are focused toward the development of a new version of the protocol, viable for IoT usage. The work presented in this chapter is in line with these efforts, by tackling the aforementioned weaknesses of mDNS/DNS-SD.

3 mDNS/DNS-SD SERVICE DISCOVERY

mDNS and DNS-SD form the core SDP in Bonjour, a Zero-configuration implementation by Apple. The protocol consists of two components: a communication protocol defined by mDNS and a SD and service description protocol defined by DNS-SD.

mDNS is an extension over the unicast DNS [39,40] for name resolution. Names can refer to addresses, as in classic DNS, or to services, using DNS-SD. The primary purpose of mDNS is to enhance name resolution in a local area network. Therefore, mDNS exclusively resolves host names ending with the .local top-level domain. The packet structure in mDNS is similar to the one defined in the standard DNS protocol. The main difference between the two comes in the message exchange protocols: DNS assumes hierarchically organized servers, and unicast messaging between clients and servers. Contrarily, mDNS foresees a distributed environment, where resolution information is stored locally on each device within a small network, and each device directly answers to incoming name resolution queries. In that sense, every participating device acts as both a server and a client. mDNS uses multicast messaging to efficiently distribute both queries and responses to all devices within the network. mDNS packets are sent to/from the reserved multicast addresses 224.0.0.251 [Internet Protocol version 4 (IPv4)] and ff02::fb IPv6, using User Datagram Protocol (UDP) port number 5353.

DNS-SD is a standardized protocol for describing and resolving services using DNS RR. DNS-SD defines how an SC can leverage standard DNS queries to discover service instances within a logical domain using the service type as selection criteria. DNS-SD describes service instances using SRV, TXT, PTR, and A/AAAA RRs (Fig. 4.2). The SRV and TXT RRs have the same structured name, in the form "<Name>.<Type>.<Domain>." The first part of the name is a unique identifier of the service instance. The service type is formed by concatenating the application protocol and transport protocol used for accessing the service instance. Lastly, the domain defines the scope of the service instance.

SRV RRs, besides the name of the service instance, contain the port number for accessing the service instance, the priority and weight parameters to discriminate between service instances of the same type, and the host name of the SP where the service instance is stored. The host name can be resolved to an IPv4/IPv6 address using A/AAAA RRs.

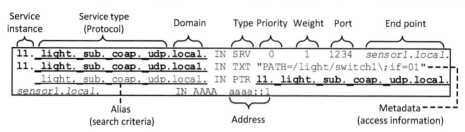

FIGURE 4.2 DNS-SD Description of a Light Sensor Service

The four resource records are connected through the name of the service.

The TXT RR contains additional service metadata in the form of [key]:[value] pairs. The exact content depends on the protocol used, and can include a URI path for a specific resource, invocation parameters, more specific service descriptions, etc. The maximum size of the TXT RR is 1300 bytes.

PTR RRs have a name in the form "<Type>.<Domain>." They provide a mapping between a service type and a specific service instance. As explained in the following section, they are the key RRs used by client to discover services. Therefore, queries sent by SCs contain PTR RRs of the wanted service types. From the return PTR RRs, the SCs can discover the names of the service instances. On selecting one or more service instances, the SCs can resolve them, by querying for the SRV and TXT RRs. Finally, a service instance is completely resolved with A/AAAA RRs, which provide a mapping between the host name of the SP and its IPv4/IPv6 address.

DNS-SD can be used with the existing DNS infrastructure, or in combination with mDNS. When used with the existing, unicast DNS infrastructure, it enables SD within existing DNS scopes. However, in combination with mDNS, it enables plug and play functionality of services within a local broadcast domain. Since there is no need to configure separate DNS servers, autoconfiguration is easier. Furthermore, adding new devices to the network is trivial, since the new devices can discover and advertise network services independently.

3.1 OPERATIONAL MODES

A client discovers a service instance using mDNS/DNS-SD, by retrieving the four RRs associated with the given service instance. DNS-SD can be used in two modes to enable SD: proactive and reactive. In proactive mode, SPs periodically advertise hosted services, by multicasting their descriptions to the network. A SC then needs to listen for these advertisements, and match them against the selection criteria. Obviously, in this mode, there is a large trade-off between overhead traffic and speed of discovery. In large networks, with many service instances, this approach is inappropriate.

In reactive mode, the SC initiates the discovery process by multicasting a query with the selection criteria. The query consists of one or more PTR RRs, with the wanted service types as name. Then, on receipt, each SP checks whether locally it has a PTR RR with the given name, and if so, sends it back. mDNS specifies that the response can be either unicast or multicast. Furthermore, the SP can also include additional RRs in the response, in case the SC asks for them later. If only the PTR RRs is returned, then the SC has to additionally query for SRV, TXT, and A/AAAA RRs. Both options are illustrated in Fig. 4.3.

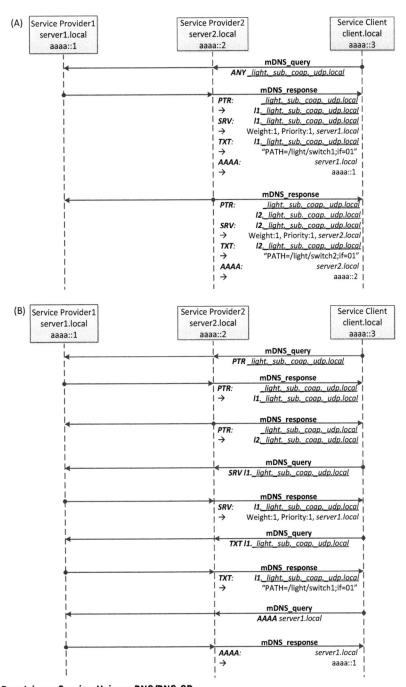

FIGURE 4.3 Resolving a Service Using mDNS/DNS-SD

First, the resolver needs to find service instances of the requested service, provided by PTR RR. Then, the actual service is resolved, through SRV and TXT RRs. Finally, the host providing the service is resolved. The four RRs can be distributed independently, in separate packets (A), or can be packed into one larger packet (B). *Note*: All messages are multicast and all nodes belong to the same broadcast domain.

The flexibility of mDNS/DNS-SD imposes some optimization problems: whether responses to queries should be unicast or multicast, and whether they should include complete descriptions (ie, send complete answers), or only concrete answers to queries (ie, send short answers). An additional complexity is the maximum available payload by lower layers. For example, using DNS name compression, four RRs for one service description occupy around 158 bytes. If IEEE 802.15.4 is used, this DNS packet would be fragmented in at least two frames. We investigate these trade-offs in the next section.

3.2 STRATEGIES FOR RESPONDING TO QUERIES

In order to compare the different strategies when responding to queries in mDNS/DNS-SD, we develop an analytical model. Consider a network of n nodes, connected in an ad hoc fashion. In the set of nodes, there is one SC $c = v_c$, $c \in \{1, \dots, n\}$, which sends a single query, and a set of SPs $P = \{p_{i,j}\}$, $| P | = k$, $k \leq n$, that matches the query. A description of a service consists of $r \geq 1$ RRs, where each RR fits in one frame. If all s RRs are bundled in one DNS packet and compressed using DNS name compression, the resulting packet is fragmented in $f \geq 1$ frames. To abstract from the propagation method and the network topology, we assume that the number of frames generated in the network for a single unicast/multicast frame is U/M, respectively.

Therefore, in the first strategy, the SC first sends a multicast query, for which all SPs in P respond. Then, the SC selects one of them, and further resolves it, using $r - 1$ queries. The total number of frames in the network is as follows:

$$C_s = r \cdot M + k \cdot R + (r-1) \cdot R \tag{4.1}$$

where the response R can be either M or U.

In the second strategy, on the transmission of the initial query, all SPs immediately respond with all RRs for a given service. Then, the total number of frames in the network is as follows:

$$C_f = M + k \cdot f \cdot R \tag{4.2}$$

If we combine these two formulas, we get

$$C_f \leq C_s \leftrightarrow k \leq \left(\frac{M}{R} + 1 \right) \cdot \frac{r-1}{f-1} \tag{4.3}$$

The given equation shows that the optimal strategy for responding to queries depends on the number of responses, the effect of the compression method, and weight of the different messaging implementations. We discuss these trade-offs in the following sections.

3.2.1 Multicast versus unicast responses

In both strategies, the decision whether to use multicast or unicast responses depends on the ratio M/U. However, their relationship is more delicate than it seems. At first thought, unicast responses should require less or as many frames for delivery in the network as multicast responses. However, an additional factor to consider is the cost of discovering a route between the SP and the SC. If this route has been established beforehand, then since multicast transmissions have a wider scope, $M \geq U$, and sending a unicast response is preferable. However, if such a route is not known, discovering it is commonly done with an additional multicast query, and a unicast response with the route [41]. As a result, using multicast responses would create less traffic in the network than using unicast responses ($M < U$).

In LLNs, a route is usually established and maintained only toward the sinks in the network. Therefore, if the SC is a sink, unicast responses are recommended. However, if other nodes in the network are SCs, then multicast responses are preferable.

3.2.2 Short versus complete responses

If we assume that the optimal response strategy is used, that is, $R = \min(M, U)$, then since $M/U \geq 1$, from Eq. (4.3) we can conclude that if $f = 1$, then sending complete responses is always optimal, and this greatly reduces the number of frames associated with a SD. However, if $f > 1$, the optimal responding strategy depends on the ratio M/U and r/f, as in Eq. (4.3).

Typically, $r = 4$ and $f = 2$ (Fig. 4.2). Then, with multicast responses, sending all RRs generates less traffic than single RRs if $k \leq 6$. Similarly, when unicast responses are used in large networks, where the multicast messaging generates a lot more traffic than unicast, complete answers are again preferable.

3.3 PROBLEMS IN mDNS/DNS-SD FOR IoT

Since mDNS/DNS-SD had been designed for SD in local area networks of high-capacity devices, it is not directly applicable to networks of resource-constrained devices. We identify the following problems:

1. *Code size*: Due to memory limitations alone, it is impossible to reuse existing mDNS/DNS-SD implementations. Table 4.2 shows the memory profile of several popular implementations. From the table, it is clear that the original Bonjour implementation cannot fit on resource-constrained devices. Even the Arduino port, which is already a feature-limited implementation of mDNS/DNS-SD, is too large to fit on small factor devices.
2. *Energy consumption*: As shown in the previous section, the mDNS/DNS-SD protocol heavily uses multicast messaging. As a result, almost all messages in the network reach all nodes, which consumes significant energy. Furthermore, SPs need to be online when SCs try to locate them. This imposes an always-on requirement for SPs, which is unfeasible for battery-powered devices.
3. *Context-aware queries*: In the current standard, the only available selection criterion is the service type, while additional descriptions of services can be added as part of TXT records. Therefore, in order to select a service with a specific context, as location, an SC must first gather the descriptions of all services of the same type, and then select the most appropriate one. In large networks, with many service instances with the same type, this generates a lot of traffic.
4. *Packet size*: Due to the expressive nature of the protocol, RRs are relatively large, and require packet fragmentation from lower layers. Such fragmentation is generally unwanted in LLNs, as end-to-end delivery of fragmented packets is difficult. As shown in the previous section, fitting all RRs for a single service description would significantly reduce the overhead traffic associated with SD.

We have solved the first problem by implementing a subset of the protocol for resource-constrained devices (Table 4.4). The implementation contains only the basic features of the protocol – an engine and API for publishing and discovering services. The implementation is open source, and is comparable with the proprietary implementation described in Ref. [17]. For the second problem, we describe a proxy architecture, based on Ref. [57], explored in Section 4. For the third problem, we develop an extension of the protocol, which allows queries to include context, as described in Section 5. Finally, we present ideas for solving the last problem in Section 6.

Table 4.4 Code and Memory Footprint of Different mDNS/DNS-SD Implementations

Implementation	Code	Memory
Bonjour by Apple	500KB[a]	/
Ethernet Bonjour for Arduino	14KB	/
uBonjour for Contiki [17]	7.69KB	0.4KB
mDNS/DNS-SD for Contiki[b]	6.51KB	0.7KB

mDNS/DNS-SD, Multicast Domain Name System with DNS-Based Service Discovery.
[a]*Based on the size mDNSResponser.exe on 64-bit platforms. Memory information is unavailable.*
[b]*Available at https://github.com/mstolikj/contiki*

4 PROXY SUPPORT FOR SLEEPING NODES

One of the drawbacks of practical implementations of mDNS/DNS-SD is that SPs are assumed to be constantly online. This comes from the distributed nature of the protocol: if an SP is not online when a query for one of its services arrives, it will not be able to respond to it; thus its services will be undetectable. Additionally, in order to be able to quickly adapt to network changes, such as SPs leaving from the network, services are advertised with relatively short time-to-live intervals (2 min on local area networks). Frequent messaging is an unwanted feature for low-power networks due to the limited battery life of the participating devices. As illustrated in Ref. [42], this behavior introduces a trade-off between signaling frequency, that is, increased traffic in the network, and the risk of discovering nonexisting services.

LLNs often include devices low radio duty cycles, or so-called Sleeping Service Providers (SSPs). SSPs regularly turn off their radios, in order to reduce energy consumption. Therefore, during these offline periods their services are undetectable via the current mDNS/DNS-SD protocol. One approach to facilitate SSPs is to introduce Proxy Servers (PSs) on high-capacity (nonbattery powered) devices. Then, PSs can take over discovery functionality from SSPs. With PSs, we create an overlay network, with SD taking place between SCs and high-capacity devices, and SSPs limited to only registering services with a PS. This enables a more flexible deployment approach, and the creation of a mixed duty cycled/nonduty cycled network.

Of course, PSs can be used for more tasks besides SD. For instance, CoAP relies on PSs to provide translation of CoAP messages to HTTP messages, along with caching support. However, due to the separation of layers, in this chapter we consider only proxy support for SD, and leave the option to extend PSs with additional capabilities as future work.

Within mDNS/DNS-SD, a key factor to include support for PSs is to develop a protocol for delegation of service descriptions from SSPs to PSs. We consider two implementations of such a protocol: active and passive proxy delegation protocols. The distinction between the two protocols is based on the role of the SSP in the proxy selection phase. Both approaches are explained in the forthcoming sections.

4.1 ACTIVE PROXY DELEGATION PROTOCOL

In the active proxy delegation protocol, the SSP is the driving party during the delegation process. Proxy functionality is another service in the network, with multiple PSs hosting different proxy service instances. The SSP initiates the protocol by first searching for a PS. After the SSP has selected a suitable PS instance, it registers with it. Depending on the messaging used, the registration can be preceded by route discovery. Then, on receiving a response from the PS, the SSP begins its sleep cycle or tries to register to a different PS. The registration protocol itself is outside of the mDNS/DNS-SD specification and varies between implementations. In this work, we use the Bonjour implementation by Apple, summarized in Fig. 4.4.

The Bonjour active proxy delegation protocol intends PSs to be located on fixed infrastructure devices as wireless routers, TV boxes, or servers. PSs advertise their service using the _sleep-proxy._udp service type. By reverse engineering the Bonjour protocol we discovered that an SSP registers with a PS using the Dynamic DNS Update (DDNS) protocol [43]. The registration message is a unicast DDNS packet, which contains all RRs that should be hosted at the proxy, and an EDNS0 RR that specifies the lease time of the delegation [44], and ownership information of the SSP [45]. The ownership information is used to transfer the Medium Access Control (MAC) addresses of the SSP to the PS. In the Bonjour implementation, this address is used both to intercept messages destined for the SSP while it is not available and for waking up SSPs using the Wireless Multimedia Extension of 802.11e for wireless SSPs, or using wake-on-LAN packets for wired SSPs. Since we focus on LLNs, using different protocol stacks, we have not implemented these features. The PS server always returns a unicast response, which informs the SSP whether the delegation request was accepted.

4.2 PASSIVE PROXY DELEGATION PROTOCOL

The active proxy delegation protocol requires several messages to be exchanged before the delegation takes place. In order to reduce traffic, we propose a new protocol, called the passive proxy delegation protocol, where the registration request is embedded in the service advertisement. As shown in Fig. 4.5, the SSP adds a parameter in the TXT RR of its service description that signals the request to treat the service advertisement as a registration request. The PS that decides to serve the request acknowledges the delegation by resending the service advertisement. The distinction between advertisements originating from the SSP and cached advertisements from the PS is done using the Authoritative Answer (AA) bit. The purpose of the retransmission is twofold: (1) the SSP knows that someone has handled its request and can start its sleep cycle; and (2) other PSs know that they need not process that advertisement. The protocol can be further optimized by retransmitting only the first SRV RR in order avoid unnecessary distribution of large (fragmented) messages. The SRV RR is unique to the SSP and can be undoubtedly interpreted by the SSP and by other PSs.

Since all messages in the passive proxy delegation protocol are multicast, the proxy delegation is transparent and visible to all nodes in the local network scope. However, this is not an additional privacy concern, as all service advertisements in mDNS/DNS-SD are already transparent to all nodes in the local network scope.

For additional functionality, the PS needs to know the duration of the sleep cycle of the SSP. With the active proxy delegation protocol, this information is sent directly to the PS within the registration message, in the form of the lease time. With the passive proxy delegation protocol, it has to be added within the service advertisement. This information, along with the registration request, can be added as additional parameters within the TXT description of the service advertisement (Fig. 4.6). Similarly, other required parameters, such as the MAC address, can be transferred. The compact nature of these two descriptions is of paramount importance since large service advertisements can lead to packet fragmentation.

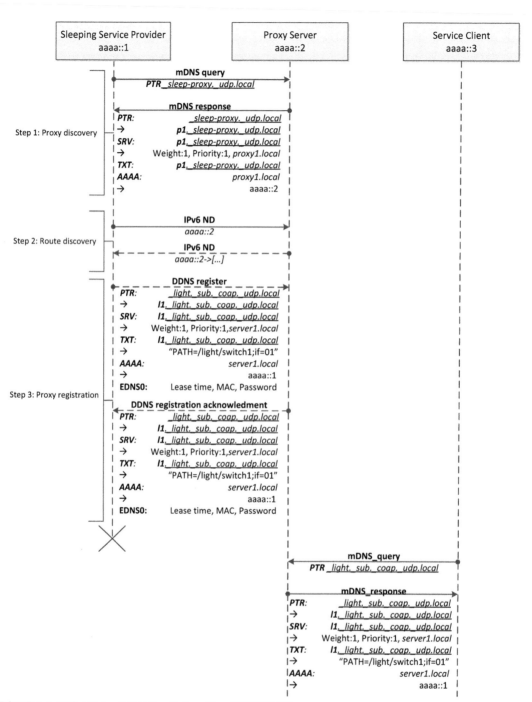

FIGURE 4.4 Active Proxy Delegation Protocol in mDNS/DNS-SD

The service provider selects a proxy service and registers with it. Afterwards, the proxy server responds on behalf of the service provider. *Note*: Full lines portray multicast messages; dashed lines portray unicast messages.

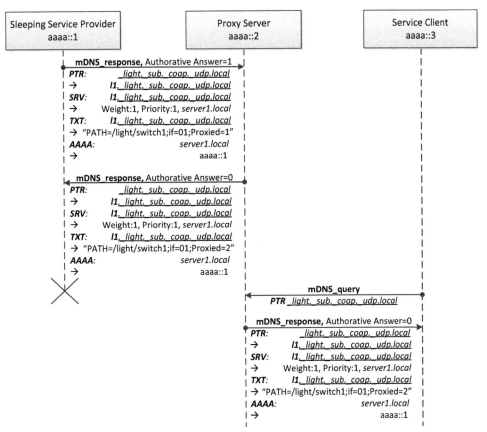

FIGURE 4.5 Passive Proxy Delegation Protocol in mDNS/DNS-SD

The service provider indicates that it wants to be served by a proxy server in the service advertisement. This request is processed by the proxy server, and from there on, it starts responding on behalf of the service provider. The request is acknowledged by resending the advertisement. All messages are multicast.

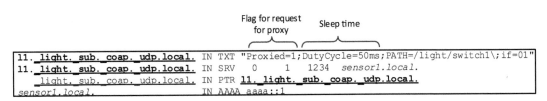

FIGURE 4.6 Embedded Registration Request for the Passive Proxy Delegation Protocol

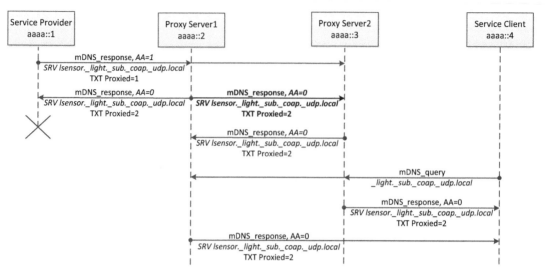

FIGURE 4.7 Possible Double Proxy Registration Using the Passive Proxy Delegation Protocol in mDNS/DNS-SD

If the advertisement resent by the first proxy server does not reach the other proxy server [marked in bold (red in the web version)], another proxy will register the same service. As a result, queries for the service will be responded to twice.

4.3 RELIABILITY

The active proxy delegation protocol does not significantly suffer if advertisement messages are lost. Simply, the proxy registration phase has to be repeated, delaying the start of sleeping time of the SSP. On the other hand, message loss in the passive proxy delegation protocol can lead to double registrations, as shown in Fig. 4.7. While this state is not functionally wrong, it leads to distribution of unnecessary messages in the network. A possible solution to this problem is to use the start-up conflict resolution protocol defined in the mDNS specification, which would resolve the proxy assignment between two or more competing PSs.

4.4 EVALUATION

We compare the performance of the active and passive proxy delegation protocols analytically and by simulating a set of nodes. We are interested in memory footprint, the delay, and energy consumption for registering a service with a PS.

4.4.1 Memory footprint

We implement the active and passive proxy delegation protocols in the Contiki operating system [46], on top of our own mDNS/DNS-SD implementation. The size of the compiled modules, compiled using the msp430-gcc (GCC) 4.5.3 compiler for the Crossbow TelosB nodes, is shown in Table 4.5.

Both the client and server components of the passive proxy delegation protocol are smaller than the corresponding components for the active proxy delegation protocol. This difference is due to the additional complexity required for implementing the DDNS protocol for the proxy registration. Even though the packet format is similar in DDNS and mDNS, additional resources are used for establishing the connection for the DDNS update protocol.

Table 4.5 Code and Memory Footprint of Different Components for Proxy Registration, in Addition to the mDNS/DNS-SD Implementation (Bytes)

Component	Proxy Protocol	Code	Memory
Service provider	Active	1.484	66
Proxy server	Active	1.268	452
Service provider	Passive	1.136	90
Proxy server	Passive	902	432

mDNS/DNS-SD, Multicast Domain Name System with DNS-Based Service Discovery.

4.4.2 Empirical evaluation

The active proxy delegation protocol requires three multicast frames for discovering the PS. After the description of the PS has been found, a route between the SSP and the PS is required. In a single-hop IPv6 network, the route discovery is realized through the Neighbor Discovery Protocol [47] by sending a neighbor solicitation, which fits in one frame. If the neighbor has been previously known and its reachability has to be verified, a unicast frame is sent. Otherwise, another multicast frame is used. The response of the solicitation comes back in the form of a unicast neighbor advertisement. Finally, the registration protocol requires three unicast frames for sending the registration and three unicast frames for the response.

In a single-hop network, a multicast transmission is essentially a broadcast. ContikiMAC implements broadcast traffic by repeating the message over the entire wake-up interval. On the other hand, unicast traffic is implemented by repeating the message until an acknowledgment is received, or an entire wake-up interval completes. Therefore, the time for sending one broadcast frame (t_b) is equal to the duration of the wake-up interval (w), plus any additional back-off by the Carrier Sense Multiple Access with Collision Avoidance (CSMA/CA) protocol (t_o). The back-off is proportional to w. The time for sending a unicast frame (t_u) equals the duration of the CSMA/CA back-off plus the transmission time. In the best case, such as when bursts of frames are sent, the sender and the receiver are in sync. Then, the frame would be transmitted once, and the receiver would immediately acknowledge it (t_{min}). In the worst case, the transmission time is equal to w. Therefore, the delay for completing the active proxy registration protocol can be estimated as follows:

$$D_{ap} = t_{sd} + t_{nd} + t_{rg} = 3(t_b + t_o) + [t_o + t_b + t_o + t_u] + 2(t_o + t_u + 2t_{min})$$

where t_{sd} is the time to discover the PS, t_{nd} is the time to discover a route, and t_{rg} is the time to perform the registration.

The passive proxy delegation protocol uses only multicast frames. Both the initial service advertisement and the repeated service advertisement are sent via two multicast frames. The delay is then estimated as follows: $D_{pp} \sim 4(t_o + t_b)$.

We can simplify the model by assuming that there is no message loss and there is no interfering traffic. In this way we prevent any retransmissions by the CSMA/CA protocol and we limit the duration of the back-off to at most w. Both of these factors influence the delay, as verified by the simulations.

The lower bound of both proxy protocols with the simplified model can then be found by assuming that no back-off takes place (ie, $t_o \sim 0$) and that the sender and the receiver are in perfect sync ($t_u \sim t_{min}$). Similarly, the worst case would then be if the back-off is maximum ($t_o = w$) and that the sender and the receiver are off-sync by w during the first unicast transmission.

Table 4.6 Upper and Lower Bounds on the Proxy Registration Protocols and the Measured Simulated Results for an 802.15.4 Channel With No Loss and Background Traffic

Wake-Up Frequency (Hz)	w (ms)	Proxy Type	Lower Bound (s)	Upper Bound (s)	Simulation Average (s)
4	250	Active	1.03	3.52	2.13
		Passive	0.75	2.00	0.98
8	125	Active	0.53	1.77	1.20
		Passive	0.38	1.00	0.52
16	62.5	Active	0.28	0.89	0.67
		Passive	0.19	0.50	0.29

Therefore, the bounds of the delay of the active and passive proxy delegation protocols are as follows: $D_{ap} \in [4w+7t_{min}; 14w+4t_{min}]; D_{pp} \in [3w+t_{min}; 8w]$. The calculated bounds are shown in Table 4.6.

As expected, due to the smaller number of messages exchanged, the passive proxy delegation protocol finishes much faster than the active proxy delegation protocol. The different wake-up intervals only increase the gap between the two protocols. Furthermore, due to the predictable behavior of the broadcasting protocol in a single-hop network, it is easier to estimate the delay of the passive proxy delegation protocol.

The model becomes more complex in a multihop network, due to the uneven load between multicast and unicast messages. The performance then depends on the implementation of the underlying multicast protocol and the role of packet fragmentation, which is outside the scope of this chapter.

4.4.3 Simulation

We simulate a single-hop scenario in Cooja [48], a cross-level simulator for the Contiki operating system. Cooja internally uses the MSPsim device emulator for cycle accurate Crossbow TelosB emulation, as well as a symbol accurate emulation of the CC2420 radio chip. The test network consists of three Crossbow TelosB nodes—an SSP, a PS, and a dummy node, all located in the same single-hop 6LoWPAN network (Fig. 4.8). The SSP advertises one service, described using four RRs: PTR, SRV, TXT, and AAAA. All four RRs are stored in a single DNS packet, which is then fragmented into two 802.15.4 frames for transport. All nodes use the CSMA/CA MAC protocol together with the Contiki-MAC [49] Radio Duty Cycling (RDC) Protocol with wake-up frequencies of 4, 8, and 16 Hz, which results in wake-up intervals of 250, 125, and 62.5 ms, accordingly.

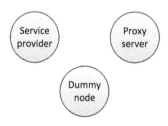

FIGURE 4.8 Test Scenario

All nodes are within the same broadcast domain.

For radio propagation, we use a constant loss rate model, with varying packet delivery ratio between 80% and 100%, at 2% increments. All charts show the mean values of 1000 runs, and the error bars correspond to the 95% confidence interval of the mean. The simulation starts with all nodes being online. At second 5, the SSP advertises its service. At second 7, the SSP starts delegating the service description to the PS using one of the previously described protocols. After the delegation finishes, the SSP turns off its radio, and the simulation is stopped. The dummy node does not participate in the proxy delegation protocol, but it overhears all traffic in the network. We use it to show the impact of the protocols to nodes in the vicinity.

Due to packet loss, the proxy delegation may not succeed in its first iteration. Therefore, we implement repetitions of individual stages of the delegation protocols based on the expiry of fixed time-outs. The time-out for completion of step 1 from the active proxy delegation protocol and the entire passive proxy delegation protocol is set at five times the RDC wake-up interval. The repetition of step 2 of the active proxy delegation protocol is dictated by the neighbor discovery protocol, and the time-out is fixed at 10 s. Finally, the time-out of the entire step 3 is set at 2 s. This should be enough to capture any retransmissions of unicast frames by the CSMA/CA protocol. We compare the performance of the active (*ap*) and passive proxy (*pp*) delegation protocol. To verify the impact of the neighbor discovery protocol (Step 2 of the active proxy delegation protocol from Fig. 4.4), we also implement an optimized version of the active proxy delegation protocol (*ap-hole*), where the neighbor discovery stage is skipped. In this optimized version, the link layer address of the PS is generated from the last 8 bytes of the IPv6 address, present in the AAAA RR of the _sleep-proxy service description.

Fig. 4.9 shows the delay, that is, the time elapsed from starting the proxy delegation protocol until its completion. As expected, increasing the wake-up interval results in much higher delays. With all

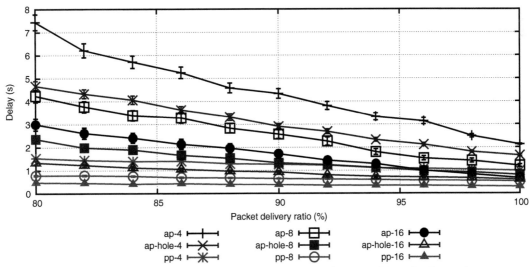

FIGURE 4.9 Required Time (Delay) for Completing the Active and Passive Proxy Delegation Protocol, as Measured by the Service Provider

ap and pp refer to the active/passive proxy delegation protocol, with 4-, 8-, and 16-Hz wake-up intervals for the ContikiMAC Radio Duty Cycling Protocol. ap-hole refers to the optimized version of the active proxy delegation protocol.

three different wake-up parameters, due to the smaller number of messages, the passive proxy delegation protocol finishes much faster than the active proxy delegation protocol. The improvements vary depending on the RDC wake-up interval and the packet loss, from twofold to sixfold. The optimized active proxy delegation protocol also converges faster, up to twofold. The difference between the two versions grows as the packet delivery ratio drops. Still, even the optimized version requires more time to complete than the passive proxy delegation protocol.

Fig. 4.10 shows the energy usage of the SSP and the PS during the proxy delegation. These measurements are profiled using the software power profiler Powertrace [50]. Contrarily to expectations, increasing the wake-up interval actually increases the energy usage for the PS and the SSP. This behavior is due to the RDC Protocol: ContikiMAC makes senders do more work than receivers during transmissions.

The energy usage of the passive proxy delegation protocol is lower compared to the active proxy delegation protocol, although the differences are not as high as with the delay. We attribute this behavior to the efficiency of the ContikiMAC protocol. Furthermore, in the active proxy delegation protocol, if a message loss occurs during the neighbor discovery phase, the sender will be silent for the entire time-out period, which introduces a large delay, but not much energy usage. This is visible in the energy usage of the active proxy delegation protocol and the optimized version with a 4-Hz wake-up interval. Even though the difference between the two in terms of delay is large, they consume similar amounts of energy at the PS side.

Finally, Fig. 4.10C shows the average energy usage of the dummy node in the network. The energy usage is measured during the entire simulation, and includes the mDNS initialization stage, where every node advertises its host name and address. The passive proxy delegation protocol requires less energy due to the smaller number of messages during both the initialization phase and the registration phase. The passive proxy is silent during the initialization phase, while the active proxy advertises the _sleep-proxy service. Surprisingly, the neighbor discovery does not significantly impact the active proxy delegation protocol in this aspect. Namely, the energy footprint of the active proxy delegation protocol is close to the optimized version.

4.4.4 Summary

From the evaluation, we can conclude that in a single-hop network, the passive proxy delegation protocol has better performance when compared to the active proxy delegation protocol. This is evident in memory usage, delay, and energy consumption. For a multihop network, the choice between the two protocols depends on the network topology, the implementation of the multicast forwarding algorithm, and whether a route between the SSP and PS is known beforehand. We leave this as future work.

5 SUPPORT FOR CONTEXT QUERIES

The DNS-SD protocol was designed for discovering all service instances of a given type within a certain logical domain. The protocol assumes that within the given logical domain, either all service instances of the same type provide the same functionality and selecting any of them is enough or an end user can select one. In the former case, when a single service needs to be selected, the priority/weight

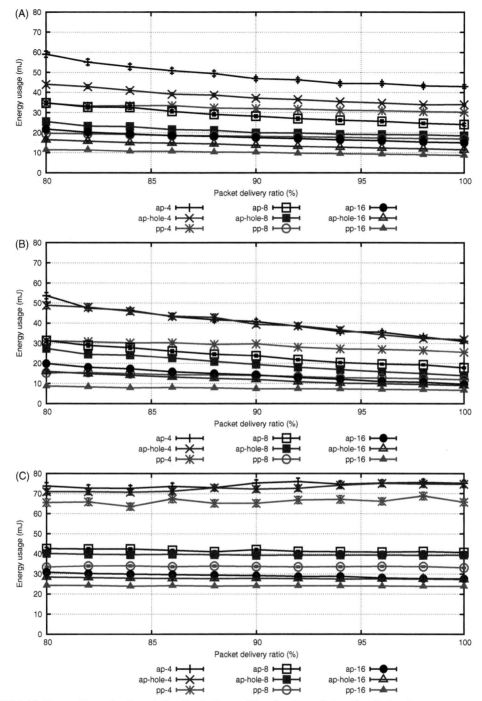

FIGURE 4.10 Energy Usage for Completing the Active and Passive Proxy Delegation Protocols

(A) Sleeping service provider; (B) proxy server; (C) overhearing node. ap and pp refer to the active/passive proxy delegation protocol, with 4-, 8-, and 16-Hz wake-up intervals for the ContikiMAC Radio Duty Cycling Protocol. ap-hole refers to the optimized version of the active proxy delegation protocol. The energy usage of the overhearing node (C) is measured for the entire simulation duration.

fields of the SRV records play an important role. In the latter case, the end user can select a preferred service instance based on the service instance name or based on the human-readable description, added as payload in the TXT RR. However, none of these methods can be added within the queries sent out by SCs, that is, they cannot be used to reduce the number of responses. Moreover, the interpretation of the TXT RR is outside of the DNS-SD specification.

In the IoT domain, services of the same type will be available in abundance. As a result, queries for common service types will result in many responses, which poses a serious burden to the network. Moreover, in Machine-to-Machine (M2M) communication, as between sensors and actuators, the selection of a service has to be done automatically, using information present in service descriptions. In such scenarios, the logical domain criteria are not enough to discriminate between services of the same type. Therefore, the DNS-SD protocol has to be adapted to provide means for stricter selection of services based on criteria present in queries.

Currently, two approaches have been proposed. The first approach, described in Ref. [51], customizes DNS-SD functionality for building control, where services are selected based on location, general type, and subtype. The location information is added in each service description as part of the domain name, while the protocol is added via the type/subtype options. For instance, a light switch located in floor 1, office 1 in the TU/e MetaForum building would have a PTR RR with the name _OnOff_light._sub._bc._udp.o01. f1.mf.tue.nl. In the remainder of the text, we refer to this method as the building-control approach.

The second approach, described in Ref. [52], uses TXT RRs inside queries as a way to impose constraints for service selection. The TXT record contains key–value pairs of entries that specify features required by the services. Then, every SP responds only if the TXT RRs associated with its service contains all entries in the query TXT RR. For example, for the same service as the previous approach, the SC would send a query with two RRs: one PTR, with _light._sub._bc._udp.tue.nl as name, and one TXT, with entries: *type="OnOff," building="mf," floor="1," office="01."* Of course, all location entries can be collapsed into one, as in the building-control approach.

Both approaches are an improvement over the current DNS-SD standard. However, neither is ideal. While the former approach lacks flexibility and is tightly coupled with location as the primary discriminator, the latter approach imposes additional overhead in the size of the queries. Therefore, we propose a mixture of the two approaches: a service description model based on context tags, added as names in PTR RRs.

5.1 CONTEXT TAG DESCRIPTORS

We define that any given service instance can be associated with a set of context tags. A context tag is an atomic descriptor of one context property. We assume an inclusive model: the association of a context tag with a service instance represents that the service instance has that specific context, while the lack of a context tag represents that the associated service instance does not have that specific context. The assignment of the context tags to a service can be done at commissioning time, or at dynamically, at runtime. The exact protocol for this assignment is outside the scope of this work.

SD then consists of sending one or more queries, listing a combination of wanted/unwanted context tags. As a response, a set of services is expected, whose set of tags satisfies the queries. We use Boolean logic to express the queries, using conjunction ($\wedge$), disjunction ($\vee$), and negation ($\neg$) operators on context tags. The discovery model then consists of a:

• set of service instances S;
• set of tags T;

- mapping function $\pi{:}S \to P(T)$ that describes which context tags are associated with a given service;
- set of queries, defined by the grammar $Q := t \mid Q \wedge Q \mid Q \vee Q \mid \neg Q$, with atoms $t \in T$;
- SD function: $Q \to P(S)$, the mapping between queries and a set of matching service instances.

The discovery function is then defined as follows:

$$\sigma(t) = \{s \in S \mid t \in \pi(s)\}$$
$$\sigma(p \wedge q) = \sigma(p) \cap \sigma(q)$$
$$\sigma(p \vee q) = \sigma(p) \cup \sigma(q)$$
$$\sigma(\neg t) = \{s \in S \mid t \notin \pi(s)\}$$
$$\sigma(\neg(p \wedge q)) = \sigma(\neg p) \cap \sigma(\neg q)$$
$$\sigma(\neg(p \vee q)) = \sigma(\neg p) \cup \sigma(\neg q)$$

In Ref. [53], we explore several options how the context model can be implemented as service selection criteria using DNS-SD, with emphasis on the trade-off imposed by added expressive value (ie, more specific queries) and increased complexity. For brevity, here we describe only the most promising one: complete predicates as a selection criterion.

5.1.1 Predicates as selection criteria

With this approach, we encode the entire predicate (σ) inside the PTR name, and use it as criterion for service selection. The predicate is parsed and evaluated for each service instance individually. If the given service instance satisfies the predicate, a response is sent back to the client.

To preserve compatibility with DNS-SD, all context tags and operators have to be in human-readable form (ie, ASCII or UTF characters). Therefore, we use the asterisk (*), dot (.), and hyphen symbol (-) for disjunction, conjunction, and negation operator, respectively. For simplification purposes, we store the predicate in Disjunctive Normal Form (DNF). This enables us to break long formulas in separate DNS RRs. Finally, after the predicate, we add the service type and the logical domain, again to preserve backward compatibility.

An obvious weakness of this approach is that it requires more complex behavior on the SP side. Namely, the SP has to be able to parse and evaluate a Boolean predicate. Therefore, we advise against using nested terms, and utilizing only simple predicates.

An additional problem is posed by lengthy predicates. The DNS-SD specification limits domain names to 253 characters. Therefore, long predicates with many atoms would have to be broken in multiple RRs, which results in additional communication overhead and processing.

5.1.2 Comparison

To compare the three approaches, we use the following scenario. Assume a network of one client and two SPs, with features as shown in Table 4.7. The client wants to discover a light switch for blue lights in office 1.

For the building-control approach, one additional PTR RR needs to be created, with the location included in the domain name. The color description is then part of the TXT RR. The client first needs to discover all light switches for office 1, and then from the responses, check the TXT RRs to find the service instance for blue lights. The complete message exchange is shown in Fig. 4.11.

Table 4.7 Description of a Test Scenario

Property	Light Switch 1	Light Switch 2
Service type	_light._sub._coap._udp.<domain>	
Location	MetaForum, Floor 3, Office 1	
Color	Blue	Red
Resource path	/light/swtich1	/light/swtich2

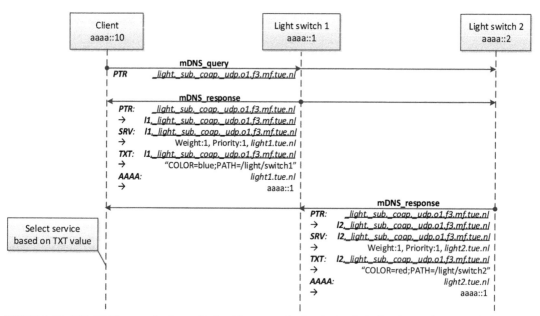

FIGURE 4.11 DNS-SD Message Exchange During Discovery of a Particular Light Service, Using the Building Control Approach

When TXT RRs are used as part of the query, all context features are encoded inside the TXT RR. The client sends the TXT RR with the wanted features together with the PTR RR as a query. Only one response is expected, as the query is very specific. The complete message exchange is shown in Fig. 4.12.

Finally, with our approach, only one PTR RR is sent as a query, containing all requested contexts. As previously, one response is expected. The complete message exchange is shown in Fig. 4.13. Note that in this case, the client does not learn any new context about the SP. If this is needed, it can be transferred in the TXT description of the service, as in the other two approaches.

This example shows the design trade-offs in the three approaches. On the one hand, the building-control approach is simple to implement. However, it does not scale well and it lacks flexibility. As shown, discriminating responses based on a property other than location and service type has to be done by the client. Furthermore, the granularity of the location has to be known in advance, as separate

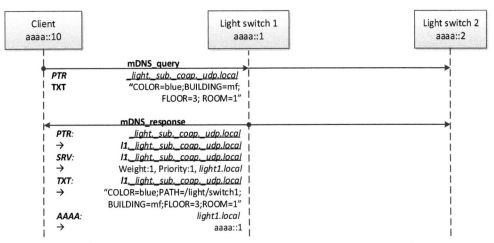

FIGURE 4.12 DNS-SD Message Exchange During Discovery of a Particular Light Service, Using TXT Records as Part of Queries

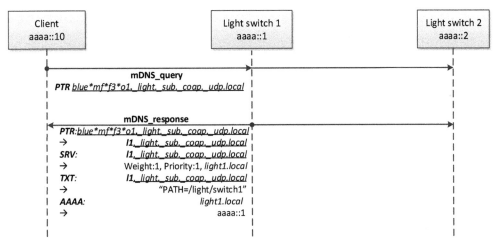

FIGURE 4.13 DNS-SD Message Exchange During Discovery of a Particular Light Service, Using Predicates Inside PTR RRs as Part of Queries

PTR RRs need to be created for any particular location context (eg, finding all light switches in a building/floor/office requires three different PTR RRs).

On the other hand, the latter two approaches enable more flexible filtering of services during the query phase, at the cost of additional complexity at the SP side. Namely, in the TXT query approach, the SP would have to process both PTR host names for matching and entries in TXT RRs and match them with any local services. Similarly, in the predicate PTR approach, the SP would have to both parse the predicate and evaluate it for every local service. However, we believe that the benefits of using such approaches, in early reduction in the number of expected responses, outweigh the implementation issues. A relieving circumstance is that in a fully distributed environment, the number of local services per SP would not be large, so the processing required is not significant.

6 FUTURE WORK: PACKET SIZE

Table 4.8 shows the minimum size of the DNS-SD RRs associated with a single service description. It is obvious that even in the best case, with human-unreadable single-letter service/host names, all four RRs are larger than the maximum available payload size in IEEE 802.15.4 with 6LoWPAN, and require fragmentation.

In Ref. [19], a compression scheme called Adjustable DNS Message Compression (ADMC) Enhanced is proposed, which aims to reduce the combined size of DNS-SD packets using four techniques. In the first step, RRs are combined to use the existing DNS name compression method. Then, the CLASS and TTL fields of all four RRs are collapsed into one. This is a reasonable assumption, since they usually hold the same value for all RRs associated with the same service. Next, the address in the AAAA record is compressed, by linking it to the address assumed to be present in the 6LoWPAN packet. Finally, 6LoWPAN compression (IP header compression/Next-header compression) is applied, yielding a DNS packet that is potentially less than 60 bytes and fits in one 6LoWPAN frame. However, this approach relies on having a relatively small service description, which is contradictory to the context-aware query support described in the previous section.

Table 4.8 Size of DNS Resource Records

RR Type	PTR	SRV	TXT	A/AAAA
Name	query name[a] (1) >14B	service name 2B (ptr->(2))	service name 2B (ptr->(2))	host name 2B (ptr->(3))
Type	2B			
Class	2B			
TTL	4B			
Length	2B			
Priority	–	2B	–	–
Weight	–	2B	–	–
Port	–	2B	–	–
Data field	service name[b] (2) >5B[d]	host name[c] (3) >5B[e]	Arbitrary field 0-1300B	IPv4/6 Address 4/16B
Total	>29B	>23B	>12B	16/28B

RR, Resource Record.
[a] <query name>.<tcp/udp>.local.
[b] <service name>.<tcp/udp>.local.
[c] <host name>.local.
[d] <service name> + ptr → (1).
[e] <host name> + ptr → (1).

Therefore, we believe that additional savings can be reached if service/host names are encoded in binary format inside of RRs. While DNS-SD advocates the use of human-readable names, this is not a necessity in M2M communication, as in the IoT. Furthermore, the encoding/decoding will be hidden by the DNS-SD API, which is responsible for translating RRs to their full, human-readable form.

A similar approach was recently taken for reducing the size of payloads for a network management interface for constrained devices using CoAP [54]. The authors use Concise Binary Object Representation (CBOR) [55] for encoding JavaScript Object Notation (JSON) [56] objects as part of the CoAP payload. In addition, commonly used strings, as paths as part of URIs, are first hashed to generate a 32-bit identifier, and registered at all parties. Then every reference to a known string is replaced with the hash. A similar approach is viable for DNS-SD, where the service types, context tags, host names, and their combination can be exchanged in hashed form. In the best case, all names can be replaced with 4-byte hashes or 2-byte DNS compression lookups, which is enough to avoid packet fragmentation.

7 CONCLUSIONS

In this chapter we analyzed the applicability of mDNS/DNS-SD protocol for SD in the IoT. We identified several key problems in the existing version of the protocol. First, existing implementations are too large to implement on resource-constrained devices. Second, the protocol requires SPs to be constantly online, which is unsuitable for battery-powered devices. Third, the selection capabilities of the protocol are limited to a service type. In a large network with many SPs of the same type, additional selection criteria are needed in order to reduce traffic load. Lastly, the expressive nature of the protocol, with service described with long, human-readable domain names, results in large packets for discovery. In LLNs, due to the limited capacity of the media, large packets are usually fragmented into several smaller ones, and their delivery can be problematic.

To resolve the first problem, we developed a small implementation of the protocol, with a limited set of features. Then, for the second protocol, we proposed the introduction of PSs, which take over SD functionality from SPs. We presented two protocols for the delegation mechanism between the SP and the PS. In the first, active proxy delegation protocol, the SP actively searches and selects the PS for delegation. In the second, passive proxy delegation protocol, the SP only signals its intention that it requires a proxy service: the PS picks up and processes the intention as a registration request. Both protocols were implemented in the Contiki operating system.

The simulations show that in a single-hop network, the passive proxy delegation protocol converges faster, requires less energy, and is smaller in code size when compared to the active proxy protocol. The improvements vary depending on the MAC protocol behavior and the loss in the wireless medium. With a sender-initiated RDC layer, the passive proxy delegation protocol can finish on average up to six times faster than the active proxy delegation protocol, consuming half of the energy.

Next, we proposed an extension of the protocol, which enables services to be described using additional context labels. With the extension, clients can then search for services that satisfy a predicated which contains a set of tags, connected with Boolean operators. As a result, clients can discover very specific service instances among a group of many services of the same type.

As future work, we present ideas for reducing the size of packets associated with SD. This reduction would enable more efficient operation of the protocol in large LLNs. Furthermore, we plan to evaluate the behavior of both protocols in a multihop network. The underlying multicast protocol then plays an important role, and emphasizes the trade-off between multicast and unicast traffic.

ACKNOWLEDGMENT
This work is supported in part by the Dutch P08 SenSafety Project, as part of the COMMIT program.

REFERENCES

[1] Deakin M, Al Waer H. From intelligent to smart cities. J Intell Buildings Int Intell Cities Smart Cities 2011;3.

[2] Cheshire S, Krochmal M. RFC 6762: multicast DNS, <http://www.ietf.org/rfc/rfc6762.txt>; 2013.

[3] Cheshire S, Krochmal M. RFC 6763: DNS-based service discovery, <http://www.ietf.org/rfc/rfc6763.txt>; 2013.

[4] Perianu RM, Hartel P, Scholten H. A classification of service discovery protocols. Technical report TR-CTIT-05-25. Enschede, The Netherlands: University of Twente; 2005.

[5] Zhu F, Mutka MW, Ni LM. Service discovery in pervasive computing environments. IEEE Pervasive Comput 2005;4(4):81–90.

[6] Presser A, et al. Upnp device architecture 1.1, <http://upnp.org/specs/arch/UPnP-arch-DeviceArchitecture-v1.1.pdf>; 2008.

[7] Waldo J. The JINI specifications. 2nd ed. Boston, MA: Addison-Wesley Longman Publishing; 2000.

[8] Guttman E, Perkins C, Veizades J, Day M. Service Location Protocol, version 2, <http://tools.ietf.org/rfc/rfc2608.txt>; 1999.

[9] Bluetooth Special Interest Group. Core version 4.0; 2012.

[10] Nidd M. Service discovery in DEAPspace. Personal Commun 2001;8(4):39–45.

[11] Helal S, Desai N, Verma V, Lee C. Konark – a service discovery and delivery protocol for ad-hoc networks. In: Wireless communications and networking, WCNC, vol. 3; 2003. p. 2107–13.

[12] Schiele G, Becker C, Rothermel K. Energy-efficient cluster based service discovery for ubiquitous computing. In: Workshop on ACM SIGOPS, EW11. ACM; 2004.

[13] Buford J, Burg B, Celebi E, Frankl PG. Sleeper: a power-conserving service discovery protocol. In: Conference on mobile and ubiquitous systems; 2006. p. 1–9.

[14] Ozcelebi T, Lukkien JJ, Bosman R, Uzun O. Discovery, monitoring and management in smart spaces composed of low capacity nodes. IEEE Trans Consumer Electron 2010;56(2):570–8.

[15] Anwar FM, Raza MT, Yoo S-W, Kim K-H. ENUM based service discovery architecture for 6LoWPAN. In: Wireless communications and networking conference, WCNC; 2010. p. 1–6.

[16] Butt TA, Phillips I, Guan L, Oikonomou G. TRENDY: an adaptive and context-aware service discovery protocol for 6LoWPANs. In: Workshop on web of things, WoT; 2012.

[17] Klauck R, Kirsche M. Bonjour Contiki: a case study of a DNS-based discovery service for the Internet of things. In: Conference on ad-hoc, mobile, and wireless networks, ADHOC-NOW; 2012. p. 316–29.

[18] Schoonwalder J, Tsou T, Sarikaya B. Protocol profiles for constrained devices. In: Workshop on interconnecting smart objects with the Internet; 2011.

[19] Klauck R, Kirsche M. Enhanced DNS message compression – optimizing mDNS/DNS-SD for the use in 6LoWPANs. In: Workshop on sensor networks and systems for pervasive computing, PerSeNS; 2013.

[20] Bosman R, Lukkien JJ, Verhoeven R. Gateway architectures for service oriented application-level gateways. IEEE Trans Consumer Electron 2011;57(2):453–61.

[21] da Silva Campos B, Nakamura EF, Figueiredo CMS, Rodrigues JJPC. On the design of upnp gateways for service discovery in wireless sensor networks. In: IEEE symposium on computers and communications, ISCC; 2011. p. 719–22.

[22] Guttman E, Perkins C, Veizades J, Day M. RFC 2608: Service Location Protocol, version 2, <http://www.ietf.org/rfc/rfc2608.txt>; June 1999 [updated by RFC 3224].

[23] Kim KH, Baig WA, Yoo SW, Park SD, Mukhtar H. Simple Service Location Protocol (SSLP) for 6LoWPAN, <https://tools.ietf.org/html/draft-daniel-6lowpan-sslp>; 2010.

[24] Chaudhry SA, Jung WD, Akbar AH, Kim K-H. Proxy-based service discovery and network selection in 6LoWPAN. In: High performance computing and communications. Lecture notes in computer science, vol. 4208; 2006. p. 525–34.

[25] Chauhdary SH, Cui MY, Kim JH, Bashir AK, Park M-S. A context-aware service discovery consideration in 6LoWPAN. In: Conference on convergence and hybrid information technology. Washington, DC: IEEE Computer Society; 2008. p. 21–6.

[26] Marin-Perianu R, Scholten H, Havinga P, Hartel P. Energy-efficient cluster-based service discovery in wireless sensor networks. In: Conference on local computer networks (LCN); 2006. p. 931–8.

[27] Raza MT, Anwar FM, Yoo S-W, Kim K-H. FESP: fast and energy efficient service provisioning in 6LoWPAN. In: Symposium on personal indoor and mobile radio communications, PIMRC; 2010. p. 2575–80.

[28] Shelby Z, Hartke K, Bormann C. Constrained Application Protocol (CoAP), <http://tools.ietf.org/html/draft-ietf-core-coap>; 2013.

[29] Wei Q, Jin Z. Service discovery for internet of things: a context-awareness perspective. In: Fourth Asia-Pacific symposium on Internetware (Internetware); 2012.

[30] Alam S, Noll J. A semantic enhanced service proxy framework for Internet of things. In: IEEE/ACM conference on green computing and communications and conference on cyber, physical and social computing; 2010.

[31] Paganelli F, Parlanti D. A DHT-based discovery service for the Internet of things. Journal of Computer Networks and Communications 2012;2012. 11 pages.

[32] Cirani S, Davoli L, Ferrari G, Léone R, Medagliani P, Picone M, Veltri L. A scalable and self-configuring architecture for service discovery in the Internet of things. IEEE Internet of Things Journal 2014;1(5):508–21.

[33] Jara AJ, Lopez P, Fernandez D, Castillo JF, Zamora MA, Skarmeta AF. Mobile digcovery: a global service discovery for the Internet of things. In: Advanced information networking and applications workshops; 2013.

[34] Germain J. The importance of openness to the internet of things, <http://www.linuxinsider.com/story/81024.html>; November 2014 [online; posted November 9, 2014].

[35] Marineau-Mes S. Why the internet of things needs 'curated openness', <https://gigaom.com/2012/09/21/rim-internet-of-things-mobilize-2012/>; September 2012 [online].

[36] Sublett J. Open to new things, <http://www.oracle.com/us/corporate/profit/big-ideas/070914-jsublett-2244448.html>; August 2014 [online].

[37] Yoon BK. The internet of things needs openness and industry collaboration to succeed, <http://www.samsung.com/us/news/newsRead.do?news_seq=24395>; January 2015 [online].

[38] Butt TA. Provision of adaptive and context-aware service discovery for the Internet of things. PhD thesis. Loughborough University; 2014.

[39] Mockapetris P. RFC 1034: domain names – concepts and facilities, <http://www.ietf.org/rfc/rfc1034.txt>; November 1987.

[40] Mockapetris P. RFC 1035: domain names – implementation and specification, <http://www.ietf.org/rfc/rfc1035.txt>; November 1987.

[41] Winter T, Thubert P, Brandt A, Hui J, Kelsey R, Levis P, Pister K, Struik R, Vasseur JP, Alexander R. RFC 6550: RPL: IPv6 routing protocol for low-power and lossy networks; <http://www.ietf.org/rfc/rfc6550.txt>; 2012.

[42] Tjiong M, Lukkien JJ. On the false-positive and false-negative behavior of a soft-state signaling protocol. In: Conference on advanced information networking and applications (AINA); 2009. p. 971–9.

[43] Damas J, Graff M, Vixie P. RFC 6891: extension mechanisms for DNS (EDNS(0)), <http://www.ietf.org/rfc/rfc6891.txt>; April 2013.

[44] Cheshire S, Krochmal M. Dynamic DNS update leases; 2006.

[45] Cheshire S, Krochmal M. Edns0 owner option; 2009.

[46] Dunkels A, Grönvall B, Voigt T. Contiki – a lightweight and flexible operating system for tiny networked sensors. In: Workshop on embedded networked sensors (Emnets-I); 2004.

[47] Narten T, Nordmark E, Simpson WA, Soliman H. RFC 4861: neighbor discovery for IP version 6 (IPv6), <http://www.ietf.org/rfc/rfc4861.txt>; September 2007.
[48] Osterlind F, Dunkels A, Eriksson J, Finne N, Voigt T. Cross-level sensor network simulation with COOJA. In: Conference on local computer networks (LCN); 2006. p. 641–8.
[49] Dunkels A. The ContikiMAC Radio Duty Cycling Protocol. Technical report, SICS T2011:13; 2011.
[50] Dunkels A. Powertrace: network-level power profiling for low-power wireless networks. Technical report, SICS T2011:15; 2011.
[51] van der Stok P, Lynn K. CoAP utilization for building control, <http://tools.ietf.org/html/draft-vanderstok-core-bc>; 2012.
[52] Aggarwal A. Optimizing DNS-SD query using TXT records, <http://tools.ietf.org/html/draft-aggarwal-dnssd-optimize-query>; 2014.
[53] Buchina N. Extending service discovery protocols with support for context information. Master thesis. Eindhoven University of Technology; 2014.
[54] van der Stok P, Greevenbosch B, Bierman A, Schoenwaelder J, Sehgal A. CoAP management interface, <http://tools.ietf.org/html/draft-vanderstok-core-comi>; 2015.
[55] Bormann C, Hoffman P. RFC 7049: concise binary object representation (CBOR), <http://www.ietf.org/rfc/rfc7049.txt>; October 2012.
[56] Bray T. RFC 7159: the JavaScript Object Notation (JSON) data interchange format, <http://www.ietf.org/rfc/rfc7159.txt>; March 2014.
[57] Stolikj M, Verhoeven R, Cuijpers PJL, Lukkien JJ. Proxy support for service discovery using mDNS/DNS-SD in low power networks. In: IEEE 15th international symposium on a world of wireless, mobile and multimedia networks (WoWMoM); June 2014. p. 1–6.

A SURVEY ON ENABLING WIRELESS LOCAL AREA NETWORK TECHNOLOGIES FOR SMART CITIES

5

N. Omheni*, M.S. Obaidat†, F. Zarai*

*Department of Telecommunications, ENET'COM, University of Sfax, Sfax, Tunisia;
†Department of Computer and Information, Fordham University, Bronx, NY, United States

1 INTRODUCTION

IEEE 802.11 Wireless Local Area Networks (WLANs) became day after day the first tool for accessing broadband Internet all over the world. Slowly, their position as the leading carrier of wireless data traffic, is clearly observed. Statistics show that in the United States, the United Kingdom, Germany, Canada, South Korea, and Japan Wi-Fi has taken part with about 73% of the whole wireless traffic on Android smartphone in 2013; this percentage was 67% in 2012 [1]. This explosion may be thanks to the appearance of a broad range of customer devices, increasing network coverage, continuing technological evolution, and the development of worldwide standards. In the next, we give a view of paradigm change pushed by the evolution of the Wi-Fi technologies. The image, at its center, detains the initiative that Wi-Fi technology, which was in the beginning developed as an Ethernet cable substitute, has become a vital wireless technology all over the world, and will go on developing to stay pace with technical expansion and spectrum availability. In 1997, the first IEEE 802.11 standard was released by the IEEE 802.11 Working Group (WG), characterizing the two layers, Medium Access Control (MAC) and Physical (PHY). Earlier retained versions were IEEE 802.11a, b, g, and n [2] that operate at 2.4- and 5-GHz unlicensed bands. At its early phase, throughput is the main Wi-Fi challenge. After the first data rates defined by the original (1–2 Mbps), several important progresses have been made to increase throughput. In 1999, the publication of the amendment 802.11a was the first attempt to improve throughput. A novel mechanism named Orthogonal Frequency Division Multiplexing (OFDM) that enhances data rate was proposed. In OFDM, a radio signal is divided into multiple subsignals that are sent at the same time at different subcarriers. Afterwards, the IEEE 802.11n standard defines a data rate up to 600 Mbps over 10 times of 802.11a's 54 Mbps. The main engines of these important enhancements are the implementation of many cutting-edge techniques of the time, such as frame aggregation, channel bonding, and Multiple-Input Multiple-Output (MIMO).

In order to achieve prolonged development, novelty, and vitality, Wi-Fi is expected to become more flexible and supple in dealing with its increasing and expanded use in a mixture of scenarios such as data rate and coverage, indoor and outdoor, individual and expert. For that reason, IEEE 802.11 WG and Wi-Fi Alliances (WFA) are discussing and publishing a number of advanced amendments that can be largely grouped into three wide classes: throughput enhancements, long-range extensions, and better ease of use.

Smart Cities and Homes. http://dx.doi.org/10.1016/B978-0-12-803454-5.00005-5

2 DEVELOPMENT OF IEEE 802.11

802.11 group was initiated in 1990 and the IEEE 802.11 standard defines the WLAN was born in 1997. The original standard has defined three PHY layers for the same MAC layer corresponding to three types of 802.11 products:

- IEEE 802.11 Frequency Hopping Spread Spectrum (FHSS), which uses the spread spectrum technique based on frequency hopping;
- IEEE 802.11 Direct Sequence Spread Spectrum (DSSS), which also uses the spectrum spreading technique but on a direct sequence;
- IEEE 802.11 Infrared (IR), infrared type.

The IEEE 802.11 FHSS and IEEE 802.11 DSSS networks are wireless networks transmitting radio in the Industrial, Scientific, and Medical (ISM) band. Given their characteristics, these three types of products are not directly compatible. Thus, an IEEE 802.11 network interface card cannot communicate with an IEEE 802.11 DSSS NIC, and vice versa. Similarly, IEEE 802.11 IR can interact with an IEEE 802.11 FHSS or IEEE 802.11 DSSS.

The IEEE 802.11 standard has not remained static, and many improvements have been made to the original standard. These improvements now continue. Three new PHY layers have been added to the IEEE 802.11b, IEEE 802.11a, and IEEE 802.11g standards:

- IEEE 802.11b or Wi-Fi uses the same ISM band as IEEE 802.11 but with rates up to 11 Mbps. IEEE 802.11b is actually an improvement over IEEE 802.11 DSSS. Thus, a feature of IEEE 802.11b is to remain compatible with IEEE 802.11 DSSS.
- IEEE 802.11a or Wi-Fi 5 uses a new band, called unlicensed-National Information Infrastructure (U-NII) band located around 5 GHz. The flow rate of IEEE 802.11a can reach 54 Mbps, but losing compatibility with 802.11 DSSS and FHSS and 802.11b, due to the use of a different band.
- IEEE 802.11g uses the ISM band but with speeds up to 20 Mbps. This standard uses the OFDM waveform of 802.11a. But unlike IEEE 802.11a, IEEE 802.11g is compatible with IEEE 802.11b and 802.11 DSSS.
- IEEE 802.11n is an evolution of 802.11g that incorporates MIMO technique.

2.1 PROTOCOL STACK

The 802.11 standard describes the first two layers of the Open Systems Interconnection (OSI) model, namely, PHY layer and the data link layer. The latter is further subdivided into two sublayers, the Logical Link Control (LLC) and MAC. Fig. 5.1 shows the architecture of the model proposed by the WG 802.11 compared to the OSI model.

The PHY layer of the 802.11 standard is the crossing point between the MAC layer and the support that allows exchanging information. Each PHY layer 802.11/a/b/g is divided into two sublayers:

- Physical Medium Dependent (PMD), which manages the encoding of data and performs modulation;
- Physical Layer Convergence Protocol (PLCP), which allows the listening of the carrier and provides Clear Channel Assessment (CCA) to the MAC layer indicating that the channel is free.

OSI Layer 2	802.11 Logical Link Control (LLC)					
Data Link Layer	802.11 Medium Access Control (MAC)					
OSI Layer 1 Physical Layer (PHY)	FHSS	DSSS	IR	Wi-Fi 802.11b	Wi-Fi 802.11g	Wi-Fi 802.11a

FIGURE 5.1 Layered Model of IEEE 802.11

2.2 FREQUENCY BANDS

The standard IEEE 802.11 uses frequencies in bands without a license. It uses free bands, which do not require authorization from a regulatory agency. Two unlicensed bands are used in 802.11a/b/g:

- ISM band
- U-NII band

2.2.1 ISM band

The ISM band used in 802.11/b/g corresponds to a frequency band around 2.4 GHz with a bandwidth of 83.5 MHz (2.4–2.4835 MHz). This band is recognized by key regulatory agencies such as the Federal Communications Commission (FCC) in the United States, European Telecommunications Standards Institute (ETSI) in Europe, and French Telecommunications Regulatory Authority (ART) in France. The released bandwidth differs according to the country (see Table 5.1).

2.2.2 U-NII band

The unlicensed band U-NII is located around 5 GHz. It provides a bandwidth of 300 MHz (greater than that of the ISM band, which is equal to 83.5 MHz). This band is not continuous but it is divided into three separate subbands of 100 MHz. In each subband, the permitted emission power is different. The first and second subbands are for indoor transmissions. The third subband relates to outdoor transmissions. As for the ISM band, the availability of these three bands depends on the geographic area. The United States uses all subbands, Europe uses only the first two, and Japan uses the first. Organizations responsible for regulating the use of radio frequencies are as follows: the ETSI in Europe, Kensa-Kentei Kyokai in Japan, and the FCC in the United States.

Table 5.1 ISM Frequency Allocation by Country	
Country	**Frequency Band (GHz)**
United States	2.400–2.485
Europe	2.400–2.4835
Japan	2.471–2.497
France	2.4465–2.4835
ISM, Industrial, Scientific, and Medical.	

2.3 PHYSICAL LAYER

As indicated earlier, the original 802.11 standard has defined three Basic PHY layers: FHSS, DSSS, and IR.

FHSS means a spreading band technique based on frequency hopping in which the ISM 2.4-GHz band is divided into 79 channels, each of 1-MHz wideband. To transmit data, the sender and the receiver agree on a specific sequence of jumps to be carried out on these 79 subchannels. FHSS layer defines 3 sets of 26 sequences, in total 78 possible sequences of jumps. The data transmission is done through a subchannel hopping another jump that occurs every 300 ms, in a predefined sequence. The latter is defined optimally in order to minimize the collision probability between multiple and simultaneous transmissions. If a station does not know the hopping sequence of channels, it cannot retrieve its data.

As the FHSS, DSSS divides the ISM band into subbands. However, the division here is 14 channels of 20 MHz each. The transmission is done only on a given channel. The bandwidth of the ISM band is equal to 83.5 MHz.

As well as these 802.11b utilizes a new PHY layer called High-Rate DSSS. 802.11a and 802.11g standards utilize OFDM. This technique can significantly increase the global debit of the network. A mixture of OFDM modulation techniques is recapitulated in Table 5.2. IEEE 802.11n uses OFDM and MIMO techniques together. The frequency band of most of IEEE 802.11 extensions is 2.4 GHz with 14 separate channels.

802.11a version uses a set of channels varying from 36 to 161, although it uses a fixed channel center frequency equal to 5 GHz. In the United States the number of nonoverlapping channels is equal to 12. This number is 19 channels in Europe [3], while the number of nonoverlapping channels for IEEE 802.11b is limited to 3 only.

2.4 MEDIUM ACCESS CONTROL LAYER

The MAC layer defines how a user obtains a channel to transmit when needed. It uses primitives provided by the PHY layer. It also proposes a standard interface to the LLC layer that can use all data transmission capabilities without knowing the specifics.

Table 5.2 OFDM PHY Layer Modulation Techniques

Data Rate (Mbps)	Modulation Coding Rate Coded Bits/Sub	Coding Rate	Coded Bits/ Subcarrier	Code Bits/ OFDM Symbol	Data Bits/ OFDM Symbol
6	BPSK	1/2	1	48	24
9	BPSK	3/4	1	48	36
12	QPSK	1/2	2	96	48
18	QPSK	3/4	2	96	72
24	16QAM	1/2	4	192	96
36	16QAM	3/4	4	192	144
48	64QAM	2/3	6	288	192
54	64QAM	3/4	6	288	216

BPSK, binary phase shift keying; *OFDM*, Orthogonal Frequency Division Multiplexing; *PHY*, Physical; *QPSK*, quadrature phase shift keying.

There are two access control functions: Distributed Coordinated Function (DCF) and Point Coordinated Function (PCF). The use of the PCF is optional and therefore it is little or not implemented in 802.11 hardware. The PCF consists of a centralized resource management. The access point (AP) orders transmissions and distributes the right to use medium.

DCF uses a distributed algorithm to manage access to channel. This algorithm uses the Carrier sense multiple access with collision avoidance (CSMA/CA) mechanism. It is completed by a random delay pulling mechanism before transmission (random backoff). Each station performs this algorithm locally to determine when it will begin its transmission. The CSMA/CA method is based on a carrier detection function to determine if the medium is busy or not.

In the contention-based mechanism named DCF if more than two stations try simultaneously to transmit, collision takes place. To avoid this kind of problem CSMA/CA can result in wrong medium information. This phenomenon is known as Hidden Node Problem where collision cannot be detected. If stations cannot communicate, the AP appeals to a Request to Send (RTS)/Clear to Send (CTS) mechanism.

2.5 IEEE 802.11 FAMILY AND DERIVED AMENDMENTS

2.5.1 IEEE 802.11g

This version is operating in the 2.4-GHz band and the offered throughput is similar to 802.11a standard. The highest range of IEEE802.11g devices is a little larger than that of IEEE802.11b. However, the range in which users can reach full data rate speed (54 Mbps) is less than that of 802.11b.

2.5.2 IEEE 802.11e

IEEE 802.11e is a version of the IEEE 802.11 standard that introduces improvements in terms of Quality of Service (QoS) at the sublayer MAC of the OSI data link layer. This amendment was approved on September 22, 2005, and published on November 11 of the same year. IEEE 802.11e provides enhancements to the QoS plan for the transportation of voice, audio, and video through WLAN connection.

2.5.3 IEEE 802.11d

IEEE 802.11d specifies a mechanism based on the parameters of transmission related to airwaves signals to set up a mobile station respecting the regulations (regarding frequency ranges and authorized powers) through specific geographic territories and traversed policies.

2.5.4 IEEE 802.11f

The 802.11f is a proposal for the vendors of APs for better interoperability of different manufacturers' products. It offers the Inter-Access Point Protocol, a roaming protocol, that allows a mobile user to switch seamlessly AP while traveling.

2.5.5 IEEE 802.11h

IEEE 802.11h is proposed to harmonize the standard IEEE 802.11a with regulatory requirements of the European Community on radio transmissions in the 5-GHz frequency band and energy savings. This amendment provides dynamic frequency selection mechanisms and control of the power transmission.

2.5.6 IEEE 802.11i

IEEE 802.11i deals with enhancing the security of exchanges in local computer network using a wireless connection. It introduces a robust security network (RSN) algorithm with improvements over

Wired Equivalent Privacy (WEP) security mode recommended by the IEEE 802.11 standard. This amendment increased the authentication and encryption methods.

2.5.7 IEEE 802.11k
This version was designed for radio resource management. This amendment is proposed to improve the way of traffic distribution inside a wireless network. In a wireless environment, all clients try to connect with the AP that offers the best signal. This procedure can cause degradation of network performance because the majority of demands will be on a single AP. With IEEE 802.11k, if the AP having the best signal is overloaded, a wireless station will be connected to another underutilized AP.

2.5.8 IEEE 802.11j
This standard deals with the convergence of IEEE 802.11 American standard and European Hiperlan 2. The 5-GHz frequency band is used for both standards.

2.5.9 IEEE 802.11p
To communicate from vehicle to vehicle in a platoon, American Society for Testing and Materials (ASTM) adopted in 2002 a wireless standard called Dedicated Short-Range Communication (DSRC). In 2003, the IEEE WG has taken over the work to define a new standard dedicated to intercommunications vehicles, called wireless access in vehicular environments (WAVE) and also known as IEEE 802.11p. This standard uses the multichannel concept to ensure communications for security applications and other services of Intelligent Transport.

2.5.10 IEEE 802.11u
This standard deals with interworking with external networks. It allows higher layer functionalities to supply overall end-to-end solutions. The key goals of this standard are assisting network discovery and selection, enabling information exchange with external networks, and supporting emergency services.

2.5.11 IEEE 802.11v
This amendment was adopted in February 2011. It describes network terminal management standards: reporting, channel management, conflict management, and interference filtering service traffic.

2.5.12 IEEE 802.11r
The IEEE 802.11r, named Fast Basic Service Set Transition, was published in 2008. It allows a wireless device that moves to stay connected. 802.11r has been launched to try to alleviate the additional burden imposed by the security and QoS during handover.

2.5.13 IEEE 802.11s
IEEE 802.11s, known as the network Mesh [Wireless Mesh Networks (WMNs)] and derived from military research, allows connected entities to create a WMN in a dynamic manner. IEEE 802.11s extends the 802.11 MAC standard by introducing an architecture and a protocol that enable both unicast and broadcast transmissions by the use of radio-aware metrics.

2.5.14 IEEE 802.11w
IEEE 802.11w is the standard that defines management frames protected for the family of IEEE 802.11 standard. The aim of this amendment was to improve safety by ensuring the confidentiality of data

contained in the management frames, via mechanisms that ensure data integrity and authenticity of data origin and anti-reexecution protection.

2.5.15 IEEE 802.11n

IEEE 802.11n, ratified in September 2009, achieves a maximum throughput of up to 450 Mbps on each of the working frequency bands (2.4 and 5 GHz). It improves the previous standards: IEEE 802.11a for the frequency band of 5 GHz, and IEEE 802.11b and IEEE 802.11g for the frequency band of 2.4 GHz. The purpose of this amendment is to increase the range and throughput of Wi-Fi networks and promises 300 Mbps (theoretical), and does not exclude the possibility to do better. To achieve this, 802.11n uses the MIMO technique. A radio signal transmitted between two points is supposed to go straight. In fact, there is nothing because the waves are reflected everywhere on walls, ceilings, furniture, etc., and the receiver does not receive a single signal but several offset signals depending on the length of each trip. MIMO will take advantage of these multipaths, considered rather disastrous in radio networks, to optimize the transmission between two points. The principle is to use multiple transmit antennas and multiple antennas for reception. It is not necessary that there is an equal amount on each side. Antennas transmit the same signals, out of phase so that the power transmitted at each time is maximal.

3 SMART CITY SOLUTIONS

Smart City solution consists of smart industry, smart government, and smart life. It mixes city management, company development, and way of people life in a ubiquitous Information and Communication Technologies (ICT)-based system through the full use of cloud computing, communications network, and information cooperation. Smart city supports stable urban progress and best possible resource exploitation, improves urban intelligence coverage, and boosts operational effectiveness and citizen approval. Fig. 5.2 shows an example of a smart city solution.

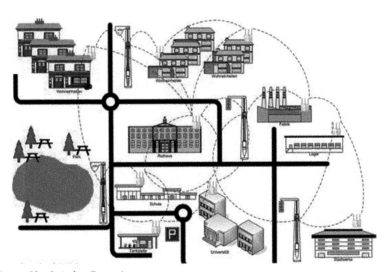

FIGURE 5.2 A Smart City Solution Example

Smart industry encourages a city's permanent economic development as good as digital and intelligent advances to draw tactical investment and complicated companies to its high-tech squares. A good example of the smart industry solution can be Smart Park that offers professional IT services and cloud-based data centers to assist enterprises to build up businesses and urban economy.

As the core of urban managing, a public government can find solutions to the imposed challenges from public safety, emergency handling, urban transport, energy consumption, and green safety. Smart Government makes use of a cloud-based data core to store and share information, and cooperates with various sectors to enhance resource utilization and optimize governmental efficiency.

Smart life makes easy smart addition among intelligent applications and interactive systems for real-time communication, education, and healthcare. Smart life picks up a city's service point and citizen fulfillment, rising quality of life. Fig. 5.3 summarizes all smart city solutions.

3.1 SMART CITIES BASED ON WLAN USE CASES

The study of the sub–1-GHz license-exempt bands [aside from TV white spaces (TVWS)] by the IEEE 802 LAN/MAN Standards Committee in 2010 showed that this spectrum is highly promising for outdoor WLAN transmission. Due to the shortage of the existing spectrum, using broadband is not allowed particularly in 802.11n and 802.11ac. However, the new modulation techniques and coding schemes (MCS), proposed in the two amendments 802.11ac and 802.11ah, can offer a rate that can

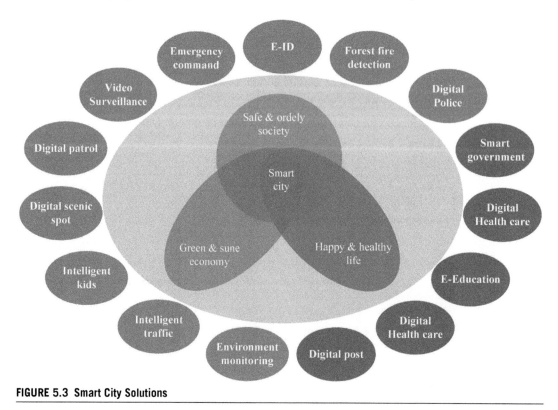

FIGURE 5.3 Smart City Solutions

exceed 300 Mbps in a good channel condition. In parallel, sub–1-GHz license-exempt bands allow to have better propagation behavior in outdoor situation than original for Wi-Fi 2.4- and 5-GHz bands, which extend transmission range over 1 km in normal conditions.

3.1.1 Smart sensors and meters

Smart sensors and meters using efficient modulation and coding schemes with good propagation characteristics and moderately narrow bands permit to introduce a new ability for sensor networks, which excels Bluetooth and ZigBee in coverage, throughput, and power consumption. The majority of considered use cases are geared toward sensing applications [4] such as:

- smart grids;
- smart meters (water, gas, and power consumption);
- automation of industrial practice (pharmacy, iron, steel, and petroleum refinement);
- agricultural and environmental monitoring (forest fire detection, pollution, humidity, temperature, water level, etc.);
- older care system (fall discovery, pill bottle check);
- indoor healthcare and fitness room.

In these use cases, an AP periodically sends out small packets to hundreds of equipment (sensors/actors). Many stations competing for the channel can result in collisions. Wide transmission range results in elevated interframe spaces and high overhead. The real throughput in these use cases does not exceed 1 Mbps. Another limit is the power consumption, as sensors are frequently battery powered. These limits must be carefully studied by standardization institutions in the future [5].

3.1.2 Extended-range hotspot

The characteristics of the sub–1 GHz such as the high data rates and the extended transmission ranges make this band very suitable for traffic off-loading inside mobile networks and for expanding hotspot range. Although 802.11n and 802.11ac throughput is equal to or even elevated than the new mobile network rates [Long-Term Evolution (LTE)], it can be utilized for off-loading in outdoor cases owing to short transmission range. On the contrary, the extension of the transmission range in 802.11ah will give a greater value, particularly in countries with wide existing S1G channel, for example, the USA [6]. In this direction, the 802.11 ah must supply at least one operation mode capable to attain a greatest aggregate multistation throughput of 20 Mbps at the PHY layer.

3.1.3 Backhaul aggregation

The final use case to make network technology for smart cities concerns backhaul link connecting IEEE 802.15.4g devices and distant servers [7]. This type of device is extensively used in industry and it is characterized by low energy consumption and very low data rates.

IEEE 802.15.4g routers collect information from devices and transmit them to the servers through 802.11ac links. These links may also be used to transmit surveillance videos obtained from autonomous cameras. Fig. 5.4 shows an enabling networking technology use case for smart cities.

3.2 ENABLING WLAN TECHNOLOGIES FOR SMART CITIES

Next generations of WLANs have evolved by incorporating the latest technological advances in the field telecommunication. The IEEE 802.11n standard chooses Single-user MIMO (SU-MIMO) technology,

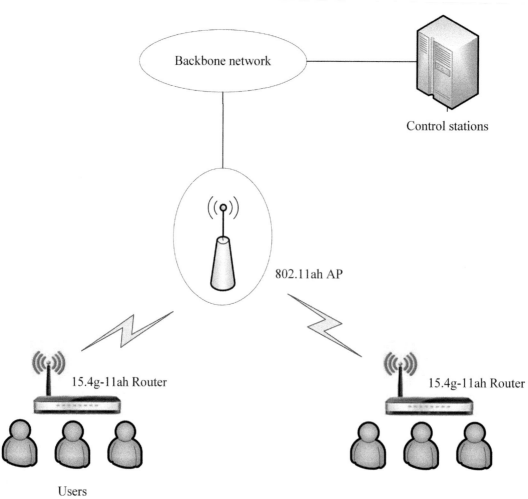

FIGURE 5.4 Enabling Networking Technology Use Case for Smart Cities

packet aggregation, and channel bonding. Those methods were more expanded in the IEEE 802.11ac standard that was published in 2013 to reach data rates about 7 Gbps. In addition, upcoming standards such as the IEEE 802.11af and the IEEE 802.11ah are expected to supply new WLAN application scenarios including long-range communication solutions, cognitive radio, advanced power-saving methods, and maintenance for Machine-to-Machine (M2M) equipment.

3.2.1 IEEE 802.11ad

The IEEE 802.11ad, called also WiGig, is a pretty new standard that was published in December 2012. Its requirement includes a "fast session transfer" mechanism [3]. The IEEE 802.11ad offers the capacity to change bands guaranteeing that active devices are regularly best connected in order to supply good concert and range criteria. Many clients in a wide deployment can keep top data rate performance,

Table 5.3 IEEE 802.11ad Key Features	
Parameter	**Description**
Operating frequency range	ISM band: 60 GHz
Maximum data rate	7 Gbps
Typical distances	1–10 m
Modulation	Single carrier and OFDM
Antenna technology	Operates beamforming
ISM, Industrial, Scientific, and Medical; *OFDM*, Orthogonal Frequency Division Multiplexing.	

without causing interference with one another or having to normally gash bandwidth similar to the legacy frequency bands [8]. The IEEE 802.11ad uses frequencies in the millimeter-range microwave Wi-Fi and provides radio coverage of a few meters. The goal of this very small range is to exchange a high volume information, for example, HD video transfers.

The amendment applies a MAC layer standard that is common with existing 802.11 standards to allow seamless session handoff between 802.11 WLAN using the 2.4- and 5-GHz bands and those working in the 60-GHz bands. On the other hand, new specifications related to 802.11ad MAC layer have been proposed such as synchronization, authentication, association, and channel access required for the 60-GHz band operation. The key features of this standard are outlined in Table 5.3.

The 802.11ad system operates in the ISM band and frequencies are between 57 and 66 GHz according to geographic location as shown in Table 5.4. The PHY layer of the 802.11ad supports three modulation techniques: spread spectrum modulation, single carrier (SC) modulation, and OFDM modulation. The used OFDM modulation allows achieving high throughput while maintaining excellent resistance alongside multiple path propagation.

The key signals specified by the 802.11ad PHY amendment are as follows [9]:

- *Control PHY*: This signal utilizes code spreading, differential encoding, and binary phase shift keying (BPSK) modulation to provide control. The Control PHY allows elevated levels of error detection and correction with a comparatively low debit.
- *SC PHY*: The modulation techniques used for this type of signal are quadrature phase shift keying (QPSK), BPSK, and 16-quadrature amplitude modulation (16QAM) on a concealed carrier

Table 5.4 60-GHz Global Frequency Allocation	
Region	**Allocation (GHz)**
European Union	57.00–66.00
USA and Canada	57.05–64.00
South Korea	57.00–64.00
Japan	59.00–66.00
Australia	59.4–62.90

positioned on the middle of channel. The symbol rate of this signal has been set at 1.76 Gsym/s and a mixture of error coding types is defined.

- *OFDM PHY*: This signal involves using multicarrier modulation technique to supply elevated modulation concentrations and elevated throughput.
- *Spread QPSK*: This 802.11ad signal engages in OFDM carrier pairs with which information are modulated. A maximum separation between the two carriers is needed in order to get better strength of the signal in case of selective fading of frequencies.
- *Low Power SC PHY*: This signal involves the use of a SC to reduce the energy utilization. It is planned for little battery equipment that is not clever to maintain the processing necessary for the OFDM design.

3.2.2 IEEE 802.11ae

The 802.11ae amendment has mainly two new mechanisms for processing management frames: a new mechanism for the supple prioritization of management frames and a new signaling algorithm for the exchanging frame prioritization policies. The prioritization mechanism is entitled the QoS management frame (QMF) service. QMF is a policy that gives a mapping between the management frame types and subtypes and the enhanced distributed channel access (EDCA) mechanism where the QoS is supported by introducing different access categories (ACs; EDCA ACs). In this way, all management frames are transmitted in an AC as specified by the existing QMF policy (Fig. 5.5).

This amendment defines a default QMF policy. But usually the QMF policies are flexible, and they can be recognized and updated using a signaling protocol defined in 802.11ae. So, this aspect permits the QMF service to be adjusted to vendor application necessities. The procedure of the signaling protocol runs according to the type of network: mesh or infrastructure. Here, it should be noted that the use of QMF is not allowed in an independent basic service set (IBSS, ad hoc) network by the IEEE 802.11ae. In the infrastructure mode, QMF policy is defined by the AP for the entire basic service set (BSS). In the mesh network, a mesh station (MS) can set the QMF policy with another MS on a per-link basis. The QMF policy can be broadcast into existing frames like beacons or using new dedicated frames as mentioned in Table 5.5.

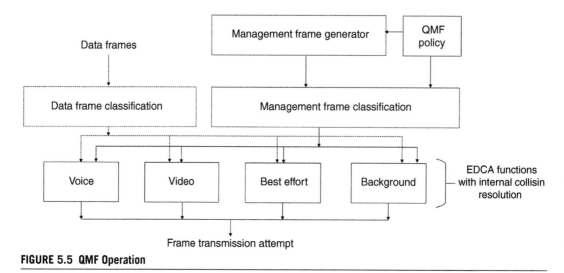

FIGURE 5.5 QMF Operation

Table 5.5 Default QMF Policy (Omitted Frames Are Assigned to BE) and Policy Dissemination Frames

Type of Frame	Description	QMF Access Category	Dissemination of QMF Policies	
			Infrastructure BSS	Mesh BSS
(Re) Association Request/ Response	Handover between APs	VO	Yes (in responses)	No
Probe Request (individually addressed)	Scanning initialization (unicast)	VO	No	No
Probe Response	Scanning result	BE	Yes	No
Beacon, ATIM, Disassociation, Authentication, Deauthentication	Network maintenance	VO	Yes (beacon)	No
Channel switch announcement	Initialization of channel switching	VO	No	No
Extended channel switch announcement	Initialization of extended channel switching	VO	No	No
QoS frames	QoS signaling (eg, TSPEC exchange)	VO	No	No
Measurement pilot	Basic scanning information	VO	No	No
Tunneled Direct-Link Setup Discovery Response	Part of direct-link setup	VO	No	No
Fast BSS Transition	Prehandover setup to speed up the handover process	VO	No	No
High-Throughput frames	Support for data rates greater than 100 Mbits/s	VO	No	No
Security Association Query frames	Procedure for robust management frame protection	VO	No	No
QMF Policy and QMF Change Policy	Dedicated frames for the dissemination of QMF policies	BE	Yes	Yes
Hybrid Wireless Mesh Protocol Mesh Path Selection	Path selection in mesh BSS	VO	No	No
Congestion Control	Congestion information dissemination in Mesh BSS	VO	No	No
Self-Protected frames	Management of security associations	VI	No	No
Disablement of Dynamic Station Enablement	Related to the operation in the 3650- to 3700-MHz band in the United States	VO	No	No

AP, access point; *BSS*, basic service set; *QMF*, QoS management frame; *QoS*, Quality of Service; *ATIM*, announcement traffic indication map; *TSPEC*, traffic specification, *BE*, Best Effort.

3.2.3 IEEE 802.11ac

The aim of this standard is to get better Wi-Fi customer practice by offering considerably higher data rates for current application fields, and to permit devices operation below 6 GHz with distribution of various data flows. With throughput more than 1 Gbps and numerous innovative characteristics, data rates and application offered by IEEE 802.11ac assure to be like those of current wired networks. The 802.11ae amendment enhances the existing 802.11n version to attain the Very High Throughput (VHT). To achieve this purpose, the 802.11ac has added a number of optional factors to those that are obligatory. The optional factors are modulation, channel bandwidth, and number of spatial streams.

802.11ac devices are operating in the 5-GHz band. The alternative to limit handling in this frequency band is principally forced by the wider channel bandwidth requirements for 802.11ac. As the bandwidth augments, channel layout becomes challenging, mainly in the crowded 2.4-GHz band. As in the case of the relatively large 5-GHz band, devices will require to adjust regular radio tuning abilities to exploit the offered resources.

As well as the 20- and 40-MHz bands used by the majority of 802.11n devices today, the 802.11ac amendment specifications define a compulsory 80-MHz channel band. The main advantage of this larger bandwidth is to double the PHY layer rate above that of 802.11n at insignificant charge boost for the chipset producer and to support new applications. Another optional 160-MHz channel bandwidth is specified in the amendment, which can be also contiguous or noncontiguous (80 + 80 MHz).

The advantage of 160-MHz PHY, compared with 40/80-MHz transmissions, is the reduction of the requirement complexity that lets devices reach gigabits per second wireless throughput and support new applications. On the other hand, 160-MHz bandwidth in the 5-GHz spectrum is not available all over the world, and implementations to support these specifications will be probably expensive. Fig. 5.6 and Table 5.6 illustrate the spectral mask specifications for 802.11ac devices.

The modulation technique used in 802.11ac applies the 802.11n OFDM. Specially, the two 802.11n and 802.11ac technologies involve device supporting QPSK, BPSK, 16QAM, and 64QAM modulation.

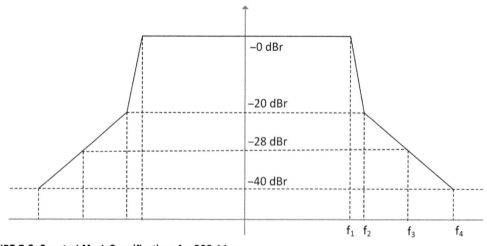

FIGURE 5.6 Spectral Mask Specifications for 802.11ac

Table 5.6 Spectral Mask for 20, 40, 80 and 160 MHz

Channel Size (MHz)	f_1 (MHz)	f_2 (MHz)	f_3 (MHz)	f_4 (MHz)
20	9	11	20	30
40	19	21	40	60
80	39	41	80	120
160	79	81	160	240

On the other hand, two main dissimilarities can be noticed compared to the IEE 802.11n standard. Primarily, the 802.11ac comprises an approved constellation mapping improvement [optional 256QAM (3/4 and 5/6 coding rates)]. The 256QAM modulation was defined as an optional mode, as opposed to an obligatory mode. Second, the number of defined MCS indices is really minimized.

Many cases of 802.11ac scenarios between an AP and another 802.11ac user device can be defined as illustrated in Table 5.7.

3.2.4 IEEE 802.11ah

IEEE 802.11ah is a promising WLAN standard that introduces a WLAN system operating at sub–1-GHz license-exempt bandwidths. The draft standard IEEE 802.11ah-D1.0 [10] was published in October 2013 and the work will be completed by 2016 [11].

Table 5.7 Examples of 802.11ac Configurations

Configuration	Typical Client (STA) Form Factor	PHY Link Rate	Aggregate Capacity
1-Antenna AP, l-antenna STA, 80 MHz	Mobile Phone, Mobile Entertainment Device	433 Mbps	433 Mbps
2-Antenna AP, 2-antenna STA, 80 MHz	Tablet, Laptop, Networked Game Console	867 Mbps	867 Mbps
1-Antenna AP, l-antenna STA, 160 MHz	Mobile Phone, Mobile Entertainment Device	867 Mbps	867 Mbps
2-Antenna AP, 2-antenna STA, 160 MHz	Tablet, Laptop, Networked Game Console	1.73 Gbps	1.73 Gbps
4-antenna AP, four 1-antenna STAs, 160 MHz (MU-MIMO)	Mobile Phone, Mobile Entertainment Device	867 Mbps to each STA	3.47 Gbps
8-Antenna AP, 160 MHz (MU-MIMO) • One 4-antenna STA • One 2-antenna STA • Two 1-antenna STA	TV Set-Top Box, Tablet, Laptop, Networked Game Console, Mobile Phone	3.47 Gbps to 4-antenna STA 1.73 Gbps to 2-antenna STA 867 Mbps to 1-antenna STA	6.93 Gbps
8-Antenna AP, four 2-antenna STAs, 160 MHz (MU-MIMO)	TV Set-Top Box, Tablet, Laptop, PC	1.73 Gbps to each STA	6.93 Gbps

AP, access point; *MU-MIMO*, Multiuser Multiple-Input Multiple-Output; *PHY*, Physical; *STA*, Station

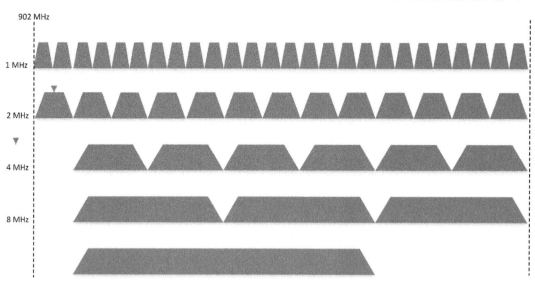

FIGURE 5.7 US Channelization

The IEEE 802.11ah PHY layer used the same basis as the 11ac standard and is adapted to available S1G bandwidth. The channel bandwidths supported by the 11ah are 10 times narrower than those in .11ac. These channels are 1, 2, 4, 8, and 16 MHz, where just 1- and 2-MHz channels are obligatory. After numerous S1G regulation studies in many countries, this amendment was confronted by the difficulty that the offered bandwidths for S1G ISM transmission differ from one country to another; the present draft of the standard defines which channels shall be used in the following countries: the United States, Japan, Europe, South Korea, China, and Singapore. Fig. 5.7 shows channelization opportunities for the United States [12]. PHY properties are similar to those of the .11ac: the use of OFDM technique, MIMO, and Downlink Multiuser MIMO (MU-MIMO).

The amendment defines some standard data rates for a variety of bandwidths and MCSs as shown in Table 5.8. They can be enhanced by minimizing the OFDM symbol period and applying numerous spatial streams.

In legacy 802.11 infrastructure networks, the MAC header length that contains three MAC addresses is 30 bytes. The field Frame Check Sequence (FCS) adds 4 bytes. Therefore, for a 100-byte payload (messages), MAC header overhead is above 30%. For shorter messages, the overhead is still considerable.

Several limitations have been mentioned in the existing WLAN versions regarding power saving especially when the number of devices in the network increases. One of the challenges is long length of the beacon frame because of the extreme length of the partial virtual bitmap in TIM (Traffic Indication Map) IE (Information Element). Additionally, if the quantity of the buffered traffic is too important to be admitted inside a beacon interval, some power-saving machines unavoidably stay in a wake status to achieve the receptions of their buffered data.

To cope with these limitations, the 802.11ah amendment introduces a new technique called TIM and page segmentation. An AP divides the whole partial virtual bitmap equivalent to one page into numerous page segments, and each beacon is in charge for carrying the buffering status of only a one page segment. Next, each power-saving station awakes at the transmission moment of the beacon that

Table 5.8 Some Regular Data Rates for Various Bandwidths and MCSs

MCS	1 MHz	2 MHz	4 MHz	8 MHz	16 MHz
MCS0	0.3	0.65	1.35	2.925	5.85
MCS1	0.6	1.30	2.70	5.850	11.70
MCS2	0.9	1.95	4.05	8.775	17.55
MCS3	1.2	2.60	5.40	11.700	23.40
MCS4	1.8	3.90	8.10	17.550	35.10
MCS5	2.4	5.20	10.80	23.400	46.80
MCS6	2.7	5.85	12.15	26.325	52.65
MCS7	3.0	6.50	13.50	29.250	58.50
MCS8	3.6	7.80	16.20	35.100	70.20
MCS9	4.0	–	18.00	39.000	78.00
MCS10	0.15	–	–	–	–

MCS, modulation techniques and coding schemes.

takes the buffering information of its segments. A new information element named segment count IE is defined to bring segmentation information. Fig. 5.8 illustrates an example of segmentation.

Moreover, the IEEE 802.11ah defines a new mechanism, named Multicast AIDs (MIDs), to multicast groups. Each station asks for MID from the AP if it fits into a multicast group. In the demand, the device must indicate its group MAC address and its preferences for listen duration. Since different stations of the same group may have diverse preferred listen periods and different power restrictions, the amendment permits to have many MIDs for each group. When the AP buffers data to a multicast group, it puts the equivalent bit in TIM so all machines of the group stay awake to receive data via the downlink; however, other stations do not squander power.

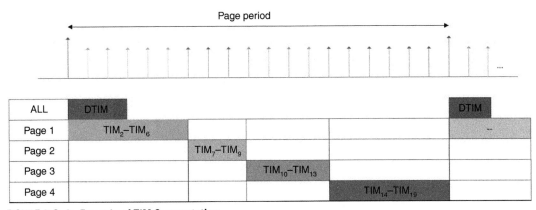

FIGURE 5.8 An Example of TIM Segmentation

3.2.5 IEEE 802.11af

IEEE 802.11af, also called White-Fi and Super Wi-Fi [13], is a wireless standard in the 802.11 family that permits WLAN operating in TVWS spectrum in the very-high-frequency (VHF) and ultrahigh-frequency (UHF) bands. The standard was accepted in February 2014 [14]. The new specifications of IEEE 802.11af are conceived to supply geolocation server access to formerly unavailable, unused, or underused frequencies to allow customers worldwide to benefit from unused or underused spectrum, based on location and time of day. IEEE 802.11af uses 6-, 7-, and 8-MHz channels, allowing thereby compatibility with existing international TV band allocations. Operation may be in one to four channels, both contiguously and in two noncontiguous blocks, offering a means for equipment to collect enough bands in a fragmented TV band spectrum to give high throughput.

TVWS is for the moment available spectrum resources in VHF and UHF bands that are initially licensed to the TV broadcasters and wireless microphones, which can be opportunistically used by unlicensed devices only if no destructive interference is forced on the licensed users. TVWS operate in 470–790 MHz in Europe and the United Kingdom, and noncontinuous 54–698 MHz in Korea and the United States, as illustrated in Fig. 5.9.

The 802.11af amendment presents the WLAN operations at TVWS to bring the so-called "Super Wi-Fi." Due to the good propagation characteristics of such low-frequency bands in contrast to 2.4- and 5-GHz bands, comprising lowered path loss and improved wall-penetrating ability, the Super Wi-Fi signal can be transmitted over longer distances than the original Wi-Fi signal. Consequently, over-the-air broadband access can be applied at minor cost by deploying 802.11af APs much less densely. IEEE 802.11af authorizes a function under the firm regulatory constraints, based on location-aware equipment and Geolocation Database. IEEE 802.11af makes use of most advanced techniques of 802.11ac such as MU-MIMO. Table 5.9 presents the main characteristics of this standard.

3.3 EXAMPLES OF SMART CITY SERVICES AND APPLICATIONS

Examples of hyperconnected smart cities are illustrated in Ref. [15] and comprise Songdo in South Korea, Masdar in the United Arab Emirates, more than dozens of cities in China, and a lot of cities in the world. In the United Kingdom, Bristol City Council is currently building a smart city in the context of the Sensor Platform for HealthCare in a Residential Environment project (SPHERE project, 2013–18) to supervise the health and well-being of residents in domicile [16]. Bristol also has a more wide smart city plan targeting home healthcare services.

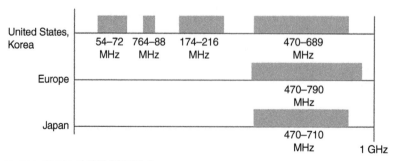

FIGURE 5.9 Frequency Bands of IEEE 802.11af

Table 5.9 IEEE 802.11af Key Features

Parameter	Description
Operating frequency range	470–510 MHz
Channel bandwidth	6 MHz
Transmission Power	20 dBm
Modulation format	BPSK
Antenna Gain	0 dBi

BPSK, binary phase shift keying.

In the United Kingdom, Love Clean Streets [17] is a new application that allows Internet-connected peoples to use their mobile phones' built-in camera and Global Positioning System (GPS) to back up and inform their local authorities about any neighborhood or environmental focuses or offenses when they are not at home. Users are able to consult later online the Love Clean Streets website to check their report status. City committees across the United Kingdom are responding on time to citizens' report request through the Love Clean Streets mobile application.

A new project called the "Internet of Everything" was launched in Barcelona [18]. The "Internet of Everything" plays the role of the backhaul around which scientific projects are being undertaken in Barcelona, rather than doing initiatives in other places in silo. Around 500 km of subversive fiber network is being established as the city performs regular repairs to its road, rail network, and other underground services, which can diminish equipment installation costs considerably.

4 REQUIREMENTS

Requirements of smart city design and achievement can be summarized in the following points:

- *Interoperability*: An obvious tendency is the interconnection of all the possible data sources inside a global infrastructure to be able to support any new service.
- *Scalability*: Networks must have the ability to accept anew services and a large number of users that is about to change.
- *Fast deployment*: The deployment of new networks and updating existing infrastructure need to be faster and easier as possible in order to encourage the quick installation of equipment.
- *Robustness*: Life in a smart city requires a number of services. Consequently, the communication should have a certain level of robustness to give guarantees in terms of availability under normal conditions and also extreme conditions.
- *Limited power consumption*: In obedience to rules posed in the field of green communications and recent activities of smart cities, installed infrastructure and devices must have a minimum impact on the environment and should be characterized by low energy consumption and to reduce their operating and management costs.

REFERENCES

[1] Roberts M. Understanding today's smartphone user: an updated and expanded analysis of data-usage patterns in six of the world's most advanced 4G LTE markets. White paper. Mobile World Congress Barcelona 2015; June 2013.

[2] IEEE 802.11-2012, part 11 wireless LAN medium access control (MAC) and physical layer (PHY) specifications, IEEE standard; March 2012.

[3] Perahia E, Cordeiro C, Minyoung P, Yang LL. IEEE 802.11 ad: defining the next generation multi-Gbps Wi-Fi. In: The 7th IEEE consumer communications and networking conference (CCNC); 2010. p. 1–5.

[4] de Vegt R. Potential compromise for 802.11ah use case document, <http://mentor.ieee.org/802.11/dcn/11/11-11-0457-00-00ah-potential compromise of-802-11ah-use-case-document.pptx>; 2011.

[5] Obaidat MS. Key enabling ICT systems for smart homes and cities: the opportunities and challenges. Keynote speech. In: Proceedings of the 2014 IEEE international conference on network infrastructure and digital content (IC-NIDC 2015), Beijing, China; September 2015.

[6] Aust S, Prasad R, Niemegeers I. IEEE 802.11ah: advantages in standards and further challenges for sub 1 GHz Wi-Fi. In: The IEEE international conference on communications (ICC); 2012. p. 6885–9.

[7] Iwaoka M. IEEE 802.11ah use case – industrial process automation, <http://mentor.ieee.org/802.11/dcn/11/11-11-0260-01-00ah-tgah-use-caseindustrial-process-automation.ppt>; 2011.

[8] Zhao F, Medard M, Hundeboll M, Ledet-Pedersen J. Comparison of analytical and measured performance results on network coding in IEEE 802.11 ad-hoc networks. In: The international symposium on network coding (NetCod); 2012. p. 43–8.

[9] Hwee-Ong E. Performance analysis of fast initial link setup for IEEE 802.11ai WLANs. In: The IEEE 23rd international symposium on personal indoor and mobile radio communications (PIMRC); 2012.

[10] IEEE P802.11ah/D1.0 draft standard for information technology – telecommunications and information exchange between systems local and metropolitan area networks – specific requirements – part 11 wireless LAN medium access control (MAC) and physical layer (PHY) specifications – amendment 6 sub 1 GHz license exempt operation; 2013.

[11] Official IEEE 802.11 working group project timelines, <http://www.ieee802.org/11/Reports/802.11Timelines.htm-tgah>; 2014.

[12] Banerjea R. US channelization, <http://mentor.ieee.org/802.11/dcn/12/11-12-0613-00-00ah-us-channelization.pptx>; 2012.

[13] Lekomtcev D, Maršálek R. Comparison of 802.11af and 802.22 standards – physical layer and cognitive functionality. Elektrorevue J 2012;3(2):12–8.

[14] Flores A, Guerra R, Knightly E, Ecclesine P. IEEE 802.11af: a standard for TV white space spectrum sharing. IEEE Commun Mag 2013;51(10):92–100.

[15] Smart cities, not only new research papers but also exciting forecast!, <http://ict4green.wordpress.com/2012/02/11/smart-cities-not-only-newresearch-papers-but-also-exiting-forecast/>; 2012.

[16] British city promotes high-tech healthcare as a smart city service, <http://smartcitiescouncil.com/article/british-city-promotes-high-tech-healthcare-smart-city-service>; 2013.

[17] Love clean streets, <http://www.lovecleanstreets.com/>; 2015.

[18] Cisco: the Internet of everything for cities, <http://www.cisco.com/web/about/ac79/docs/ps/motm/IoE-Smart-City_PoV.pdf>; 2013.

LTE AND 5G SYSTEMS

6

E. Kammoun*, F. Zarai*, M.S. Obaidat†

*Department of Telecommunications, ENET'Com, University of Sfax, Sfax, Tunisia; †Department of Computer and Information, Fordham University, Bronx, NY, United States

1 INTRODUCTION

Mobile communications systems revolutionized the way people communicate, and cooperate. During the past few decades, the telecommunications field has witnessed a marvelous evolution that impacted all aspects of our life. Since the appearance of the first generation (1G)—Analog Cellular Telephony— during the 1980s, mobile industry has witnessed a continuous evolution almost each decade. Each generation introduced new technologies and new architectures. These systems perform ubiquitously and conventionally and are spread worldwide. Thus, this work summarizes the recent cellular network evolution as shown in Fig. 6.1 and specifies the handover management in detail.

The formation of Third-Generation Partnership Project (3GPP) [1] happened in 1998, with an intention to support the development and maintenance of the radio access technologies starting from the Global System for Mobile Communications (GSM) and GSM evolved technologies such as General Packet Radio Service (GPRS) and Enhanced Data Rates for GSM Evolution (EDGE). Moreover, the 3GPP Long-Term Evolution-Advanced (LTE-A) recently emerged. There are promises to offer significant improvements over previous technologies such as Universal Mobile Telecommunications System (UMTS) and High-Speed Packet Access (HSPA) by introducing a novel physical (PHY) layer and reforming the core network (CN) [2].

Lately, the telecommunications field went beyond the existing mainstream radio access technologies cited earlier in order to introduce a new system based on the UMTS evolution entitled LTE-A that is going to promote modern concepts such as Smart Cities and the Internet of things (IoT) during the next few years. The IoT is essentially based on connecting Smart objects and aims to vigorously interconnect devices and consequently cities in the physical world alongside with the digital one. Thus this emerging concept enables a huge number of diverse and diversified end systems to exchange data using All-IP–based networking mobility.

Mobile devices such as PCs and smartphones are evolving largely since they are the most useful and suitable means of communication that are not limited by place and time. So, having a continuously connected environment and being surrounded by actively smart and intelligent devices is definitely going to revolutionize human beings' daily life in the next years with the evolution of Mobile Cloud Computing (MCC), Device-to-Device (D2D) communications, Multi–Radio Access Technologies (Multi-RATs) Heterogeneous Networks (HetNet), RFID applications, and Wireless Sensor Networks (WSN), among others. For example, WSNs enable the collection, processing, analysis, and dissemination of valuable

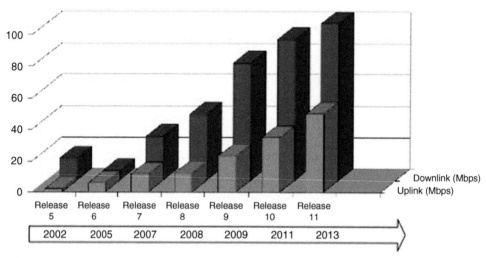

FIGURE 6.1 Cellular Network Evolutions

information, gathered in a variety of environments [3]. For all the reasons cited earlier, researchers have been recently focusing mainly on the Next-Generation Networks (NGNs) that are the fourth generation (4G) and also known as LTE-A, which are capable of overcoming the bandwidth limitation because they can significantly increase bandwidth capacity for subscribers. The 4G network is capable of providing up to 100 Mbits/s (for "LTE-A" standard) and 128 Mbits/s (for "WirelessMAN-Advanced" standard) for mobile users, whereas the current third-generation (3G) network supports a maximum of 14.4 Mbits/s [4]. It is indisputable that they can operate independently from the used heterogeneous technology as well as from the time and place limitations. It is undeniable that both LTE and the World-wide Interoperability for Microwave Access (WiMAX) are promising technologies that offer minor latency and superior transfer rates in order to bring the customers' needs to completion. In this manner, one of the major and necessary requirements is providing the most adequate and satisfactory Quality of Service (QoS) especially for real-time applications. LTE is an all-IP packet system where guaranteeing QoS is a real challenge [5]. It promises to increase the transmission capacity, able to offer the same bandwidth for further users, and provide higher data rates to the same number of users. In addition, the reduction of the interval of data transmission, known also as latency, would significantly improve the responsiveness of the network. Finally, LTE is expected to consume less energy than UMTS, especially at the terminal; its autonomy is thereby extended, despite the connection to a broadband data service.

The first phase of the LTE standardization work is to define the requirements that must be satisfied. In summary, the major goal of LTE is to improve the data services support through increased capacity, upgraded data rates, and reduced latency. In addition to these performance requirements, 3GPP also defined functional prerequisites such as the spectral flexibility and mobility with other 3GPP technologies. It is doubtless that monitoring the QoS is critical for the operator to ensure a satisfactory user experience. That is why the main purpose of LTE technology is to provide diverse servants with various QoS requirements. Thus, LTE network offers differentiated QoS mechanisms to facilitate the consideration of the various service constraints. Mobile services can be distinguished according to two main criteria, often interrelated. These service features involve a differentiated network support.

Understandably, processing a voice call will not impose the same constraints as downloading a file. In general, the real-time services (eg, voice or streaming video calls) require short transmission times, but can tolerate transmission errors. In contrast, non–real-time services (eg, a download of electronic mail or file) can tolerate speed constraints but do not tolerate transmission errors.

The next section of this chapter outlines a general overview of the cellular network evolution and specifies in particular the wireless 4G networks. Moreover, the horizon of implication of the upcoming and universal fifth-generation (5G) mobile technology is detailed. Because of the gradually growing demand for a richer communication environment that is characterized by higher data rates and mobility enabling people and things to be connected at anytime and anywhere, our main focus will be devoted on studying the 5G systems.

2 CELLULAR NETWORK EVOLUTION
2.1 OVERVIEW ON THE THIRD GENERATION

Mobile and wireless networks have experienced unprecedented growth in recent years. It was first acted on the one hand by the deployment of several successive generations of telecommunications networks mainly dedicated to telephony [second generation (2G), GSM] and more oriented toward multimedia (3G, UMTS). On the other hand, wireless Local Access Networks (LANs) have invaded everyday's life through leading standards such as Wi-Fi and Bluetooth. Moreover, the natural evolution of wireless communication systems is widely oriented now to take into account the QoS. It is inherent in solutions for telecom operators (GSM, UMTS), but only remained marginal in the first wireless network solutions (Wi-Fi or Bluetooth) for which it basically offers initial solutions that work for undemanding users. This is no longer acceptable; thus, nowadays all the new standards incorporate mechanisms to ensure the QoS to certain flows and/or certain users.

The 3G systems supported new technologies comparing to 2G systems such as Wideband Code Division Multiple Access (WCDMA) since it is the broadly chosen 3G air interface. However, 3G cannot be seen as a single unique standard because it is actually a set of standards working all together such as UMTS and HSPA+. Thus, the omnipresent technologies that are based on two parallel backbone infrastructures, one consisting of circuit-switched (CS) nodes and one of packet-oriented nodes [6], are further discussed in the following section.

2.1.1 UMTS

UMTS is a mobile system with a downlink maximum speed of 384 kbits/s and an uplink maximum speed of 128 kbits/s. It is the successor of GSM, based on the 2G networks, offering only phone calls and small amounts of data via MMS service. In fact, the UMTS, which represents a major 3G wireless technology, offers wireless Internet services at a worldwide scale, extending the scope of the 2G wireless networks from simple voice telephony to complex data applications including voice over Internet protocol (IP), videoconferencing over IP, web browsing, multimedia services, and high-speed data transfer [7]. Therefore, once the 3G system swept the market in 2003, it offered larger data amounts such as videos and a higher range of extremely advanced services, which allowed the new ubiquitous technology to overcome previous ones. A variant of the UMTS Time Division Duplex (TDD), called Time Division Synchronous Code Division Multiple Access (TD-SCDMA), is also standardized by the 3GPP. The technology operates on a bandwidth of 1.28 MHz, and is primarily deployed in China. UMTS has one major development that we present briefly in Section 2.1.2.

2.1.2 HSPA+

HSPA technology evolution, commercially known as 3G+, was introduced to remedy the limits of the Release 99. Thus, HSPA [8] is actually the amalgamation of two mobile telephony protocols: High-Speed Downlink Packet Access (HSDPA) and High-Speed Uplink Packet Access (HSUPA) that expand and enhance the existing WCDMA protocols' performance. These developments have been defined by the 3GPP Release 5 (2002) and Release 6 (2005) with the Enhanced Dedicated Channel (E-DCH), also known as HSUPA. In standardization forums, WCDMA has emerged as the most widely adopted 3G air interface technology for mobile communications. Since the first 1999 release of the WCDMA-based universal terrestrial radio access network (UTRAN), new features have been added such as high-speed HSDPA in Release 5 and HSUPA in Release 6 [9].

Latency can be defined in general as the response time taken by the system to a user request, and is a key factor in the perception of user's data services. In 2007, another feature of HSPA+ in the 3GPPP Release 7 introduced for the first time the Multiple-Input Multiple-Output (MIMO) technique, which allows the use of up to two antennas: one for transmission and the other for reception, also known as 2×2 MIMO for downlink (DL) transmission. Release 11 was introduced in 2013 by adding the 4×4 MIMO for the DL and 2×2 MIMO for the uplink (UL). The main innovation of HSPA concerns the passage from a switching circuit on the radio interface, where radio resources are reserved for each User Equipment (UE) for the call duration, to a packet where the base station decides dynamically the resources sharing between active UEs. The dynamic resource allocation is performed by the scheduling function, depending in particular on the instantaneous quality of the radio channel of each UE, its QoS constraints, and the overall system efficiency. Packet switching (PS), thereby, optimizes the use of radio resources for data services. The outstanding 3G capabilities improvement was perfectly obvious; however, researchers did not stop upgrading the existing communication systems in order to cope with the tremendous demand. Thus, 4G mobile networks were introduced and are discussed in the following section.

2.2 4G WIRELESS NETWORK

To satisfy the ever-increasing demand for mobile broadband communications, the International Mobile Telecommunications-Advanced (IMT-A) standards have been ratified by the International Telecommunications Union (ITU) in November 2010 and the 4G wireless communication systems are currently being deployed worldwide [10]. The tremendous prosperity of the Internet and its correlated services incited users' and industrialists' expectations to crave for larger bandwidths and faster communications. Wireless devices have been exponentially and obviously growing; thus, urgent users' demand for new data-intensive services has been incredibly increasing. Yet, the prevailing 2G/3G networks were not designed to cope with this explosion in data consumption. Although sufficient for many uses, their design limits their potential for the needs of tomorrow's forthcoming concepts such as Smart Cities and IoT. These actual technologies are currently struggling with the bursting needs because it is no longer capable to completely cope with the claiming demand. As a result, 3GPP recently released the release 8 (Rel-8), which can be considered as dividing line between both 3G and 4G.

Actually, the LTE network, which is the successor to the UMTS mobile radio standard, is one of the omnipresent concepts in mobile industry designed in order to meet the requirements of true 4G system as defined by the ITU. Its introduction addresses the exponential growth of mobile data traffic, which is doubled in the past few years in global average. 4G systems promise to offer not only IP-based services, higher speed and superior capacity but also requisite to cope with the Scalability challenge, which is basically defined as the capability to handle expanding users' number and services' variety. Besides, one of the main goals of the 4G requirements is fulfilling a satisfying QoS in order to be able to coordinate between the various HetNets. Thus, the 3GPPP LTE, also known as 3.9G, has already been introduced to the market as the 4G. It is well presented in the following section along with its successor LTE-A.

2.2.1 4G: 3GPPP LTE

Telecommunication industry has witnessed a considerable cost reduction comparing to its first emergence, which led consumers to expect better performance versus a noticeable charges reduction. In the expeditiously changing wireless communications field, experts have recently developed one of the recent telecommunications standards, which is LTE, in order to improve data rates and decrease handover latency.

The deployment of 4G mobile broadband systems based on the 3GPP LTE radio access technology [11] is now ongoing on a broad scale. Thus, in order to efficiently satisfy escalating users' needs, the LTE radio access technology never ceased unfolding and gradually evolving. In 2010, the evolution of LTE was mainly initialized by the occurrence of the 3GPP Release 10 (Rel-10), known as LTE-A, which is actually an improvement of Rel-8 and release 9 (Rel-9) standards. Smooth transition to the 4G is accomplished by providing services using the technologies developed for 4G and the same frequency bands as for 3G. However, unlike UMTS, LTE involves two procedures of duplex data transmission: the Frequency Division Duplex (FDD) and TDD [12].

These two mobile communication technologies rely on different frequency bands and are able to aggregate and to expand the capacity of bandwidth in their entirety. On one hand, the FDD uses two distinct carriers: One for transmitting and one for receiving data, which gives more capacity to LTE networks. On the other hand, TDD uses a single frequency to transmit data asynchronously to aggregate to leverage their respective capabilities. In all cases, both techniques are based on different frequencies for the operator that exploits both. Unlike CS services, LTE is able to support packet-switched services. Moreover, the LTE architecture changed since all Radio Network Controller (RNC) functions have been moved to eNodeB, also known as the Evolved–Universal Terrestrial Radio Access Network (E-UTRAN) node B; it no longer had a central radio controller node.

These adjustments in the Radio Access Network (RAN) framework template are fundamentally caused by the indispensable requirement of supplying users with more flexibility in terms of delay and spectral efficiency. Evolved Packet Core (EPC) is the CN, all-IP–based packet. Unlike the 2G and 3G networks where the CS domain and PS are clearly distinguished in the CN, the new network only has packet domain called EPC. All Services will be offered over IP including those formerly provided by the circuit domain such as voice, videotelephony, Short Messaging Service (SMS), and all telephony services. Besides, the EPC is able to interact with the premature 2G/3G networks and CDMA2000 in case of mobility. It is possible to route traffic from the EPC to the LTE access, CDMA2000, 2G, and 3G and thus ensure the handover between these access technologies to provide seamless communications in a heterogeneous environment. The EPC supports not only Default bearer but also dedicated bearers. Indeed, when the user is connected with the EPC network, it creates a default bearer representing permanent connectivity (maintained until the user is connected to the network) without guaranteed bandwidth.

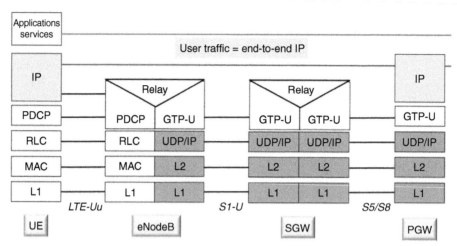

FIGURE 6.2 LTE Protocol Stack

When the user wishes to establish a call that requires a certain QoS such as voice or videotelephony call, the network may establish for the duration of the call a dedicated bearer that supports QoS required by the service flow and especially which has a guaranteed speed to emulate the circuit mode. The EPC supports packet filtering (deep packet inspection, eg, virus detection) and advanced taxation (taxation-based service flow). In fact, the LTE provides sophisticated charging mechanisms to tax the service accessed by the client based on the volume of the session, duration, event, content, etc. One of the major dissimilarities between both the 3G and the LTE paradigms concerns mobility management, since the proposed LTE technology is based on a complete different architecture as shown in Fig. 6.2.

2.2.2 5G: LTE-Advanced

While LTE deployment is evolving at a rapid pace, the term 5G has been quite an interest for both researchers and industrialists and is eventually going to swamp the market worldwide. It may sound premature to already start discussing the upcoming 5G since the LTE-A has not been deployed yet. However, the growth of user requirements has been tremendous during the past few years that forced scientists to investigate more research efforts in order to come up with faster and more advanced technology that is the 5G, which is going to be the next step in the NGN. Deploying the universal ultra-broadband network infrastructure necessitates specific aspects and characteristics as depicted in Fig. 6.3 in order to build smart cities. It is true that the LTE Rel-8, which is mainly deployed in a macrocell/microcell layout, provides improved system capacity and coverage, high peak data rates, low latency, reduced operating costs, multiantenna support, flexible bandwidth operation, and seamless integration with existing systems. Yet, the LTE-A notably upgrades the existing LTE Rel-8 using wider transition communication bandwidths, better coverage, and higher throughput leading to an excelling user experience.

Furthermore, LTE-A systems target support for DL peak spectral efficiency of 30 bits/s/Hz and UL peak spectral efficiency of 15 bits/s/Hz, and support for approximately 1.5× improved cell average and cell edge spectral efficiency over Rel-8 and Rel-9 standards [13]. They also used numerous antennas called MIMO data transmission in the PHY layer on both the transmitter and the receiver sides. This radio feature is a key enabler technique that provides rapid wireless data services, which enhances the

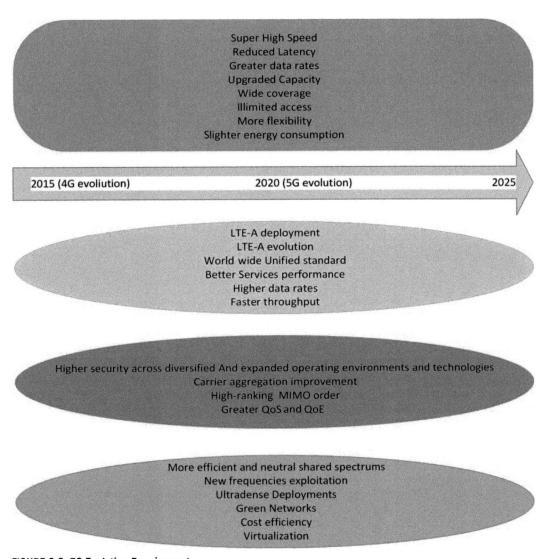

Super High Speed
Reduced Latency
Greater data rates
Upgraded Capacity
Wide coverage
Illimited access
More flexibility
Slighter energy consumption

2015 (4G evoliution)   2020 (5G evolution)   2025

LTE-A deployment
LTE-A evolution
World wide Unified standard
Better Services performance
Higher data rates
Faster throughput

Higher security across diversified And expanded operating environments and technologies
Carrier aggregation improvement
High-ranking MIMO order
Greater QoS and QoE

More efficient and neutral shared spectrums
New frequencies exploitation
Ultradense Deployments
Green Networks
Cost efficiency
Virtualization

FIGURE 6.3 5G Evolution Requirements

overall efficiency. They also added carrier aggregation technique in this version of release that enables users to download data from diversified sources simultaneously, which empowers them with higher speeds. As a matter of a fact, the Evolved Packet System (EPS) is actually standardized by the 3GPP standardization organization and has several functions such as network access control, mobility management, packet routing and transfer, and radio resource management (RRM) functions. The primary role of the policy and charging rules function (PCRF) is to provide service control without going through the UE and to dynamically manage the data sessions, which is the main difference of the LTE architecture compared to the previous ones. As shown in Fig. 6.4, it is directly connected to the PGw via Gx interface.

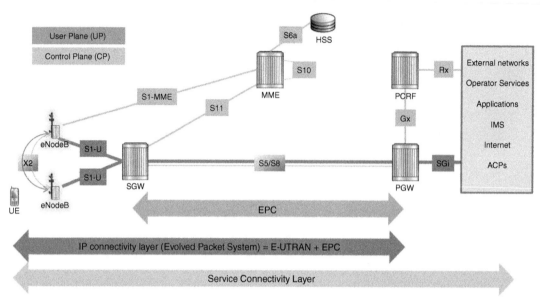

FIGURE 6.4 EPC Interfaces

Battery life is one of the most important parameters to take into account because it is more efficient to use mobiles that consume the least possible amount of battery power. Unfortunately, it is compulsory to use a transmission mode characterized with a constant power level, which cannot be offered by the Orthogonal Frequency Division Multiplexing Access (OFDMA) technique. As a result, the Single-Carrier Frequency Division Multiple Access (SC-FDMA) technique is used since it can reduce power consumption of the terminal and thus contributes to the increase in battery life. For the DL of the LTE radio links, for which there is a lower energy constraint, the OFDMA technique is basically used because it allows—for the same spectral width—higher bit rate.

So, in order to accomplish the different requirements such as a maximum spectral efficiency, minimal delay necessity, and faster data rates, the researchers eliminated the WCDMA technique to use the OFDMA that exactly fits the LTE's particular requirements in enabling multiuser access by allocating different subcarriers to various users, enhancing spectrum flexibility, on the radio side, thus upgrading the peak rates on the DL and discrete Fourier transform (DFT)-spread OFDM also acknowledged as SC-FDMA on the UL. On the core side, the LTE allowed a migration from a telephone network that is built to enable some data toward an all-IP network that is built to support converged voice video and data services over one common architecture. In this release, the 3GPPP separated it into two distinguished projects: One focusing on the radio, the needs, and the evolution that needed to happen in the radio network and the other is in the CN. Hence, the radio side is specified by the LTE and the core side is characterized by the System Architecture Evolution (SAE). The SAE as a part of the 3GPP activity focuses on the CN of a mobile network, whereas the changes in the RAN [PHY and Medium Access Control (MAC) layers] are handled in the LTE project, which focuses on the RAN [14]. Those two projects produced as an output the E-UTRAN and the EPC, which is the CN defined as a part of the SAE project. It can provide seamless handovers and high data rates. On one hand, the E-UTRAN is composed of several eNodeBs. The eNodeB is equipped with two planes: the Control Plane that is responsible for Radio Resources Control (RRC), which is charged of lower layers' configuration, and the User Plane situated between the eNodeB and the UE. It consists of the subsequent sublayers, which are MAC, Packet Data Convergence Protocol (PDCP), and Packet Radio Link Control (RLC). The RRC is charged of making Handover Decisions based on the different measurement reports of the UE and controlling these reports. It also is responsible for paging and broadcast procedures, all mobility functions, and radio bearer control, whereas each of the second layer's sublayers has its own function. For instance, the MAC sublayer is responsible for transporting format selection and for traffic volume measurement. The PHY layer transfers physical channels to MAC that transports it as logical channels to the RLC layer. The PDCP exists on both control plane and user plane. It takes charge of decryption process and the ciphering process in both sides. However, it only manages the integrity protection for control plane and the transfer of its data. It offers header compression and retransmission along in a handover procedure for the user plane as shown in Fig. 6.5.

Handover Management, Macrodiversity and Encryption and Radio channel coding, IP header compression and encryption of user data streams, RRM, Location Management, Measurement and measurement reporting configuration for mobility and scheduling, and Traffic Management are involved in the radio access protocols in the E-UTRAN access stratum. These are functionalities that are involved in the E-UTRAN. On the other hand, the EPC is composed of a Mobility Management Element

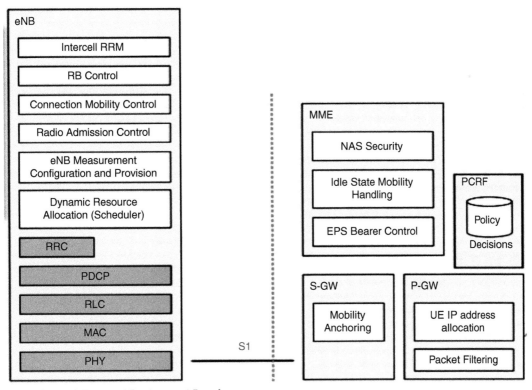

FIGURE 6.5 EPC Elements' Fundamental Functions

(MME), a Serving Gateway (SGw), a Packet Data Network (PDN) Gateway (PGw), and a PCRF. It enables functions such as giving access to external IP networks. When the user shifts from one cell to another, it is primordial that the EPC supplies a smoothly continuous mobility at the third layer through various access technologies such as WiMAX and Wireless Local Area Network (WLAN). It is also defined in RFC 6459 as an evolution of the 3GPP GPRS system characterized by a higher-data-rate, lower-latency, packet-optimized system that supports multiple RATs. The EPS permits numerous mobility protocol usage in accordance with the used access technology. That is why, from the user through the radio network through the EPC toward a destination IP network, that whole end-to-end architecture is the EPS. It is important to note that only eNodeB form the E-UTRAN. The MME, the Home Subscriber Server (HSS), the SGw, and the PGw all together form the EPC that is combined to E-UTRAN to make an EPS. As shown in Fig. 6.6, the LTE protocol stack is composed of four main components, which are the UE, eNodeB, SGw, and PGw. The PGw is the gateway of the CN, whereas the SGw operates as the gateway for the EPC.

The UE refers to the device controlled by the wireless network. An important concept is the separation of the user plane and the control plane because it enables the eNodeB to connect and communicate with two nodes from the CN. As shown in Fig. 6.6, the LTE radio interface known as Uu interface is both a user and a control plane component. It enables data transfer between the eNodeB and the UEs and provides radio access to the UEs. The various protocols and functions required for the control operations, and the data transfer of the Uu interface are implemented in the eNodeB. Note also that in LTE there is an important difference from the previous UMTS architecture, which is adding the eNodeB. In the online mode, the eNodeB can configure a radio signal level threshold above which the UE is not obliged to perform measurements on neighboring LTE frequencies or on any other systems, even if the intervals for these measurements are activated. This threshold is called s-Measure, as in the standby mode.

The role of this threshold for the operator is to limit UE consumption while simplifying the configuration step. The s-Measure value can also be adapted to UE activity. An active UE will need good radio conditions for a satisfactory QoS and a good continuity of service. However, a lower value may be used for a UE little or none active for which battery consumption must be minimized [UE in Discontinuous Reception (DRX) mode, eg]. Besides, it tunnels data between the UE and the IP network or service that the user is trying to access via S1-U interface. An interface in this terminology refers to a logical connection between two points. Depending on the interface, different layers might be defined for it.

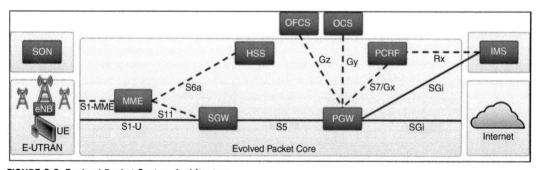

FIGURE 6.6 Evolved Packet System Architecture

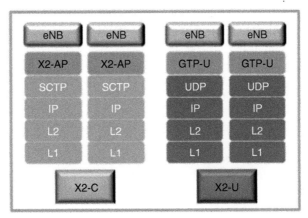

FIGURE 6.7 X2 Protocol Stack

So, for example, the S6a interface between an MME and a HSS is defined with a whole protocol stack for it, which is going to be over IPv4, IPv6, etc. In order to tunnel user data from eNodeB to the SGw, a protocol called GPRS Tunneling Protocol (GTP-U) is actually used to transport data in any IPv4 or IPv6 packet formats to the user plane. In addition, the User Datagram Protocol (UDP) is also charged of transferring data to the user. It is—as detailed in RFC 768—a leading telecommunication protocol used by the Internet. It is part of the transport layer of the Transmission Control Protocol (TCP)/IP protocol stack. The role of this protocol is to allow a very simple way of data transmission between two entities, each defined by an IP address and a port number. The backbone network protocols UDP/IP are used in control signaling and routing data.

In case of sending packets toward destination network, it is compulsory then to send them from the UE to the SGw. When moving through the network, the LTE architecture is quite simple because it allows peer-to-peer connections since each eNodeB is able to communicate with another eNodeB and effectively forwards traffic to them. Unlike UMTS or HSPA+, it reduces latency delays and enhances round-trip time (RTT) and thus packet loss. Thus, it is possible to say that X2 interface plays the RNC's role since cell towers were not able in the previous architectures to directly communicate with each other as depicted in Fig. 6.7.

They actually had to communicate with an RNC or a Base Station Controller (BSC) in order to be able to communicate with the other cell tower. SGw is the serving gateway that acts as a signaling intermediary between the PGw and the eNodeB. The main purpose of the SGw is, instead, to route and forward user data packets among LTE nodes, and to manage handover among LTE and other 3GPP technologies. The PGw allows the LTE network interconnection universally, supporting connectivity between the various UEs and the external PDNs. As its name indicates, in LTE, the MME is in charge of the mobility management function. The MME keeps track of UEs that are registered on the network and access as a gatekeeper, handling user requests for network access, and setting up and tearing down data sessions. Other main functionalities of MME include Authentication, Non-Access Stratum Protocol (NAS) mobility management, handover, SMS, and voice support. The MME communicates with the HSS to get security information. Hence, the subscriber needs to be authenticated and traffic needs to be anticipated. It handles the authentication initiation of users on the network, as well as basic authorization. The data passes then to the HSS to get security that will interrogate the authentication, authorization, and accounting (AAA) server. HSS is essentially a common database containing subscribers' information that can be queried and demanded by the MMEs in order to decide which services are allowable or permitted. HSS also has the primary key information for all SIM cards in a mobile network. Moreover, it controls the users' location in the network; thus, HSS is responsible for informing the old MME of data related to the MME, networks, and SIM.

Consequently, the HSS can be seen as the Home Location Register (HLR) since it has a similar role in previous architectures. The MME's role is to select what SGw and PGw will be used for a given session. Actually, one or more SGw will serve a given group of eNodeBs for User Plane Data (UPD). A single UE can be served by one SGw at any time. SGw reserves instruction from the MME to set up and tear down sessions for a UE. It also handles user IP Packets between P-Gw and eNodeB. That's why a SGw is an IP router with support for a mobile-specific tunneling protocol GTP. The PGw provides access to PDNs. If a UE has multiple data sessions to multiple PDNs, the UE can be connected to multiple PGws. However, the UE will still be served by only one SGw.

As a conclusion, the development of the 5G is expected to overcome the previous architectures' issues and to offer higher data rates, reduced energy consumption, better performance, deeper user coverage, faster throughput, and lower latency. Since reducing latency is one of the most demanding goals to be achieved in LTE-A, choosing to use terminal on idle or active state is required in order to lower signaling and control plane latency. But for now, location management and handover management are still two main issues when it comes to terminal mobility management. The latter will be discussed briefly next.

3 HANDOVER MANAGEMENT

Handover can be defined as the process that prevents an ongoing communication from getting interrupted as the mobile equipment changes its attachment point. As a mobile phone hands off from one cell to another among HetNets, some disruptions may occur in the active communication due to packet losses and delays.

It is usually initiated once the signal quality degrades and deteriorates or once one cell crosses its boundaries. An important aspect that should be further processed in the context of cognitive radio is the treatment and the collection of information. Indeed, this aspect is necessary for a complete perception of the network's external environment in order to enable it to make effective decisions accurately and quickly for better radio resources management. Typical RRM intends to provide networks with enabling mechanisms that offer well-informed call admission control (CAC), reduced power consumption, and minimized traffic session disruption.

3.1 HANDOVER STEPS

One of the fundamental and essential features in the cellular communication systems is the handover. It is one of the fundamental keys that prevent users from disruptions during communication in order to get a guaranteed QoS. In case a handover is foreseeable, it must be prepared earlier than planned; thus the handover latency and packet loss will be reduced. A handover is basically composed of three phases that are cells and networks discovery and finally handover execution.

3.1.1 Cells and networks discovery

During this initial phase, neighbor networks and resources discovery (Bandwidth, frequency, channel number, etc.) are parts of handoff preparation stage since they are required in gathering and collecting related data in case the mobile needs to move from one cell to another. The gathered information, by either the network or the Mobile Terminal (MT), provides multiple metrics for the user about the network and the terminal such as security, recognized quality, available network coverage, bit error rate, and network interfaces ready for use. Thus, the handover preparation phase basically aims to minimize blocking probability and to reduce the communication's suspension time. During the preparation phase, the UE gathers all the required information about the candidate networks to which it can easily hand off. In order to make the right decision, the most suitable network must be carefully selected based on some parameters such as user preferences, energy preferences, and obtainable bandwidth. Therefore, a handover is triggered based on the gathered refined handover measurements. In the LTE case, the preparation phase can be carried out between two eNodeBs via the X2 interface if it exists, or, by default, through the MME via the S1 interface. In both cases, however, the procedure on the radio interface is identical. Using the S1 interface for the handover is necessary when the operator cannot implement some of X2 interface between some eNodeBs.

Nevertheless, preparation time and data transfer can be longer since the messages pass through the MME and S1 and cross two interfaces (between eNodeB and MME source, and then between MME and eNodeB target). If the target cell belongs to the same eNodeB, it undertakes no preparation procedure.

In the preparation phase, the eNodeB source provides, among others, the following information to the target eNodeB:

- the global identifier of the target cell Enhanced Cell Global Identifier (EGCI), allowing the target cell identification unambiguously;
- the handover cause (eg, radio conditions, reducing the load, and resources optimizing);
- security settings such as implemented algorithms, the eNB key (KeNB) and the short identifier of MAC (MAC-I);
- list and description of the E-UTRAN Radio Access Bearer (E-RAB) to configure;
- RRC context of the UE, which describes the radio configuration of the RRC connection to the source cell, the parameters of the DRX cycle in case they were used;
- information on mobility in the UE's history, informing the target eNodeB of the list of the last 16 cells seen by the UE, and, for example, recurring movement between cells ("Ping Pong").

The target eNodeB, on receipt of the message X2 Application Protocol (X2AP) Handover Request, performs admission control that checks whether it has the radio resources to accommodate the UE and, in particular, the active E-RAB on the source cell. If it is able to establish at least one of these E-RABs, the eNodeB should respond positively to the source eNodeB indicating whether one or several E-RABs are capable to be maintained. It includes in its response the RRC message destined to the UE and that will be sent by the source eNodeB when ordering the handover. This message contains the configuration that the UE should apply when accessing the target cell, including radio bearers associated with maintained E-RAB. The target eNodeB also returns to the source eNodeB, for each E-RAB, the end point of the GTP tunnel between two eNodeB, if a transfer of the source eNodeB to the target eNodeB of DL data (received by the source eNodeB from the SGw) was requested by the source eNodeB in the Handover Request message.

3.1.2 Handover decision and network decision

During the system discovery phase, the required information to identify the need for a handover is collected. Both the network and the UE collaborate in order to collect the appropriate data. It is essential to eventually choose the most appropriate point of attachment. Hence, the data gathered on the previous step allow the equipment user to figure out accessible networks in the handover decision phase and determine which one is mostly and principally convenient for each and every functioning application. Thus, after analyzing the network environment, the mobile has to identify whether a handover is indeed needed in order to initiate it afterwards by preparing the serving cell to target the destination cell in which the UE will eventually move to after taking the accurate measurements. Undeniably, handover decisions made without taking into account the neighboring access networks may result in catastrophic handover performance because they can simply be wrong and lead to choose the defective network or even the same point of attachment. Right after this stage, the handover must be executed.

3.1.3 Handover execution

Right after releasing the previous connections, a new connection's establishment is preceded. The handover execution phase cannot be started unless the UE receives a confirmation that the destination network resources are ready and apt to adopt the ongoing communication by providing a performing communication quality. As a matter of fact, in order to obtain a seamless and lossless handover with high-quality communications, two essential metrics are used to evaluate the handoff performance, which are the handover latency and the QoS. The latter criteria should not exceed few hundred milliseconds. The further is derived from both the source and the target schemes should be approximately identical. Thus, and in order to boost the cited metrics, handover process must be upgraded by integrating mechanisms for uninterrupted and seamless services and higher-speed applications. Eventually, the venerable resources are discharged and a new connection is built and established at this final phase. After receiving the response message of the Handover Request, the eNodeB source triggers the handover by sending a RRC message to the UE, which indicates in particular:

- the target cell [frequency, if different, and its physical cell identifier (PCI)];
- its Cell Radio Network Temporary Identity (C-RNTI) in that cell;
- security settings (eg. algorithms) enabling it to derive the new encryption keys and RRC integrity.

On receiving this message, the UE must immediately attempt to switch to the target cell, even if it could not acknowledge the RRC message reception [Automatic Repeat reQuest (ARQ) or Hybrid Automatic Repeat reQuest (HARQ) acknowledgments/RLC]. It resets the MAC layer and proceeds to reestablish its RLC and PDCP layers. The RRC layer configures the PHY Layer, MAC, RLC layers, and PDCP according to the parameters provided by the target eNodeB and transmitted by the eNodeB source in the RRC Connection Reconfiguration message. The UE then derives the new KeNB, either from the current Key Access Security Management Entity (KASME) key (means the one used for the calculation of the current eNodeB key) or from the new KASME key if the NAS security proceeding was performed. The eNodeB indicates to the UE which of the two mechanisms to use for this derivation. The UE then conducts random access on the random-access channel (RACH) of the target cell and, if successful, sends the RRC Connection Reconfiguration Complete message to the eNodeB, which finishes the signaling procedure. The access to the RACH can be achieved with a dedicated preamble, if the target cell was provided to the source cell in the preparation phase. This method has the advantage of removing the risk of collision with other UE preambles, thus increasing the chances of success of the procedure and tending to reduce the overall time. Finally, the UE lifts periodic measurements enabled on the source cell and removes the configuration of the measurement intervals used for interfrequencies or intersystem measurements. Registration process is always done right after IP assignment. In this final phase, the new base station allocates the new resources. This operation is known as channel assignment, which can be seen as an important element of resource management and CAC.

3.2 HANDOVER DELAY

Handover delay can be described as the time taken in order to discover the new subnet the Mobile Node (MN) is going to move to, the establishment of a recent care-of-address, and the period of time designated to notify the home agent along with its correspondents about the MN's movement toward a new location. Actually, the handover delay involves three delays, which are as follows: the authentication, the IP assignment, and the registration delays. The handover delay highly persists, thus remaining unacceptable for real-time applications. In order to improve handover delay, we can allocate and classify the various requests and provide each one with a specific priority. The highest priority must be given to the most compelling and urgent request. Hence, in order to maintain an effective ongoing communication, a seamless mobility mechanism characterized by a low handover delay is quite mandatory. Therefore, using LTE-A that is an all-IP–based cellular network not only decreases latency but also increases users' data rates. So, once the handover latency is decreased, the QoS is increasingly guaranteed. Actually, the QoS is what everyone seeks these days. People never stopped deploying efforts on improving their quality of life; thus a recent concept was established: smart city. It is featured with the exploitation of human resources, information and communication technology (ICT) infrastructure, social capital, economic development, and environmental sustainability. The novel rising approach is illustrated in detail in the next section.

4 EMPIRICAL CASE: SMART CITIES

Cities are powerful engines of economic growth, fueled by numerous interpersonal and high levels of expertise. Since it is estimated that 70% of the world population will live in cities by 2050, sustainable urbanization is becoming a major political issue for governments worldwide. Thus, following the technological development, and the urban population growth, the concept of Smart City appeared. A city can be seen as smart once it affords continuous economic development with reasonable management of natural resources when investing in human and social capitals, modernized ICT infrastructure, etc. Thus, the upcoming 5G network systems must be suitable and targeted to provide the envisioned scenarios of smart cities in the context of this universal perspective during the next era. The deployment of the 5G networks is expected to overcome the previous network problems by eliminating spectral efficiency, latency, throughput, and mobility problems. Also, in intelligent cities, human intervention is going to be as reduced as possible since the network infrastructure and its deployment is semiautomatic or even automatic.

Thanks to the explosion of IoT, the number of connected devices has increased. Moreover, a seamless uninterrupted guaranteed connectivity should be provided 24/7 with a deeper coverage even in unordinary forbidding conditions. In this way, the elementary radical infrastructure for structuring the emergent smart cities will incite mobile network depiction, and promote their capability requirements and capacity performance to their extremes. This emergent smart city concept is specified in detail in the following subsection.

4.1 CONCEPT CHARACTERIZATION

Since there is still no precise customary universal definition of the term "Smart City," researchers are agreed on considering a city as Intelligent and sustainable in case it uses innovative ICT and other means to improve the quality of life, the efficiency of urban management and urban services, as well as competitiveness while respecting the needs of present and future generations in the economic, social, and environmental fields.

The concept of smart cities knows a marvelous success that continues unabated. ICTs have a crucial role to play here, allowing greater efficiency in environmental terms, in all sectors of the industry, and encouraging innovation in areas such as intelligent transportation systems (ITS) and "smart" water, energy, and waste management.

The technical characteristics of a smart city are being able to manage it in real time, power management that requires a smart grid and an optical fiber, Web and Wi-Fi free access, a seamless mobility among mobile networks at any time and any place, a multimodal transport system, the use of smart cards and geolocation to provide all digital citizenship services, with the possibility of contactless payments, and a municipal control room for the integration of services (e-commerce, police, fire, hydro, health care, e-government, etc.).

5G systems are going to offer the basic infrastructure for the smart cities' design where extremely high data volume and sensitive real-time applications are simultaneously processing. Thus, latency higher than few milliseconds would not be tolerated anymore. Over 20 billion connected devices are interacting and performing simultaneously. The IoT concept unfolds the Machine-to-Machine (M2M) and D2D services to allow houses, smart grid, transportation systems, cities, and governments to be connected. In this way, citizens will be able to control their devices remotely on a daily basis.

Thus, inhabitants will benefit from smart energy management, intelligent mobility, home security guarantee, power consumption reduction, smart metering surveillance such as smart grids, and city automation such as traffic monitoring, smart parking assistance, air pollution control, natural disasters' early estimation, smart shopping, and smart educational systems. That kind of information enables the city citizens to be more aware of their environment, which leads them to manage it in a more efficient way, to develop new projects, to increase the economic growth, and to encourage costly operations.

4.2 CURRENT AND FUTURE CHALLENGES

Humanity's future will be brighter once characterized with a better quality of life by offering more innovative services and enabling higher interaction and interconnection of urbanizing populations within a smart city. Actually, the population's perplexing increase and the urban areas' challenges, are a real nightmare for city planners and engineers. Moreover, excessive centralization of people in urban areas will cause environmental problems such as deforestation and agriculture lands' absence, environmental deterioration, and an excessive pressure on natural resources such as land, water, and energy. Besides, social issues such as unemployment and pressure on housing have become serious challenges. One of the major dares in smart cities is the extensive use of ICT.

Cities must be aware of the tremendous data volume that is going to be exchanged daily within devices and people all over the world. Therefore, the currently deployed infrastructure is no longer able to cope with such evolutionary concept. A well-managed coordination of service delivery and communicating information among the different entities with a wider broadband and a higher speed must exist. According to Ref. [16], the forthcoming LTE-A standardization process should update the concept of small cells and the integration of broadband personal communications with M2M and D2D communications will represent a key challenge to be addressed by future 4G standards.

4.3 INTERNET OF THINGS

It can be seen as the fundamental progress of the prevailing Internet into a pervasive network of codependent interconnected objects. Urban IoTs, in fact, are designed to support the Smart City vision, which aims at exploiting the most advanced communication technologies to support added-value services for the administration of the city and for the citizens. In order to achieve that, the IoT must coherently combine various heterogeneous systems via the utilization of ICTs to maintain a seamless and smooth urban environment performance. WSN that can be deployed to constantly supervise bridges, remotely control city lightning, and monitor roadways in order to obtain a more practically overall efficiency and safety.

The automation of several applications such as recycling, picking up trash, electricity service, and water provisions will effortlessly simplify humans' tasks, expeditiously reduce cost, and conveniently guarantee their safety. However, no one could deny that with the city growth and the citizens' numbers' explosion, the number of connected objects will exponentially increase. The use of the Internet will tremendously increase, which means more resources exploitation leading to network's congestion. People will no longer be restricted to portable devices. Thus, the utilization of these resources must be as wise and efficient as possible.

4.4 MOBILITY INTEGRATION

Smart connectivity with existing networks and context-aware computation using network resources is an indispensable part of IoT [15]. Since 4G-LTE, Wi-Fi, and WiMAX have been widely and broadly expanded, the progression toward universal information and communication systems is unquestionable and even noticeable. However, to profitably utilize the IoT concept, ICT systems must go beyond the traditional use of portable devices and smartphones, and unfold into new pervasive communication systems that goes in line with time by embedding quickness, intelligence, and autonomy into those everyday use wireless sensors and portable devices found in the smart city environment. In order to attain a unified city management, a ubiquitous mobility process must be deployed.

Thus, mobility and smart city are two complementary concepts. Since today's current solutions are basically made for voice coverage and are characterized by less-efficient costs, they cannot possibly line with increased demands required in the future. Thereby, the recent Rel-10 LTE-A objective is to cope with the challenges presented by the ever-growing use of smart wireless devices. These intelligent gadgets necessitate extravagantly more tremendous spectral resources than familiar and conventional cellular telephones. While contemplating our daily life, one can easily notify how our world has been going into wider and deeper changes during the past decade. From deploying High Definition (HD) surveillance cameras city-wide for security purposes to extending freely broadly high-speed public Wi-Fi connectivity access, quotidian quality of life has profoundly changed. As a result, small cells combined with HetNets are part of the proposed solutions. In the LTE-A perspective, a small cell is a low-power and low-cost radio base station, whose primary design target is to provide superior cellular coverage in residential, enterprise, and hotspot outdoor environments [16]. Actually, they will contribute in the progress of the present infrastructure of the mobile broadband in order to enable purchaser of the finest quality of experience (QoE). HetNets are the efficient solution to cope with the smart city challenges since they efficiently upgrade the used spectrum's accessibility. By adopting a HetNet topology, LTE-A seeks to provide a uniform broadband experience to users anywhere in the network in a cost-effective manner [17]. Thus, it is expected that the networked society will have a set of intelligent cities with heavy use of mobile networks that could process and deal with the huge amount of connected devices.

5 CONCLUSIONS

In summary, this chapter has first enumerated cellular network evolutions, starting with a review of 3G and 4G and opening to the universal upcoming worldwide 5G. Since LTE-A is supposed to be a high-capacity–low-latency wireless system, we presented a handover management analysis, in which we tackled the main steps of a handoff process and the handover delay issue. We finally focused on an empirical case, specifically the imminent smart cities' case, which is expected to address serious challenges in the urban areas and big cities as well as solve these in a cost-effective and comfortable manner.

REFERENCES

[1] Rahnema M. Overview of the GSM system and protocol architecture. IEEE Commun Mag 1993;31(4):92–100.
[2] Ikuno JC, Wrulich M, Rupp M. System level simulation of LTE networks. In: 71st IEEE vehicular technology conference (VTC 2010 Spring), Taipei; May 1–5, 2010.
[3] Yick J, Mukherjee B, Ghosal D. Wireless sensor network survey. Comput Netw 2008;52(12):2292–330.
[4] Dinh HT, Lee C, Niyato D, Wang P. A survey of mobile cloud computing: architecture, applications, and approaches. Wireless Commun Mobile Comput 2011;13(18):1587–611.
[5] Zaki Y, Weerawardane T, Görg C, Timm-Giel A. Multi-QoS-aware fair scheduling for LTE. In: 73rd IEEE vehicular technology conference (VTC Spring), Yokohama; May 1–5, 2011.
[6] Khan AH, Qadeer MA, Ansari JA, Waheed S. 4G as a next generation wireless network. In: International conference on future computer and communication, Kuala Lumpur; April 2009. p. 334–8.

[7] Zarai F, Boudriga N, Obaidat MS. WLAN-UMTS integration: architecture, seamless handoff, and simulation analysis. The Society for Modeling and Simulation International (SCS) 2006;82(6):413–24.

[8] Shah S. UMTS: high speed packet access (HSPA) technology. In: IEEE international networking and communications conference, INCC, Lahore; May 2008. p. 2.

[9] Mogensen P, Na W, Kovács IZ, Frederiksen F, Pokhariyal A, Pedersen KI, Kolding T, Hugl K, Kuusela M. LTE capacity compared to the Shannon bound. In: 65th IEEE vehicular technology conference, VTC 2007 Spring, Dublin; April 2007. p. 1234–8.

[10] Hossain E, Rasti M, Tabassum H, Abdelnasser A. Evolution towards 5G multi-tier cellular wireless networks: an interference management perspective. IEEE Wireless Commun 2014;21(3):118–27.

[11] Astley D, Dahlman E, Fodor J, Parkvall S, Sachs J. LTE release 12 and beyond. IEEE Commun Mag 2013;51(7):154–60.

[12] 3GPP Technical Specification Group Radio Access Network. Base station (BS) radio transmission and reception. 3GPP TS 36.104 releases 8, 9, 10; 2008, 2009, 2010.

[13] 3GPP TR 36.913. Technical Specification Group Radio Access Network; requirements for further advancements for evolved universal terrestrial radio access (E-UTRA), 2008.

[14] Pontes A, Dos Passos Silva D, Jailton J, Rodrigues O, Dias KL. Handover management in integrated WLAN and mobile WiMAX networks. IEEE Wireless Commun 2008;15(5):86–95.

[15] Gubbi J, Buyya R, Marusic S, Palaniswami M. Internet of things (IoT): a vision, architectural elements, and future directions. Future Gen Comput Syst 2013;29(7):1645–60.

[16] Cimmino A, Pecorella T, Fantacci R, Granelli F, Rahman TF, Sacchi C, Carlini C, Harsh P. The role of small cell technology in future smart city applications. Trans Emerg Telecommun Technol 2014;25(1):11–20.

[17] Bjerke BA. LTE-Advanced and the evolution of LTE deployments. IEEE Wireless Commun 2011;18(5):4–5.

CARS AS A MAIN ICT RESOURCE OF SMART CITIES

7

F. Hagenauer*, F. Dressler*, O. Altintas[†], C. Sommer*

*Heinz Nixdorf Institute, Paderborn University, Paderborn, Germany;
[†]Toyota InfoTechnology Center, Tokyo, Japan

1 INTRODUCTION

Information and Communication Technology (ICT) is a foundation of today's and tomorrow's smart cities [1] and research activities span such diverse areas as smart buildings, smart urban planning, and smart healthcare. Fig. 7.1 gives an overview of many of these activities, categorizing them roughly as pertaining to smart cities' *Infrastructure*, their *Administration*, and (more directly) smart cities' *Inhabitants*.

Smart sensing systems and dynamic wireless network infrastructures make future smart cities possible—from enabling applications such as environment monitoring, to a smart grid, and to smart mobility. In addition, they improve survivability in emergency situations. Technology-wise such applications demand robust and fault-tolerant wireless communication and widespread sensing capabilities, that is, a tremendous amount of data. Yet, even today, operators of infrastructure are struggling to keep up with the increasing data demand, as evidenced by a massive push toward newer technologies and more bandwidth for moving data.

We see a way out in the use of cars as a main ICT resource of smart cities. We believe that, by acting in two capacities, cars will play a major role in smart cities of tomorrow [2]: First, while driving they can provide access to the services of a powerful vehicular network – in essence acting as mobile base stations. They can be accessed by people in their homes or waiting at the curb, by devices in smart buildings, or by sensors deployed throughout the smart city. Second, cars themselves form a network with incredible processing and storage capacities, as well as sensing capabilities beyond any possible roadside deployment. Thus, cars can not only offer access to services but also offer services of their own. Further, by also including parked cars, both storage and processing capabilities can (to some degree) be made persistent in time and space.

Such a network formed by cars is different than typical Internet-based solutions that are readily available only in a certain geographic context and during a certain time interval. On the contrary, cars are ubiquitous. This holds even in critical emergency situations such as after disasters such as hurricanes or tsunamis [3]. Either way, these networked cars will play a big role in future cities by enabling a multitude of services in an architecture that we term *Car4ICT*.

Some first activities toward turning vehicles into a larger-scale networked system are already taking place [4]. Such architectures, however, commonly require high market penetration rates of the system, which is a problem in early market introduction—and even more so after disasters. Further, current

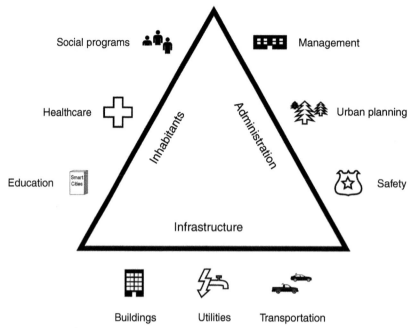

FIGURE 7.1 The Different Building Blocks of Smart Cities

approaches often focus on only a single application. Our architecture, in contrast, is able to support a huge variety of new applications both independent of and in conjunction with infrastructure.

Fig. 7.2 outlines our concept. Cars are collecting data from smart sensors along the road or they connect to smart buildings. They also provide services such as storage, processing power, or access to their sensors. They relay, distribute, and store data and connect people to this data or to the Internet. As a consequence of this design, our architecture is extensible and flexible. It can be used as a base for many applications such as distributed processing, monitoring, and sensing.

As Fig. 7.3 outlines, the same system can just as well be envisioned to provide communication support in case of a disaster. Many large cities are built in areas where disasters (eg, earthquakes, hurricanes) can cause widespread damage—also to infrastructure, particularly to installed communication infrastructure. If such an event occurs, our architecture can help emergency services by providing communication. Fig. 7.3 shows an early concept how Car4ICT can support the rescue staff in case of a disaster. Inside the disaster area people can connect to cars via ad hoc Wi-Fi (or other communication means which need no infrastructure) and send messages to nearby cars. Any such car can then take messages to an evacuation site where other vehicles might be waiting. If no other cars are in range during the drive to the site, cars transport the messages in a store-carry-forward manner. If other cars are in range, messages can be sent to them via short-range communication [eg, Dedicated Short-Range Communication (DSRC)] in a multihop fashion. Finally, there might be a car that is close enough to a working cellular base station and sends the data to the Internet, delivering the messages to people outside the disaster zone. Such a scenario works with Car4ICT because the whole concept is envisioned to be able to work without any external infrastructure – although it might make use of infrastructure if and where available.

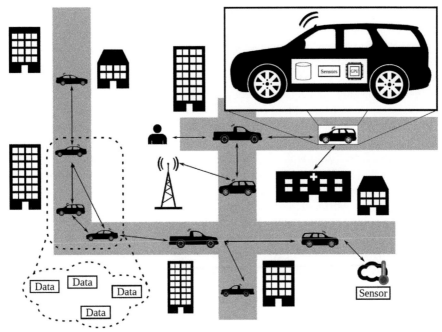

FIGURE 7.2 Concept of Providing ICT Resources in Smart Cities Using Mobile and Parked Vehicles

In the following, we describe the envisioned *Car4ICT* architecture and sample use cases. In particular, we motivate the use of cars as a main ICT resource in smart cities – in a network architecture going beyond current information-centric networking (ICN) systems (based on Ref. [5]). We present a technique for the simulative performance evaluation of such a complex system (based on Ref. [6]) and conclude with a presentation of basic performance results that outline the capabilities of our architecture.

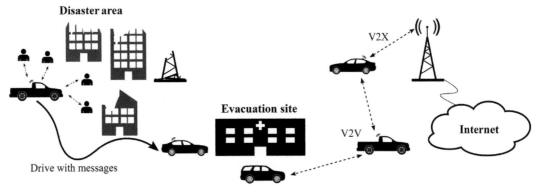

FIGURE 7.3 Concept How Car4ICT Creates an Infrastructure in Case of a Disaster and Helps the Emergency Services

2 RELATED WORK

The idea of using cars and their communication capabilities to form networks has been investigated since more than a decade [4].

In much of this work the focus was on improving traffic safety, driving efficiency, and entertainment solutions based on data exchanged between cars. Early work on data exchange via ad hoc routing in vehicular networks demonstrated that routing works only in very local contexts; thus, clustering solutions have been investigated as an enabling technology [7]. Today, many network protocols and architectures are, in essence, broadcast protocols that have been augmented to be adaptive to network conditions – first and foremost to make them congestion aware. One example is Decentralized Congestion Control (DCC) [8]. The bases of these protocols are different communication technologies, of which a multitude have been investigated in the past years. Early solutions were based on the exclusive use of cellular networks such as Universal Mobile Telecommunications System (UMTS) or Long-Term Evolution (LTE); later research moved to using Wi-Fi access points as Roadside Units (RSUs) (ie, hotspots along the streets). More recently, a dedicated short-range radio solution based on Wireless Local Area Network (WLAN) technology was developed and standardized for vehicular networks as IEEE 802.11p. Standardization of Inter-Vehicle Communication (IVC) protocols is focusing on this in particular. In the United States, one of the driving forces is the DOT, which plans rulemaking to make IEEE 802.11p-based DSRC systems mandatory for new cars [9]; consequently, a standard called IEEE WAVE is developed. In Europe, *ETSI ITS-G5* is planned to be rolled out as soon as cars are equipped with IEEE 802.11p radios. Japanese automakers have recently announced 760-MHz–based vehicle-to-vehicle and vehicle-to-roadside devices as part of an optional package offered in some 2015 model cars in Japan. In parallel to IEEE 802.11p, many other communication technologies are researched that are viable candidates for IVC in future smart cities. This includes visible light communication [10], Bluetooth [11], and millimeter-wave communications superimposed on radar signals [12]. While, traditionally, all these technologies were seen as alternative ways of transporting data, most recently, we observe a trend toward heterogeneous vehicular networking, the use of multiple communication technologies in parallel – each bringing individual strengths and compensating others' weaknesses [13]. Along with many new challenges, this heterogeneous vehicular networking also offers great opportunities. It allows us to design applications and networks in a situation-aware and intelligent way.

Investigating a heterogeneous vehicular network, however, also means that appropriate simulation tool support needs to be guaranteed. Many established simulation tools for vehicular networking research fall short of providing the needed support for such a complex coupled system. First, vehicle mobility has to be adequately modeled – particularly as far as the influence of additional information on the mobility is concerned, for example, in terms of trips, routes, or driver behavior. Second, communication according to the different technologies has to be modeled – again, with a particular focus on the interrelation of packet transmissions using one technology on another. There exist well-established tools that can model network communication and its influence on vehicles' movement patterns: *Veins*, *iTETRIS*, and *VSimRTI*. All of them have good support for mobility but are missing the tools to simulate heterogeneous networks; especially support for LTE networks is missing.

Similar to research on vehicular communication technologies, research on vehicular applications and architectures has been going on for many years now. In the most closely related proposals, cars are envisioned to become gateways for streaming multimedia content or simply providing Internet access to their occupants, or they become an enabler for emergency services. Gerla et al. propose the concept of a vehicular cloud [14], focusing on autonomous driving more than on services provisioning to users: The vehicular cloud is envisioned to provide sensor data and other information that is needed for autonomous vehicles. Other work investigated the establishing of clusters of parked vehicles to form an information hub [2]. Here, parked cars are organized using virtual coordinate-based routing concepts. Users are then able to store data inside (and retrieve data from) this network. Interdomain routing is envisioned to provide network connectivity and data management for disconnected operation. Another, already existing system is proposed by Barros [15]. It uses mesh routing between and across cars to provide Internet access to users. A similar solution coordinates delay-tolerant transfer of bulk data, for example, between data centers, via a vehicular network embedded in a centralized system [16]. In our Car4ICT architecture, we go one step further and enable a large-scale vehicular network to provide services to other cars, outside users, and even sensor systems in future smart cities.

Using Service Discovery Protocols in vehicular ad hoc networks (VANETs) has been investigated in recent years, although with no consent on a good solution. An early proposal, designed for mobile ad hoc networks (MANETs), is by Sailhan and Issarny [17]. By building a virtual overlay network they aim to reduce the load on the network. Based on network densities virtual Service Directories are established or removed. Services are described using the Web Service Description Language (WSDL) and sent following four geographic trajectories. This allows to search for a service and find one by also searching in the same four trajectories. As nodes in their evaluation randomly move with a speed of 1 m/s, it is unclear how well this works for VANETs, as cars are usually moving much faster. A protocol specifically designed for VANETs is proposed by Abrougui et al. [18]. Their system relies on infrastructure in the form of RSUs and on clustering mechanisms to reduce network load. RSUs are able to move small distances and are clustered around a service. Three types of services are supported in their approach: moving services, fixed services, and migratory ones. The last kind is fixed services staying in a certain region and is passed to other vehicles in case the current vehicle leaves the area. Lakas et al. present a service discovery solution with four different actors [19]. Instead of the usual three actors, Service Provider, Service Consumer, and Service Directory, they also see cars as mobile SDs. Requests are flooded via broadcast until a user providing the service is found or the lifetime of the message is over. Because they rely heavily on broadcast, the scalability of the proposed Service Discovery Protocol remains unknown.

For routing in vehicular networks, there exist a lot of suitable solutions for urban or rural environments. There are reactive and proactive approaches, as well as protocols focused on georouting. As evaluating and comparing them is out of the scope of this chapter, we refer to Ref. [4], which presents the best known algorithms.

The Car4ICT system will need to be able to identify available services in a location-based approach. The task of content identification is most closely related to future Internet research. At its core, a basic solution is to associate content with unique names. Here, concepts such as ICN [20] and Named Data Networking (NDN) [21] have been proposed. Applying the concepts of such ICN to vehicular networks has been proposed in Ref. [22], yet with a focus on a single-use, single-protocol network. By focusing only on IEEE 802.11p communication, the authors were able to move away from Internet Protocol (IP) and toward a new custom protocol to enable efficient ICN networking. Other architectures, by Wang et al. [23] and Grassi et al. [24], apply the concept of NDN to the vehicular domain. Both approaches, however, are targeting only a limited set of applications, optimizing the system for a specific use case and a specific communication technology. More recently, *Internames* [25] was proposed, a system that is envisioned to work with all kinds of approaches including IP, cellular, and ad hoc networks. At its core, Internames maps all names to locations as well as the needed protocol to reach them. Still, it focuses very much on one application domain. Our architecture tries to act as a base for a multitude of applications, building upon concepts described in these approaches such as the association of items with hash tags. Further, current ICN and NDN solutions commonly fall short when it comes to the specific requirements in the very dynamic and disruptive environment of vehicular networks, which we try to directly address in Car4ICT.

3 THE CAR4ICT CONCEPT

Our concept makes no further assumption than that there are some network-enabled vehicles driving on the road. Such vehicles can be envisioned to be available almost ubiquitously in future smart cities. No preinstalled infrastructure is assumed to be available (or to offer free capacity), although if it is, the network is assumed to be able to exploit it for improved performance. As discussed, this network of vehicles is designed to complement the available Wi-Fi access points, hotspots on the streets, and cellular networks based on UMTS (3G) or LTE (4G), where these are available.

Building on this assumption, we aim to make the additional network of connected vehicles available to users. One use case might be to exploit the movement of a car when there is no other connection available—thus having a car deliver data to another region simply by exploiting its movement, either to the intended destination of the data or to an area where connectivity to the destination is available.

This enables the car network to act as central components of future ICT systems of smart cities. Our concept revolves around requesting *services* or offering them to users: either smart devices or, of course, people. Potential use cases are people offering storage space, unused CPU power, or devices offering access to smart sensors. Users are envisioned to request access to these services (eg, for processing large amounts of data on the go). Service requests and offers are mediated by cars, which act as coordinator: storing service offers and matching them with service requests. Two possible methods are allowed in the architecture: reactive service fulfillment where neighbors are actively queried once a request is received and proactive exchange of service tables. Once a request and one or more offers have been matched, the Car4ICT system compiles a list of offers and returns it to the requesting entity. People or devices can then elect to use a service, exchanging data with the entity offering the service—again mediated by the Car4ICT system, which employs the network of vehicles to relay the data between the user requesting and the one offering the service. Such data relaying can also be done in multiple possible ways, from routing to pure store-carry-forward.

3.1 USE CASES

As the architecture is designed to be used both by people and by machines, various different use cases arise for Car4ICT. Consider, for example, smart sensors of future smart cities detecting a natural disaster. Because data validation and postprocessing is a resource-intensive task and because infrastructure might already be failing, they too much "might" exploit the Car4ICT system to transport large amounts of raw data to a central authority—or even to offload the tasks of postprocessing and evaluating the data to CPU resources offered by the network itself.

As another use case, consider a tourist who wants to quickly store more pictures than can be kept on his or her camera. Without having to rely on compatible infrastructure or on a data plan in a foreign country, he or she can rely on any technology (Wi-Fi or Bluetooth being the most likely ones) to connect to any Car4ICT-enabled car to access the system. The system will try to match the incoming request for secure replicated storage to an existing service. If one is found, the tourist can use the Car4ICT network to offload pictures. Back at the hotel, the tourist can query any passing Car4ICT-enabled car to download the pictures back to his or her device. Automated systems can also make use of the Car4ICT architecture, for example, a weather forecast system. To acquire the necessary input data for a certain region the system can now exploit Car4ICT to search for temperature sensors in that area. The temperature data can be offered by sensors at fixed locations in that area or cars that just happen to drive through it. Choosing now from a vast range of possible sensors, the forecast system is able to get lots of data via Car4ICT and generate the temperature forecast. Note that searching for such sensors and finally acquiring the data are two steps and allow the forecast system to decide from which sensors to get the temperature. More and more cars in recent years come equipped with so-called dashboard cameras that constantly record videos or take pictures. This provides another use case related to semilive picture sharing. Cars can offer the dashboard camera data via Car4ICT and are therefore able to provide a semilive view of certain areas. Depending on the storage and network capacities of a car, it can offer only recent pictures or also older ones to provide a view into the past. With cars offering this data it is now possible to search for a place and gather pictures from this place taken by dashboard cameras. This gives the possibility to see how the area looked recently or at some point in the past.

3.2 BASIC ARCHITECTURE

Two types of entities interact in the Car4ICT architecture. First, *members* are the nodes that are connected to form the Car4ICT network. Members are responsible for transporting all control and data messages through the network. Members will typically be cars driving in a smart city, but parked cars or temporary support can be envisioned to participate just as well. The second type of entities, *users*, relies on any nearby Car4ICT member to serve as their gateway to the Car4ICT network. They inject new service requests or offers into the network and use the network to consume services. A Car4ICT user might be a person (eg, a tourist accessing a service via their smart phone's Wi-Fi, Bluetooth, or a DSRC connection) or a machine (eg, a sensor in a building connecting wirelessly or even a subsystem of the car itself, connected by wire). The choice of which communication technology to use for interacting with a member is left up to the user. Depending on metrics such as delay, connectivity, and cost, the user might want to rely on provided ad hoc communication or on a more costly LTE data connection. While persons should be enabled to make conscious choices about the technologies they use, machines will need to be equipped with a way of automatically reasoning about the suitability of available technologies—thus enabling them to use the Car4ICT architecture in true Machine-to-Machine (M2M) fashion. To prevent abuse of the system every user must be prepared to undergo a verification step before being allowed to offer or request services via a member of the Car4ICT network. Using, for example, a Public Key Infrastructure (PKI) any member is equipped to verify a user's credentials (presented wirelessly when connecting to the network) and to reply with an individual grant.

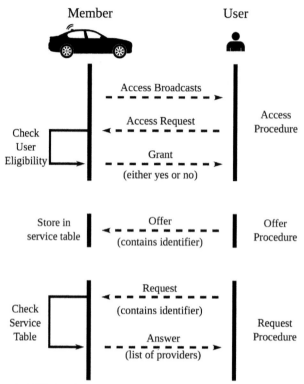

FIGURE 7.4 Entities of the Car4ICT Architecture

Fig. 7.4 illustrates this architecture: Members advertise their availability for connections by means of *access broadcasts*. On receiving such an access broadcast by a member, a user may choose to establish a connection to the Car4ICT network via this member. The user sends security credentials to the member, which are checked for the user's eligibility to access the Car4ICT network. After confirming the eligibility, the member responds with an individual access grant. This access grant can be used to prove eligibility to access the network—not just at the member that generated the grant. How and where this grant is generated is left open at this time.

A user that is authorized for access to the network is free to send offers to any Car4ICT member, or to send requests for services that will be matched with existing offers before being returned to the user. It is left to any member to decide whether a valid grant is necessary or whether this step can be skipped (eg, because the member is a well-known device in the same car connected via a trusted wired connection).

Fig. 7.5 illustrates how the entities of a Car4ICT system interact to exchange control messages. Shown are potential users requesting access from a member, the access being granted, and finally the user accessing a service offered (in this case, directly on the member the user is connected to). Also illustrated is proactive exchange of service tables between members and (via an optional cellular network) with an (optional) central server.

Our architecture can serve as the base for providing a wide array of different services, for example, data storage, distributed computing, sharing messages with a specific region, or retrieving sensor readings.

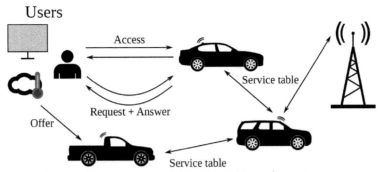

FIGURE 7.5 Illustration of Control Message Exchange in the Car4ICT Architecture

Service offers can be stored on each member of the Car4ICT network by means of service tables, which can (but need not) be shared among neighbors. This enables each member to query its local table whenever a user requests a service. If the matching of requested service and known (or discovered) service offers was successful, an answer is sent back including the ID of the service. Naturally, not each service will be known everywhere, so there are multiple steps that can be taken to improve the chances of a query's success:

- A car could forward the request to neighboring cars that then in return could check their service tables for a potential match. This method probably reduces the overall network load but might increase the delay for a service discovery request.
- Another approach would be to share the service tables proactively. In such a case most likely the response time for requests decreases but the communication overhead would increase.
- As some of the cars also have a cellular link, it could be helpful to store offers at some central server reachable via the Internet. If a car with such a connection receives a request that it cannot answer, it can query this central service table and get an answer from there. In such a case the location information of the request will play an important role.
- Mixed approaches are also possible. A car may first execute an expanding ring search using only short-range radio communication. If this remains unsuccessful, and if sufficient incentive exists, a car might then use an LTE uplink to query more distant members.

When successful, this process yields a list of users offering the requested service, as well as meta-data that indicates how these users might be reachable via the Car4ICT network. It is left up to the user to decide whether and which service(s) to use. For a person (eg, using the system for replicated data storage), the incentives to use a given service might be different than those of a machine (eg, for building a small-scale weather, pollution, or allergen map). In any case, data transport to and from the service is taken care of by the Car4ICT network, that is, by its members.

3.3 SERVICE AND NEIGHBOR TABLES

As described, the operation of the Car4ICT architecture hinges on service tables, which are kept locally on Car4ICT members and which are continuously updated.

Such updates can occur by express messages, sent via connected users. Similarly, updates can be triggered by overhearing service announcements from other members. We believe that these service announcements can be sent in a bandwidth-conserving manner by exploiting other protocols. Current standardization by the IEEE and ETSI in the United States and Europe, to give two examples, assumes the continuous exchange of Basic Safety Messages (BSMs) and Cooperative Awareness Messages (CAMs), respectively. They are employed to maintain neighbor information for road safety applications, but can just as well serve to support other applications, for example, for maintaining neighbor tables. Similar to other applications, it might be feasible to piggyback (subsets of) service tables onto CAMs—if just to make vehicles aware that a certain service is still operating. If no updates for a known service are received for a certain duration (either via piggybacked information or in response to queries), it is the task of Car4ICT members to expire its corresponding entry in their service tables. Similarly, services that are offered only in (or are relevant to) a certain geographic region need to be purged on leaving this area.

Services are identified, distinguished, and tracked using the following scheme: Each service is associated with a hash tag and metadata. The hash tag is a string of fixed length that uniquely identifies the type of service. This might be the hash of specific content being offered (eg, a file), or it might be a well-known string referring to such resources as a certain type of sensor, CPU power, or replicated storage. In addition to the hash tag, a service is annotated with metadata pertaining to, for example, geographic area or validity constraints. This scheme follows the concepts discussed in the ICN context [26] and has proven itself to be both flexible and extensible.

Table 7.1 shows an example service table, identifying multiple services by one hash tag each, along with metadata: Three different users (1, 3, and 7) are offering a certain video file, *file1*, of a given length (2 GB), at different locations (*Paderborn* and *Tokyo*). Another file, an image named *file2*, is hosted by two users (1 and 12). User 7 is also offering both computation resources and storage, each with associated metadata describing the details of, for example, the CPU and the storage offered.

Geographic position is considered to be a first-class metadata that has to be included in all messages. This can be exploited to use only services in a certain area or prevent service requests from leaving a certain area. For example, if a user wants to get a processing resource close to his or her location, the request can be already stopped if it is outside of this area – therefore reducing the load on the network.

There are multiple ways to compare identifiers with each other. An identifier I_x is denoted by the tuple (h_x, M_x), where h_x is its hash and M_x a set of its metadata elements. We identified three ways of comparing two identifiers I_1 and I_2:

Table 7.1 A Service Table Example

User	Hash Tag	Metadata		
1	hash(file1)	location = Tokyo	type = video	size = 2 GB
3	hash(file1)	location = Tokyo	type = video	size = 2 GB
7	hash(file1)	location = Paderborn	type = video	size = 2 GB
1	hash(file2)	location = Tokyo	type = image	
12	hash(file2)	location = Tokyo	type = image	size = 500 MB
7	CPU	location = Paderborn	type = ARM	
7	Storage	location = Paderborn	type = hours	size = 78 GB

- *Hash only* ($\underline{h}$): This comparison uses only the hashes; therefore $I_1 \overset{h}{=} I_2 \leftrightarrow h_1 = h_2$. It can be used, for example, if a user wants to search for storage but has no specific requirements.
- *Subset matching* ($\subseteq$): To match only a subset of an identifier with another the hashes have to be equal and all $m \in M_1$ have to be included in M_2, but not all $m \in M_2$ have to be in M_1. This can be used to search for processing power with some requirements such as the type of CPU needed. Therefore $I_1 \subseteq I_2 \leftrightarrow (I_1 \overset{h}{=} I_2) \wedge (\forall m \in M_1 : m \in M_2)$.
- *Equals* (=): This comparison compares both hashes to be the same and all $m \in M_i$ have to be in every other $M_j, j \neq i$. The comparison = can be formulated as $I_1 = I_2 \leftrightarrow (I_1 \subseteq I_2) \wedge (I_2 \subseteq I_1)$.

As the geographic position is a first-class metadata entry, an Identifier I_x can be extended to a triple (h_x, pos_x, M_x). A user can now send a request with a certain position pos_u and a radius r. If the user requests now a so-called *distance match* with the identifier I_u from the Car4ICT network, a matching identifier I_x can be matched by the earlier comparisons. Only the first one has to be modified, as the others inherit the distance check from it: $I_u \overset{h}{=} I_x \leftrightarrow (h_u = h_x) \wedge (\text{dist}(\text{pos}_u, \text{pos}_x) \leq r)$.

4 SIMULATIVE PERFORMANCE EVALUATION

To validate and evaluate the proposed complex architecture a simulation framework is needed that can accurately model all of vehicle movement, user behavior, short-range radio communication between users and members, as well as cellular communications where such infrastructure is available. We performed simulations using *Veins LTE* [6], a simulator developed with heterogeneous vehicular networking in mind. Such networks usually consist of cars communicating via multiple network technologies. For example, IEEE 802.11p-based technologies (eg, DSRC or ETSI G5) can be used to communicate with cars in close proximity, while a cellular network, for example, LTE, is used to exchange data with cars farther away or to connect to the Internet. To be able to simulate such networks we developed *Veins LTE* and published it as Open Source software. Veins LTE, as the name implies, is based on *Veins*, which itself is based on the discrete network simulator OMNeT++ and connects to the mobility simulator *SUMO*. *Veins* has a so-called feedback loop that allows the network simulator (OMNeT++) to react on events from the mobility simulator (SUMO) and vice versa. Such a feedback loop enables the simulation of scenarios that are dependent on information from the network and allows cars to change their route during the simulation (which would be not the case with trace-based simulations). While Veins already provides means to simulate IEEE 802.11p-based networks, *Veins LTE* itself adds *SimuLTE* [27] to the mix—which provides the possibility to simulate the complete LTE stack. With this addition two network technologies (IEEE 802.11p and LTE) can be simulated in great detail by using a single simulator.

We conducted an initial study of using a heterogeneous vehicular network for cooperative intersection collision avoidance according to an algorithm proposed by Tung et al. [7] to ensure that the combined simulation framework was behaving as expected. As a simple indication, Fig. 7.6 shows a sample result, the delay in the LTE uplink. As can be seen the delay increases with the frequency of sent messages. It also should be noted that the delay is shown independent of the transmission distance—this also adds longer delays for messages as more retransmissions are needed.

On top of *Veins LTE*, applications for such heterogeneous vehicular networks can be developed. Any application can decide to send the packet via a specific network stack or let a dedicated module—named *Decision Maker*—decide which network technology is better suited for the current transmission.

Uplink delay (s)
0.01
0.25
5
1
2
5
10
Message interval (Hz)

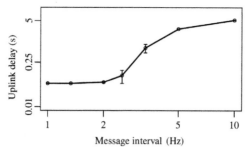

FIGURE 7.6 LTE Message Interval Versus LTE Uplink Delay (Mean and 95% Confidence Interval) for an Allocation of 100 Resource Block (RB) (20 MHz Bandwidth)

Note that both scales are logarithmic [6]

The stack can be seen in Fig. 7.7. As Car4ICT is meant to work in exactly such a heterogeneous vehicular network where different technologies can flexibly be used when and where available, this simulation environment is the perfect fit for the evaluation and the validation of Car4ICT.

We added a new *User* module to the simulator that is able to run multiple applications and has a *Connector* that is in charge of connecting the user (and its applications) to the Car4ICT network. Applications then send messages to the *Connector* that in turn forwards it, if a connection exists, to a car. On the car side we developed the *Car4ICT module*. This module has three main tasks:

1. If an offer is received, it is stored in the service table.
2. If a message arrives that is neither a request nor an offer, it should be, if possible, routed toward its destination.
3. As cars can also include applications, the *Car4ICT module* also offers an interface for such applications. Our simulation is built so applications can be easily used in cars and by users – only a single flag has to be set that defines the connection type (eg, local or wireless).

Figs. 7.8 and 7.9 shows screenshots of a running Car4ICT simulation. Both times, a user requests a service provided by another user (big dots). In the first case, shown in Fig. 7.8, service discovery remained unsuccessful. As the first car – the one the user connects to – does not have a corresponding entry in its service table, the request is sent further to other cars, but the request expires before reaching a valid destination. In the second case, shown in Fig. 7.9, a service is found and the user requesting it is

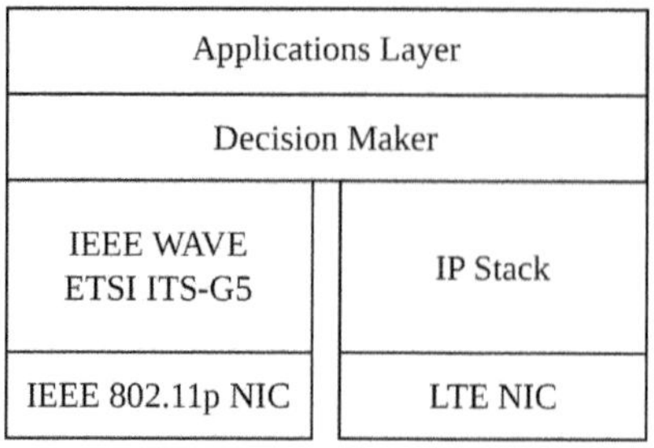

Applications Layer
Decision Maker
IEEE WAVE
ETSI ITS-G5
IP Stack
IEEE 802.11p NIC
LTE NIC

FIGURE 7.7 The Heterogeneous Stack of Veins LTE [6]

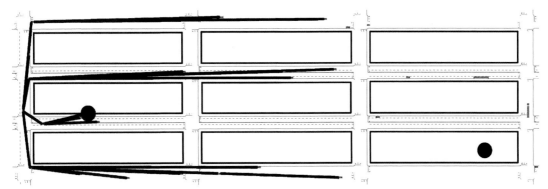

FIGURE 7.8 Simulation screenshot of an unsuccessful search for a service provider

No reached car knows how to match requester and provider (*big dots*)

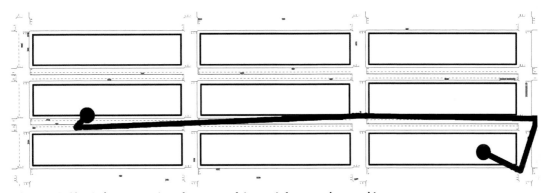

FIGURE 7.9 Simulation screenshot of a successful search for a service provider

Shown is an example path from provider to requester

informed. A thick line indicates the hops a data packet might take between the requesting user and the one offering the service. Note that there are two ways of answering to requests: *proactively* by sharing the service tables upfront and *reactively* by searching for a provider after a request cannot be answered.

5 SIMULATION STUDY

To illustrate the feasibility of the Car4ICT architecture, we conclude with an evaluation of the performance of service discovery in our Car4ICT framework under progressively worse conditions.

We used the described simulation framework, implementing the outlined approach in detail. Each member (ie, each car) in the simulation periodically announces its presence to users nearby. We opted for implementing a *proactive* Car4ICT service discovery variant, exchanging service tables in fixed time intervals. Users periodically request different services. As long as the request cannot be matched to an offer by the Car4ICT architecture, the request is retried. When the request is successfully matched and a positive response is delivered to the user, we note the delay between first try and fulfillment as the *discovery latency*.

Our implementation supports all described ways of comparing service identifiers. The two most important ones for when a user is searching for a service are $\underline{h}$ and $\subseteq$. The first one, $\underline{h}$, can be used to get an overview of all offered services of a certain kind in an area. Second, $\subseteq$ helps a user to find a service with some prerequisites. The comparison $=$ is mostly used when comparing new offered services if they do not already exist in the service table.

As the simulated traffic scenario we chose a challenging Manhattan Grid topology, populated by as much as 415 equipped cars/km^2 and down to as little as 35 equipped cars/km^2 taking random trips. Both road and building dimensions correspond to downtown Manhattan. Figs. 7.8 and 7.9 shows a small section of this scenario. Buildings next to each road shield radio transmissions across streets, thus cutting down further on the number of potential communication partners in low-density scenarios. To make the scenario more challenging, we add exactly five users connecting to a close-by vehicle and offering a service, five users requesting this service, and five more users offering unrelated services. This scenario closely mirrors the discussed use case of secure replicated storage, but with a tightly limited amount of available resources. All parameters are summarized in Table 7.2.

We illustrate results of our discovery latency evaluation in the form of an empirical cumulative distribution function (eCDF) as service requests can take anything from being instantaneously fulfilled to never being fulfilled (in increments of 2 s, the retry interval).

In Fig. 7.10 we show results for different traffic densities (measured in vehicles per square kilometer), given a service table broadcast interval of 10 s. As could be expected, traffic density has an immediate impact on discovery latency. While for high densities of 415 vehicles/km^2, it takes no more

Table 7.2 Used Simulation Parameters

Parameter	Value
Simulated area	0.7 km^2
Average number of equipped cars per square kilometer	35–415
Total number of users	15
Number of users requesting	5
Number of users offering	5
IVC technology	IEEE 802.11p
IVC maximal transmit power	10 mW
Simulation duration	80 s
Service table broadcast interval	0.1–10 s
Neighbor table entry lifetime	10 s
Service table entry lifetime	10 s
User request interval	2 s
Request time-out	30 s
IVC, Inter-Vehicle Communication.	

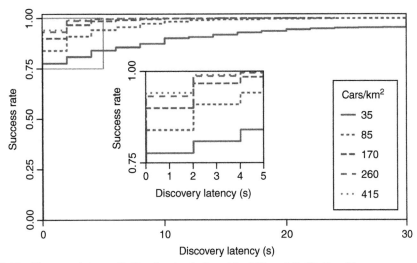

FIGURE 7.10 The Discovery Latency Until a Service Is Found for Different Traffic Densities

The service table broadcast interval was set to 10 s. The inset shows a zoom to the first 5 s

than 2 s to match 99% of service requests to a corresponding offer, at slightly lower densities of 85 vehicles/km² this fraction already drops to 90%, as can be seen in the zoomed-in part of the figure. At this density, fulfilling 99% takes as much as 12 s. At lower densities, even 30 s is not enough to fulfill 99% of requests. While this would be well in the acceptable range for M2M communication like in a distributed sensing use case, it would likely be above the tolerable threshold for a user-facing service.

We thus look to smaller service table broadcast intervals as a potential solution at low traffic densities. Fig. 7.11 illustrates our results, now using service table broadcast intervals as low as 0.1 s. We observed

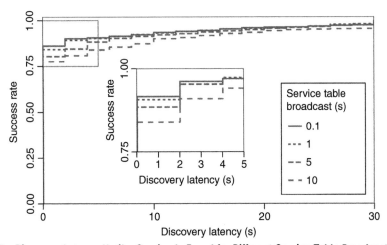

FIGURE 7.11 The Discovery Latency Until a Service Is Found for Different Service Table Broadcast Intervals

The traffic density was 35 vehicles/km²

that adjusting the service table broadcast interval from 10 to 1 s was enough to achieve a better success rate than with these lower rates. Even for medium to low discovery latency constraints, system performance can be seen to be in an acceptable range, not just for M2M services but also for user-facing services such as in the discussed secure replicated storage use case. Decreasing the interval further yielded no substantial performance gain. Thus, while an adaptive service table broadcast interval would certainly help conserve channel capacity at high vehicle densities, configuring a moderate service table broadcast interval at any traffic density yields a system that delivers good performance at both high and low traffic densities.

6 CONCLUSIONS

We outlined our idea for a novel architecture, which we term Car4ICT, that uses cars as a main ICT resource of smart cities. We showed how such an architecture can help cope with the forecasted massive data demand of smart cities – and how it can operate in disaster situations where no infrastructure might be available. Based on this architecture, a huge variety of new applications is made possible in a network architecture going beyond current ICN systems. Our concept revolves around offering and requesting services, serving smart devices just as well as people. We highlighted use cases of this architecture such as offering storage space, unused CPU power, or devices offering access to smart sensors. We presented a flexible and extensible scheme for identifying, distinguishing, and tracking services. The architecture is technology agnostic, using any of multiple available wireless technologies to connect cars to each other and with users. This necessitated a custom-built simulation framework, which we published as Open Source. Based on this framework, we illustrated the feasibility of the proposed Car4ICT architecture, concluding with an evaluation of the performance of service discovery under progressively worse conditions. We were able to show that even in such conditions simple tweaks to protocol parameters can deliver good user experience in terms of low service discovery latencies.

REFERENCES

[1] Caragliu A, Del Bo C, Nijkamp P. Smart cities in Europe. J Urban Technol 2011;18(2):65–82.
[2] Dressler F, Handle P, Sommer C. Towards a vehicular cloud – using parked vehicles as a temporary network and storage infrastructure. In: 15th ACM international symposium on mobile ad hoc networking and computing (Mobihoc 2014): ACM international workshop on wireless and mobile technologies for smart cities (WiMobCity 2014). Philadelphia, PA: ACM; August 2014. p. 11–8.
[3] Altintas O, Seki K, Kremo H, Matsumoto M, Onishi R, Tanaka H. Vehicles as information hubs during disasters: glueing Wi-Fi to TV white space to cellular networks. IEEE Intell Transportation Syst Mag 2014;6(1):68–71.
[4] Sommer C, Dressler F. Vehicular networking. Cambridge: Cambridge University Press; 2014.
[5] Altintas O, Dressler F, Hagenauer F, Matsumoto M, Sepulcre M, Sommer C. Making cars a main ICT resource in smart cities. In: 34th IEEE conference on computer communications (INFOCOM 2015), international workshop on smart cities and urban informatics (SmartCities 2015). Hong Kong: IEEE; April 2015. p. 654–9.
[6] Hagenauer F, Dressler F, Sommer C. A simulator for heterogeneous vehicular networks. In: 6th IEEE vehicular networking conference (VNC 2014), poster session. Paderborn, Germany: IEEE; December 2014. p. 185–6.
[7] Tung L-C, Mena J, Gerla M, Sommer C. A cluster based architecture for intersection collision avoidance using heterogeneous networks. In: 12th IFIP/IEEE annual Mediterranean ad hoc networking workshop (Med-Hoc-Net 2013). Ajaccio, Corsica, France: IEEE; June 2013.

[8] Vesco A, Scopigno R, Casetti C, Chiasserini C-F. Investigating the effectiveness of decentralized congestion control in vehicular networks. In: IEEE global telecommunications conference (GLOBECOM 2013). Atlanta, GA: IEEE; December 2013.

[9] Wald ML. U.S. plans car-to-car warning system. In: The New York Times; February 4, 2014. p. B3.

[10] Viriyasitavat W, Yu S-H, Tsai H-M. Channel model for visible light communications using off-the-shelf scooter taillight. In: Vehicular networking conference (VNC), 2013 IEEE. Boston, MA: IEEE; December 2013. p. 170–3.

[11] Bronzi W, Frank R, Castignani G, Engel T. Bluetooth low energy for inter-vehicular communications. In: 6th IEEE vehicular networking conference (VNC 2014). Paderborn, Germany: IEEE; December 2014.

[12] Hasch J, Topak E, Schnabel R, Zwick T, Weigel R, Waldschmidt C. Millimeter-wave technology for automotive radar sensors in the 77 GHz frequency band. IEEE Trans Microw Theory Tech 2012;60(3): 845–60.

[13] Dressler F, Hartenstein H, Altintas O, Tonguz OK. Inter-vehicle communication – quo vadis. IEEE Commun Mag 2014;52(6):170–7.

[14] Lee E, Lee E-K, Gerla M, Oh S. Vehicular cloud networking: architecture and design principles. IEEE Commun Mag 2014;52(2):148–55.

[15] Barros J. How to build vehicular networks in the real world. In: 15th ACM international symposium on mobile ad hoc networking and computing (Mobihoc 2014). Philadelphia, PA: ACM; August 2014. p. 123–4.

[16] Baron B, Spathis P, Rivano H, Dias De Amorim M, Viniotis Y, Clarke J. Software-defined vehicular backhaul. In: IFIP wireless days conference 2014. Rio de Janeiro, Brazil: IEEE; November 2014.

[17] Sailhan F, Issarny V. Scalable service discovery for MANET. In: Third international conference on pervasive computing and communications (PerCom 2005). Kauai, Hawaii: IEEE; March 2005. p. 235–44.

[18] Abrougui K, Boukerche A, Pazzi R. Design and evaluation of context-aware and location-based service discovery protocols for vehicular networks. IEEE Trans Intell Transportation Syst 2011;12(3):717–35.

[19] Lakas A, Serhani MA, Boulmalf M. A hybrid cooperative service discovery scheme for mobile services in VANET. In: 7th international conference on wireless and mobile computing, networking and communications (WiMob). Shanghai, China: IEEE; October 2011.

[20] Ahlgren B, Dannewitz C, Imbrenda C, Kutscher D, Ohlman B. A survey of information-centric networking. IEEE Commun Mag 2012;50(7):26–36.

[21] Zhang L, Afanasyev A, Burke J, Jacobson V, Claffy K, Crowley P, Papadopoulos C, Wang L, Zhang B. Named data networking. ACM SIGCOMM Comput Commun Rev 2014;44(5):66–73.

[22] Amadeo M, Campolo C, Molinaro A. CRoWN: content-centric networking in vehicular ad hoc networks. IEEE Commun Lett 2012;16(9):1380–3.

[23] Wang L, Wakikawa R, Kuntz R, Vuyyuru R, Zhang L. Data naming in vehicle-to-vehicle communications. In: 31st IEEE conference on computer communications (INFOCOM 2012): workshop on emerging design choices in name-oriented networking. Orlando, FL: IEEE; March 2012. p. 328–33.

[24] Grassi G, Pesavento D, Wang L, Pau G, Vuyyuru R, Wakikawa R, Zhang L. ACM HotMobile 2013 poster: vehicular inter-networking via named data. ACM SIGMOBILE Mobile Comput Commun Rev 2013;17(3):23–4.

[25] Melazzi NB, Detti A, Arumaithurai M, Ramakrishnan KK. Internames: a name-to-name principle for the future Internet. In: 10th international heterogeneous networking for quality, reliability, security and robustness (QShine 2014). Island of Rhodes, Greece: IEEE; August 2014. p. 146–51.

[26] Farrell S, Kutscher D, Dannewitz C, Ohlman B, Keranen A, Hallam-Baker P. Naming things with hashes. In: IETF, RFC 6920; April 2013.

[27] Virdis A, Stea G, Nardini G. SimuLTE – a modular system-level simulator for LTE/LTE-A networks based on OMNeT++. In: 4th international conference on simulation and modeling methodologies, technologies and applications (SIMULTECH 2014), Vienna; August 2014.

FROM VEHICULAR NETWORKS TO VEHICULAR CLOUDS IN SMART CITIES

8

M. Soyturk*, K.N. Muhammad*, M.N. Avcil*, B. Kantarci†, J. Matthews**

*Department of Computer Engineering, Marmara University, Istanbul, Turkey;
†Department of Electrical and Computer Engineering, Clarkson University, NY, United States;
**Department of Computer Science, Clarkson University, NY, United States

1 INTRODUCTION

Smart city and digital cities retrieve information in collaborative environment and store it to the Internet cloud. According to Yovanof et al. [1], a smart and digital city provides connected infrastructure to ensure an appropriate service quality standard for the inhabitants of the society. Smart city can also be described as an ICT-centered information city where technical infrastructure, diverse people and good governance are combined in action [2]. Giffinger et al. also included the importance of characteristics such as economy, mobility and lifestyle [3]. The study in Ref. [4] presents the role of smart city applications on government administration, modern health service, intelligent transportation, efficient utility and secure environment for public services. These applications of smart city concept motivate the role of Vehicular Ad hoc Networks (VANETs), cloud computing, and Intelligent Transportation Systems (ITS) in modern computing era.

As an inseparable component of smart cities, the concept of VANETs is a research area expanding very rapidly where the research community and the industry are working in a collaborative manner. The main aim is offering new and novel communication technologies and building infrastructures for the vehicles to communicate among themselves. Recent work in this area has resulted in established standards such as IEEE 802.11p and IEEE 1609.

VANETs are considered to be one of the challenging forms of wireless communication technologies that facilitate road efficiency and safety applications through ITS. ITS aim at the betterment of the transportation in cooperation with the Information and Communication Technologies (ICTs). Furthermore cyber-physical solutions in smart cities have enabled interaction between the physical and computational components of systems. VANETs are specialized forms of Mobile Ad hoc Networks (MANETs) where protocols performed well in MANET may not be suitable for VANET. This is due to VANETs' experiencing significant constraints in terms of node mobility, frequent topology change and varying speed of the nodes. Due to its intrinsic characteristics, enabling Quality of Service (QoS), reliability, and security in addition to the well-known network operations, for example, scalable routing has become much more challenging in VANETs. Clustering appears as a promising solution to cope with all these kind of issues as it has been proven to be in the case of MANETs.

A VANET is a useful platform for ITS. Through peer-to-peer ad hoc based communication, VANETs provide collaborative infrastructure to construct a robust information network. The real time status of a particular vehicle is retrieved through its On-Board Units (OBUs) and eventually transferred to its peers. Thus the network is not particularly dependent on roadside unit or sensors for information propagation.

With the advent of cloud computing concept, computing, communications, and storage resources are being provided as services within a shared pool of resources with rapid elasticity and pay-as-you-go fashion. Furthermore, the advancements in mobile communications reveal the potential of mobile devices to form a mobile cloud environment. Vehicular networks combined with mobile cloud computing introduces the vehicular cloud concept that enables connected vehicles to share their resources through a cloud platform. Several smart city applications can be empowered by the vehicular clouds. These applications can be listed as highway/downtown traffic monitoring, environmental monitoring, emergency assistance, disaster management, multimedia content delivery and so on. Cloud computing can facilitate a scalable system with optimum cost through on demand access to the shared infrastructure. Vehicular cloud is a unique idea to combine information from mobile nodes and various Road Side Units (RSUs), and store it in the cloud.

A vehicular cloud is formed by a cluster of vehicles to enhance unified communication and enable self-organization on the basis of network demand. The VANET technology proposed in Refs. [5,6] has discussed the concept of vehicular communication through Vehicular Cloud Computing (VCC) and Information Centric Networking (ICN). VCC provides a method of network service provisioning and ICN ensures the process of cloud centric data routing and dissemination. Through peer-to-peer connection, the nodes are internetworked for resource sharing directly in a decentralized manner. However, for the sake of computing efficiency, one vehicle might be elected as the broker that is responsible for resource allocation based on some metrics. Thus VCC can provide resource monitoring, efficiency in routing, securing the intervehicle privacy issues and virtualization among vehicles. Based on the network, it is easier to create new application on vehicular cloud. ICN provides the scope to spread the cloud contents efficiently among the vehicles. Hu et al. [7] introduced a Service-Centric Contextualized Vehicular cloud (SCCV) that is efficient and can mitigate the network overhead.

Two important issues of Vehicular Cloud Networking (VCN) are addressed in Ref. [6]: the process to quantify the value of resources and process of preventing vehicles moving freely. The corresponding paper also introduces the design principles of VCN services: data storage, sensing services and computing services.

Fig. 8.1 illustrates a sample smart city scenario where connected vehicles collaboratively share the traffic conditions to assist paramedics to hospitalize the patients who need urgent medical care. According to this application, alternate destination medical centers are crowdsourced, and alternate route-trees toward those destinations are computed through collaboratively collected road and traffic condition data. The mobile end-user device sends an inquiry to the vehicular network that reports the traffic profile to the alternate route-tree selection module along with a number of possible medical centers that the patient may be taken to. Due to the possible sudden changes of the traffic profile in an urban area, it is crucial to determine the alternate routes to predetermined destinations. Moreover, in case of a sudden change in the traffic profile, the ambulance driver should be able to switch to an alternate route toward an alternate medical center.

Besides their benefits, connected vehicle systems denoting both vehicular networks and vehicular clouds introduce several challenges. These challenges vary from clustering problems related to spatial and temporal properties of the communication medium, QoS assurance, vehicle-to-infrastructure/

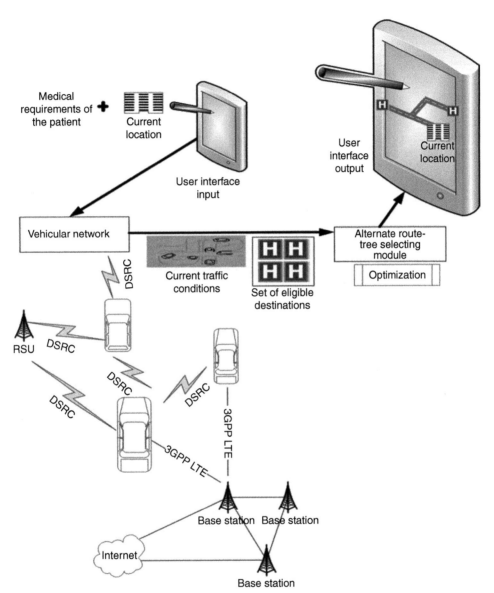

FIGURE 8.1 Sample Smart City Application in an Urban Area Aiming at Emergency Assistance for Paramedics [8]

infrastructure-to-vehicle communication challenges in VANETs to virtualization and security/privacy/trust problems in vehicular clouds. This chapter presents a comprehensive survey of the current state of the art in VANETs and vehicular clouds for smart cities. The chapter starts with a definition of the VANET and vehicular cloud architectures by surveying the existing solutions, as well as identifying research and application challenges; then a thorough study of the existing solutions for

VANET challenges in smart cities are presented and compared to each other using several criteria. This section is followed by a section where vehicular cloud solutions under dynamic and static scenarios are presented along with the challenges that are mentioned previously. The studies in this section will be grouped under the following categories: Virtualization for computation and storage services, security challenges, privacy and user experience issues and context-aware solutions for all services in vehicular cloud environments. Comparisons of the surveyed solutions are presented along with relevant arguments. The chapter is wrapped up with a thorough discussion of open issues, challenges and possible directions for the researchers who would like to pursue solutions for vehicular communications in smart cities.

2 VANET ARCHITECTURE

ITS utilize the communication technologies to connect vehicles, people, and any facility for more secure, safer, and highly mobile transportation in an urban environment [9]. Vehicular Networks are the key component of an ITS, enabling and integrating the use of various technologies, communication standards and the infrastructures [10–12]. The city is turned into a smart, connected city by the ITS with the use of vehicular networks and its infrastructure. Every region, facility, driver, passenger, and even pedestrian is envisioned to be connected to the ITS and could be made aware of local, or region of interest (ROI), or city-wide events and updates/changes to the transportation system even in real-time. Within that architecture, real-time and non-real-time information will be used and provided by the ITS and vehicular networks for safety and efficiency [13].

Since the 1980s, modern vehicles have been able to gather a massive amount of data from the electronic control units placed within them. These data has been stored on board by the vehicles but has been processed by only the manufacturers. However, there is a great demand in the industry to utilize these data for various purposes, for example, safety, efficiency, and driving comfort. Data collected by the vehicles can be used to identify and quickly locate the available parking spaces or to reduce traffic congestion. VANETs can be used to increase the efficiency and the utilization of resources, for example, saving time and reducing fuel consumption. It could be used to create smart cities for a better quality of life. Moreover, such kind of data and networking infrastructure has a great commercial value that can be used to improve competition in the market. Building such kind of cost-effective (considering cost of the radios in vehicles), distributed and decentralized networking system is the common aim of both the industry and the research community.

Vehicular networks are composed of vehicles and infrastructure units. Communication takes place between the vehicles, which is named as Vehicle-to-Vehicle (V2V) communication and between the vehicles and infrastructure points (roadside units), which is named as Vehicle-to-Infrastructure (V2I) communication. Vehicles use OBU for V2V and V2I communication.

Although VANETs are specialized forms of MANETs, compared to MANET and other wireless and mobile networks, VANETs show unique characteristics. These are [14–16]:

- *Intermittent connectivity*: Due to the high and variable speed of vehicles, the connectivity of the vehicles does not last for a long time but vehicles get connected instantly and frequently.
- *Dense versus sparse topology*: Density may vary in time and space. In urban areas, density is high and variable during the daytime and becomes very crowded in rush hours but is sparse after midnight. On the other hand, the topology is sparse in rural areas.

- *Predictable mobility pattern*: Since the vehicles are mobile and usually follow each other, it is easier to predict the mobility of the vehicles. Drivers who use the same path in their daily life make the route and mobility patterns more predictable.
- *Broadcasting and controlled flooding*: Due to the characteristics mentioned previously, constructing and maintaining routes between the communicating pairs are not feasible solutions. For the safety and nonsafety applications, beaconing and controlled flooding are accepted approaches for information dissemination.

VANETs integrate several networking technologies such as Dedicated Short Range Communications (DSRC) [17], IEEE 802.11p [18], WAVE IEEE 1609, WIMAX IEEE 802.16, and even ZigBee IEEE 802.15.4 are amongst these technologies.

2.1 VANET PROTOCOL ARCHITECTURE

After several years on standardization efforts, the Wireless Access in Vehicular Environments (WAVE) has been accepted as the system architecture for vehicular communication, and several standards have been released for short-range communication in VANETs. These standards are described in Table 8.1 [19–25]. IEEE 802.11p (IEEE Std 802.11p-2010) [18] is the physical layer standard including a set of extensions to the IEEE 802.11 standard. Upper layers include the family of IEEE 1609 standards, which relies on IEEE 802.11p (Fig. 8.2). The family of 1609 standards defines the architecture, communications model, protocols, security mechanisms, network services, multichannel operation and the

Table 8.1 The IEEE 1609 Protocol		
WAVE Standard	**Usage**	**Description**
IEEE 1609.0-2013	Architecture	Describes the WAVE architecture and services necessary for WAVE devices to communicate in a mobile vehicular environment
IEEE 1609.2-2016	Security services for applications and management messages	Covers methods for securing WAVE management and application messages. Also describes administrative functions required to support the core security functions
IEEE 1609.3-2016	Networking services	(Approved draft standard.) Provides services to WAVE devices representing higher layers in the communication stacks, including TCP/IP, and provides management and data services within WAVE devices
IEEE 1609.4-2016	Multi-channel operation	Describes various standard message formats for DSRC applications at 5.9 GHz
IEEE 1609.11-2010	Over-the-air electronic-payment data-exchange protocol for Intelligent Transportation Systems (ITS)	Specifies the application service layer and profile for Payment and Identity authentication, and Payment Data transfer. Defines a basic level of technical interoperability for electronic payment equipment, i.e. OBU and RSU using WAVE
IEEE 1609.12-2016	Identifier allocations	Specifies allocations of WAVE identifiers defined in IEEE 1609 series of standards

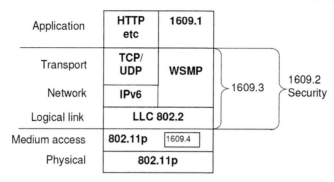

WSMP—*WAVE Short Message protocol*

FIGURE 8.2 IEEE 802.11p and IEEE 1609 Protocol Family in the Communication Protocol Stack

use of Provider Service Identifiers in the vehicular environment. It is aimed to support high-speed (up to 27 Mb/s) short-range (up to 1000 m) low-latency wireless communications [26].

There are 75 MHz of bandwidth in the 5.9 GHz (5850–5925 MHz) that have been allocated for ITS in the United States [27] and 70 MHz of bandwidth in the same spectrum has been allocated in European Union (5855-5925 MHz) [28]. Channel allocations vary in United States and European Union [28–30]. As shown in Fig. 8.3, allocated bandwidth is divided into seven channels of 10 MHz forming one control and six service channels in US and EU regulations [30]. The control channels are located in the middle and are used for control and safety messaging. The service channels are used for messaging by non-safety applications after coordination in the control channel.

2.2 VANET APPLICATIONS

VANET applications are classified in a variety of ways in the literature. Willke et al. [31] define four types of applications according to the aim of the application: (1) general information services, (2) vehicle safety information services, (3) individual motion control, and (4) group motion control. Karagiannis et al. [32] categorize applications in three categories: (1) active road safety applications,

Center frequency	5.860 Ghz	5.870 Ghz	5.880 Ghz	5.890 Ghz	5.900 Ghz	5.910 Ghz	5.920 Ghz
EU regulatory channel number	1	3	4	5	6	7	9
US regulatory channel number	172	174	176	178	180	182	184
IEEE channel number	172	174	176	178	180	182	184
EU allocation	SCH	SCH	SCH	SCH	CCH & SfCH	SCH	SCH
US allocation	SfCH	SCH	SCH	CCH	SCH	SCH	SfCH

FIGURE 8.3 Frequencies and Channels Allocated for VANET in the United States and European Union

SCH, service channel; *CCH*, control channel; *SfCH*, safety channel

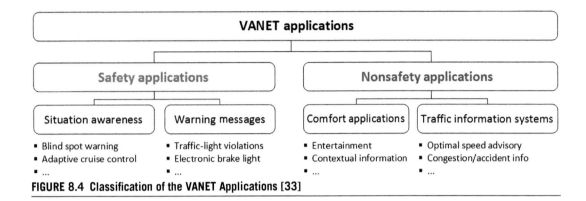

FIGURE 8.4 Classification of the VANET Applications [33]

(2) traffic efficiency and management applications, and (3) infotainment applications. One classification has been defined in Ref. [33] as shown in Fig. 8.4. In Ref. [33], applications in VANETs are generally classified as safety or non-safety applications. Safety applications are subcategorized as situation awareness applications and safety messaging applications. Non-safety applications include the applications for comfort driving, enhancing the driving process and traffic information systems, which do not present any safety or life-critical requirements.

3 VEHICULAR CLOUD INFRASTRUCTURE

Smart cities call for a new business model for vehicular communications where the vehicles can join a pool of resources and/or offer their resources as a service to others. With the advent of the cloud-computing paradigm, offloading local resources and rapidly accessing to a shared pool of resources has appeared as a more feasible solution to accelerate computing and storage services. A vehicular cloud is formed by incorporating cloud-based services into vehicular networks. Many service models such as Computing-as-a-Service (CompaaS), Storage-as-a-Service (StaaS), Network-as-a-Service (NaaS) [34], Cooperation-as-a-Service (CaaS) [35], Entertainment-as-a-Service (ENaaS), Information-as-a-Service (INaaS) [36], and Traffic Information-as-a-Service (TIaaS) can be delivered via vehicular clouds. In Ref. [37], the authors model the vehicular cloud as a data center with mobile hosts that have limited computing and/or storage capability. Migration from the conventional VANET model toward the vehicular cloud model enables the vehicular drivers to access mobile cloud resources rapidly in a pay-as-you-go fashion.

As shown in Fig. 8.5, vehicular cloud infrastructure can be either static or dynamic. In the static implementation of a vehicular cloud, the computing resources of a group of vehicles that remain fixed at a specific geographic location for a reasonable amount of time are pooled in a data center-like structure. "Data center in a parking lot" is a typical application of this kind of implementation. Indeed, the parking lot application is very similar to the static cloud data center implementation where servers are always switched on unless they are idle. The only difference between the parking lot implementation and the cloud data center is the more limited computing and storage capability of the data center in a parking lot.

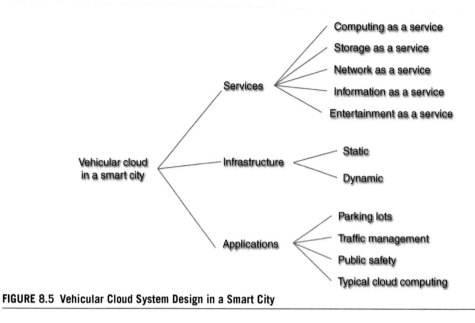

FIGURE 8.5 Vehicular Cloud System Design in a Smart City

Dynamic implementation of a vehicular cloud system can be formed by a pool of computing/storage/communication resources in mobile vehicles that are interconnected via VANET infrastructure and further linked to the Internet via RSUs. In such a dynamic implementation, incorporating the RSUs in the vehicular cloud infrastructure improves the manageability of mobile hosts (ie, computing resources in vehicles) [38]. As mentioned in Ref. [37], a vehicular cloud system should ideally utilize the underlying VANET infrastructure and minimize the involvement of RSUs.

VCC architectures consist of three segments: vehicle-unit, communication and cloud (see Fig. 8.6). The vehicle-unit segment retrieves the vehicle information through various sensors to monitor characteristics such as vehicle pressure, temperature and driver attributes [39]. This information can be transmitted to the cloud for storage. The communication segment of a vehicular cloud operates with V2V and V2I systems. V2V uses DSRC protocol [40,41] and uses Emergency Warning Messages (EWMs) to warn of abnormal road characteristics, sudden changes in direction, speed limits or major problems with individual vehicles. Vehicles propagate information to neighboring vehicles and to the cloud for storage. V2I exchanges occur through 3G, Internet or satellite communication and improve the safety standards and performance of the vehicular network [42].

The cloud layer employs data aggregation and data mining techniques on the data stored in a distributed manner. Stored data can be of the same type with the VANET data that is related to safety, entertainment or driving comfort. Hence, data stored in the cloud is analyzed in a distributed manner to retrieve information with the ultimate goal of safety and quality-of-life in the smart city. By adopting the cloud-computing concept in vehicular communications, having computing and storage resources in the vehicles will be avoided to be underutilized most of the time. Thus, a vehicle can rent its computing, storage or communication capacity to other vehicles in the network.

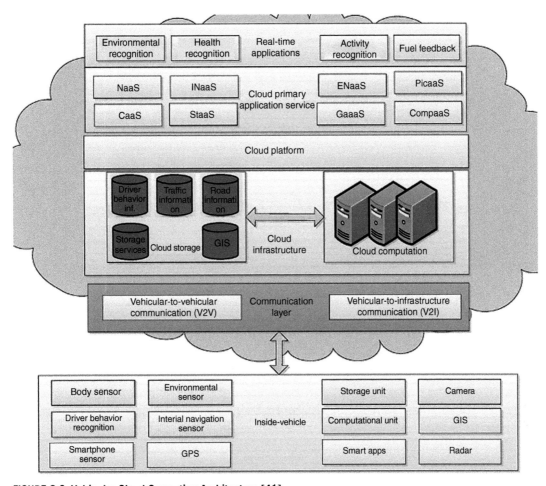

FIGURE 8.6 Vehicular Cloud Computing Architecture [41]

Primarily cloud services are applications in real-time such as NaaS, ENaaS, and StaaS. Cloud storage and cloud computation segments are infrastructure for vehicular cloud. Information that is retrieved through vehicle-unit is stored in Cloud storage. Cloud computation is used to compute data based on storage and real-time data.

In Ref. [43], VANET-based clouds are classified in three groups as shown subsequently:

- *Vehicular clouds* are formed by the VANET infrastructure, gateways and brokers. This architecture is similar to the dynamic vehicular clouds mentioned previously. The brokers are called authorized entities, and they are elected by the vehicles that join the cloud. Election of the

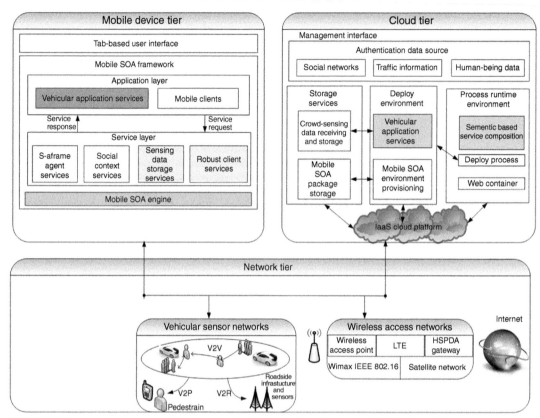

FIGURE 8.7 System Architecture of Service-Centric Contextualized Vehicular (SCCV) Cloud [7]

authorized entities also forms the boundary of the cloud as the elected authorized entities send invitation messages to other vehicular nodes within the boundary to join the cloud. The higher authorities (ie, broker cum gateway) authorize the brokers to pool resources for the cloud in case the number of participating vehicular nodes is higher than a certain threshold.

- *VANETs using clouds* is the vehicular cloud architecture where VANETs access the Internet cloud via gateways on the move. Services provided by this type of cloud are real-time traffic information, road side help and infotainment.
- *Hybrid vehicular clouds* combine the two approaches mentioned previously. Thus, a vehicular node can join a vehicular cloud as a service provider (SP) while at the same time, they can access the Internet cloud via gateways. Peer-to-Peer (P2P) file sharing and/or IaaS are good examples of hybrid vehicular clouds where vehicles rent their resources intermittently. A specific type of hybrid vehicular clouds, namely the SCCV is illustrated in Fig. 8.7 along with the corresponding system architecture.

As a vehicular cloud system is said to be a variant of a conventional cloud data center with mobile hosts (ie, intermittent on/off switching), virtualization is the key component to maximize resource utilization and isolate services provided to different users or groups. Virtual machine (VM) placement [44] is the mapping of virtual resources allocated to given service requests to physical resources. VM placement in a vehicular cloud is mostly application-driven. Real-time navigations run data mining functions, and they mostly run on road-side clouds due to enhanced computing capability. Road-side clouds can also provide VM hosting for distributed storage for video surveillance. Furthermore, downloading large files in a cooperative manner requires VM hosting in the road-side clouds. VM mapping schemes that are proposed for conventional data centers can be adopted whereas migration of VMs is a crucial issue.

In the conventional cloud data centers, VMs can be migrated between physical hosts due to several reasons such as energy saving, maintenance, efficiency, hot-spot prevention and so on [44]. In a vehicular cloud, the factors that trigger VM migration are various and mostly mobility-driven as the physical hosts rapidly change their location. Therefore, connected vehicles as physical hosts of the VMs in a vehicular cloud system experiences VM management as an ongoing challenge. In Section 5, ongoing works and preliminary research findings in the literature will be summarized along with potential applications.

4 VANET CHALLENGES AND SOLUTIONS IN SMART CITIES

VANETs have become a unique solution for implementing safety and security standards in transportation systems. The ITS not only standardizes an approach to overall road safety, but also it makes vehicle driving more comfortable and stress-free. With the combination of RSUs and smart vehicles the traffic incidents and laws can be monitored by local and centralized systems.

4.1 SMART DRIVING

The major goals of enabling ITS based smart cities are to increase traffic safety and to increase energy efficiency. Bifulco et al. [45] have addressed the 2020 ambition of Europe with efficient and sustainable energy utilization. According to the study of Kley et al. [46] one solution to achieve the goal will be introducing electric vehicles with modern ITS. In addition, Bifulco et al. mentioned the concept of smart driving which has been explored in various research projects including projects from Microsoft in San Francisco, Accenture in Amsterdam and IBM in Singapore to build smart cities. For smart transportation, Microsoft launched smart parking in 2004 in Bay Area of San Francisco. With combination of geo-referencing and Windows Azure; Microsoft research utilizes real-time data on transportation system and traffic flow [45,47].

Accenture have been collaborating in Amsterdam for smart city ITS where the initiative is to connect cars, electric bicycles, vessels and river cruisers in smart grid [48]. This smart city project was launched in 2009 with the ambition of energy solutions with smart grid. Eventually the platform trends to integrate all necessary service in single smart system and progresses for ITS [49]. IBM has tested the ITS in Singapore to monitor traffic in roadways. The system is expected to deliver accurate traffic patterns and already have been implement in Lyon (France). The algorithm considers road type, density, speed-limit, and event data to predict traffic patterns [50].

One of the interesting characteristics of VANETs is to provide real-time updates to drivers. However, VANET-equipped vehicles and smart cars are currently rate in the market. Google is collaborating with Open Automotive Alliances (OAA) [51] to bring the Android platform to vehicles. Smart parking solutions like Parker [52] and Apparcar [53] assist drivers to identify nearby parking places. Moreover, mobile application like Waze [54] can assist in tracking the traffic status in a particular region. VANETs can help avoid accidents by monitoring motion and direction change in vehicles. Awareness of upcoming road-side curves and damage can be passed from car to car to assist the drivers. Special notification for ambulance, fire and brigade vehicles can be passed to surrounding vehicles and RSUs to create space for service vehicles.

A grand challenge of VANETs is to pass this sort of application information from vehicle to vehicle and enhance the real-time collaboration of vehicular network with efficiency.

4.2 SAFETY CHALLENGES

Traffic law violations have a major impact on road safety of smart cities. Specially, speed, tailgate and fitness issues are primarily being monitored to ensure road safety standards. Moreover, unregistered and law violating vehicles can create security threats.

Existing solutions for VANET safety are primarily based on Trusted Third Parties (TTP) or proxies [55,56]. Barba et al. have introduced a protocol for VANET that is able to report traffic violation anonymously.

The protocol chooses next forwarding route randomly by maintaining privacy anonymous-ness and uses a forward probability measurement (Fig. 8.8). For privacy, the protocol operates in application layer while reporting to next hop being anonymous for privacy. The approach considers 802.11b MAC protocol and other routing protocols, for example, Ad hoc On-Demand Distance Vector (AODV), Greedy Perimeter Stateless Routing (GPSR) during communication.

4.3 MANAGEMENT CHALLENGES AND CLUSTERING SOLUTIONS

There are several clustering algorithms proposed to address the challenges of VANET clouds. Main challenge in clustering algorithms is forming more stable clusters in terms of cluster size, member node exchange, cluster head election and long duration of cluster heads. One good clustering algorithm for VANETs should also form fewer clusters to ease maintenance and stability. A Fuzzy Clustering-based Vehicular Cloud Architecture (FCVCA) has been proposed in Ref. [57] with a new clustering technique to group vehicles. In this paper, authors propose a new 3-step approach for estimation of traffic volume in a particular road segment. Initially, traffic information has been collected from different clusters with the help of the proposed clustering algorithm. With a virtual chain among clusters this information is transmitted toward the road-side cloud. The virtual chain is used to meet the connectivity demand between clusters with RSU as RSUs have limited transmission range. Later, the total traffic volume has been calculated with a generalization method from the collected data. In the simulation, the proposal has been tested with performance metrics of inter-vehicle distance, density and flow rate. Flow rate is the amount of nodes crossing particular road segment within specific time. The metrics used in the simulation environment are duration of cluster head and amount of clusters. Within the similar environmental scope, the proposed algorithm is compared with Lowest-ID [58] and MCMF [59] techniques.

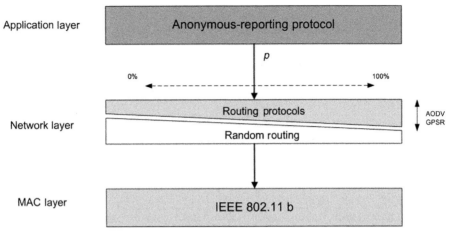

FIGURE 8.8 Barba et al. Consider Standard Protocols in Conjunction With Their Proposed Approach That is an Anonymously Reporting Protocol [55]

According to comparison results, it has been reported that the proposed approach can construct a higher number of stable clusters. Authors also measured the quality of volume estimation for the evaluation of the traffic volume estimation accuracy in their approach. While the scheme is compared with Online Learning Weighted Support-Vector Regression (OLWSVR), the proposed method performs better in terms of low, mid or high flow rates. Thus this proposed scheme has been illustrated with reduced amount of cluster formation in comparison with Lowest-ID and MCMF approaches.

Arkian et al. [60] have proposed another clustering technique to solve the resource limitation problem by cooperatively providing the resources in a vehicular cloud. The scheme focuses to create clusters with flexibility and to select a cluster head using the fuzzy logic. Fuzzy logic is a decision making process based on input membership functions that operates similarly to the human brain. To improve the efficiency of cluster head decisions, a Q-learning based SP selection has been introduced in this scheme. With Q-learning, each cluster head maintains a two-dimensional Q-table and periodically updates the Q-table to improve actions. In addition, three queuing strategies have been considered for resource allocation in Virtual cloud for efficiency, QoS and fairness. While comparing with Lowest-ID and user-oriented fuzzy logic-based clustering scheme [61], the simulation study demonstrates improvement with the proposed COHORT clustering approach. The simulation results also demonstrate that the proposed COHORT clustering approach has a significant impact on reducing service discovery delays and service consuming delays in comparison with CROWN. CROWN (discovering and consuming services within vehicular clouds) [62] is a system that enables vehicles in a VANET to search for mobile cloud servers that are moving nearby and discover their services and resources. The system uses RSUs for cloud directories to register the mobile cloud servers. Within a specific zone, RSUs distribute their registration data to vehicles to discover and act as mobile cloud server. The proposed system is later evaluated in NS2 to measure performance service discovery and service consuming delays and packet success ratio. The result of CROWN is compared with a broadcasting-based protocol. CROWN is the one of the pioneer cloud service discovery protocol proposed for vehicular clouds.

4.4 EMERGENCY AND DISASTER RECOVERY

VANETs have an important role for message propagation during emergency situations and disasters such as cyclones, earthquakes, fires, volcanos, etc to reduce loss of life and resources. It has been noted that message dissemination during a crisis is one of the key challenges for future research in this area. Alazawi et al. [63] have proposed a system using VANETs, cloud computing, and mobile technologies which is based on the transportation in a real setting. The system is able to retrieve real-time data and utilize the connectivity in between mobile and social networks with VANETs. In emergency situation, an efficient system is expected to control road traffic accordingly and able to propagate the information with available resources. Through data analysis the system can design a strategy to optimize the impact of disaster, for example, public transportation for evacuation from the point of incident. Authors have evaluated the proposed solution during disaster scenario and found the proposed system better than conventional approach for evacuation. Their future work aims to work on real-time cases and enlighten the scopes of intelligent disaster management system.

In a smart city vehicular network, emergency service solution is a new research direction. Amici et al. [64] have introduced a routing protocol where real traffic scenario is collected using 370 taxicabs every 7 s in the city of Rome. These cabs are connected throughout the city without the utilization of 3G/4G network infrastructure. Real-time analysis of this traffic data is used to identify emergency situations. An infected vehicle propagates emergency messages to other participants of the network. An infected vehicle is the uniformly distributed random vehicle at a given time where the primary messages have been delivered. Later, these emergency messages are propagated to other not-infected vehicles when these vehicles are within the coverage range of infected vehicles. Upon receiving the emergency message the vehicles become infected. Considering the average speed and wait time in real traffic scenario, the message can be propagated throughout the network.

5 VEHICULAR CLOUDS CHALLENGES AND SOLUTIONS IN SMART CITIES

As vehicular clouds are envisioned to be widely adopted in smart cities several challenges have to be solved [65]. Most of the studies identify security and trust issues as the grand challenge in vehicular clouds. Due to mobility of vehicles and intermittency of short range communication links makes the establishment of trust relations and authorization of mobile vehicular nodes more complex than in conventional VANETs [66]. Indeed, provisioning delay in such an environment is also a big concern due to the same factors as mentioned in Ref. [67]. As there are multiple SPs, privacy preserving in the intermittent contracts between the vehicular nodes and the SPs have to make sure that the need to reveal private information is minimized. Other virtualization-based challenges also remain in vehicular clouds such as VM migration in an IaaS scenario where vehicular nodes are mobile [37].

5.1 SECURITY IN VEHICULAR CLOUDS

Security challenges in VANETs and cloud computing are inherited by vehicular clouds as mentioned in Ref. [43]. These challenges have been studied in detail in Refs. [68–70]. These studies can be improved

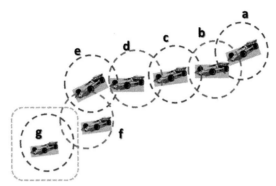

FIGURE 8.9 Geographic Security Approach Presented in Ref. [68]

The square is a securely protected area (eg, military headquarter). Only vehicle-g can access the ciphertext sent by vehicle-a

by taking the specific conditions and requirements of vehicular clouds into consideration. In Ref. [66], the authors summarize these challenges as spoofed identities, nonrepudiation, Denial of Service (DoS) attacks, and mobile authentication. Yan et al. have proposed a security framework that addresses most of these challenges [68]. The main targets for an adversary are reported as confidentiality of VMs in the vehicular cloud, integrity of the content that is stored in a distributed manner, and the availability of physical machines, resources, services and applications in the vehicular cloud. Based on these three targets, a typical attack scenario has been defined as follows: (1) the geographic location of the victim vehicle is identified and possible physical hosts (ie, vehicular nodes) in the vicinity are discovered where the victim vehicle is possibly being served, (2) the vehicular node that serves the victim vehicle is discovered upon submitting several service requests to the cloud, (3) Once the VM that is allocated to the victim is identified, services are requested on the same host, (4) Finally higher privilege is aimed to be obtained to collect assets via system leakage. This type of attack is inspired from the attacks that explore information leakage in the clouds [71]. While authenticating the nodes with high mobility, it is not viable to use well-known metrics such as ownership, knowledge and biometrics. Furthermore, Sybil-like attacks are always possible in such an information network [72].

In Ref. [68], the authors propose building trust relationships between vehicle clusters. It is essential that the behavior of a vehicle in a cluster can be monitored by all members in the same cluster. Besides, for mission critical applications in a vehicular cloud, the authors propose geographic location-based security. Thus, when a ciphertext is sent by a vehicle, only vehicles in a certain area are authorized to access the ciphertext and the corresponding decryption key. Fig. 8.9 illustrates a scenario that is presented in Ref. [68]. A group of cars communicate with the cloud that provides service to the clients at a naval base. Once a message is encrypted and sent out to the cloud, only vehicle-a can access the ciphertext and the decryption key.

As the cloud topology changes dynamically, based on the number of vehicles, security strategy of the vehicular cloud may need to be reconfigured. The idea behind this is that the higher the number of vehicles involved, the stricter the security protocols should get. Hence, the authors propose a queuing theory-based mode to predict the volume of the vehicles in the vehicular cloud. Increasing volume of the vehicles in the cloud introduces the scalability problem for the security schemes. To cope with the scalability issues, each VM is divided into sub-VMs when the number of accesses to the VM exceeds a predefined threshold. The VM allocates the resources of an incoming request to the sub-VMs in order to fulfill the load balancing requirements. A VM middleware serves like a resource broker; it caches the recent accesses and usage information. Whenever a new request arrives, the sub-VMs are allocated based on the recent usage and load balancing among the sub-VMs.

The authors in Ref. [73,74] propose a pseudonym system by introducing anonymous public keys and the public key infrastructure (PKI). Despite the efficiency of the PKI-based framework, certification of the public key may lead to latency while there exists a trade-off between the frequency of updating the public key certification and communication overhead.

5.2 PRIVACY AND USER EXPERIENCE IN VEHICULAR CLOUDS

Privacy is a major concern for cloud users due to virtualization-based vulnerabilities where sensitive information can be revealed to adversaries [75,76]. The authors in Refs. [77,78] present the benefits of mobile cloud computing in a vehicular cloud environment while presenting the privacy and security challenges of a vehicular cloud environment in detail. Existing privacy considerations for cloud systems and VANETs [69,70,73] can be adopted by vehicular clouds however special requirements of vehicular clouds have to be taken into consideration. Anastasopoulou et al. [79] use game theory-based solutions to improve privacy of cloud-based mobile apps. The proposed methodology analyzes the user interactions, and makes a compromise between the QoS in the cloud and user privacy.

Recently, Aloqaily et al. have defined privacy as a component of a function denoting user experience [80]. Moreover, the authors propose a hierarchical framework where multiple cloud providers and multiple TTPs exist. To this end, the authors formulate a weighted sum of provisioning delay, the amount of information reveal and the service cost of each TTP-cloud provider tuple. Three key factors have been considered for vehicular clouds. This solution employs a TTP between the vehicular nodes and the SPs. To ensure scalability and to cope with computation overhead, a hierarchical clustered architecture is used. Service requirements of the vehicular users determine the best TTPs.

The benefit of the QoE-based (QoE - Quality of Experience) architecture is reported as the vehicle node's ability to prioritize its preferences on latency, price and information revealed to the SP. With the adjustment of the coefficients, the vehicular nodes can be served with affordable price, by revealing less information to SP and/or low latency. The output of the negotiation can be improved if the experience of previously provisioned drivers is used as an input. Furthermore, the SP cannot detect any identifier about the vehicular nodes that request service through the vehicular cloud as minimal reveal of identifying information to the SP is without disclosure of the user identities. As the user always has the flexibility of switching between SPs, dealing with the TTPs rather than trusting the SP as the user can switch to another SP in the future.

A vehicular node initially requests service from the first available TTP within its range. Upon the vehicular node-TTP matching, the TTP negotiates with the SP on behalf of the vehicular node, and the SP delivers the service to the vehicular node through the TTP. Direct feedback of the vehicular node is used to evaluate the delay, price, and privacy offered by the SP. As the vehicular node can prioritize any of these objectives, the vehicular nodes that are associated with users who are more sensitive to revealing personal information can minimize the information revealed to the SP. It is worthwhile mentioning that in such architecture, the user credentials such as credit card/visa information are not kept with multiple SPs but within one TTP.

5.3 VIRTUALIZATION-BASED CHALLENGES

As mentioned before, a vehicular cloud can be considered as a data center with unstable physical hosts. Therefore, virtual machine management appears to be a challenging issue in vehicular clouds. VM migration may occur when a vehicle is about to go off the grid and cannot be in range of any RSUs, or when a handover occurs between two RSUs that are in range of a vehicle. To cope with this challenge, Refaat et al. have proposed a VM migration scheme in Ref. [37]. The VM migration algorithm works as follows: Out of the nearest vehicles, the source vehicular node selects a destination vehicular node based on the search criteria. If the destination vehicular node does not have sufficient available capacity to host the VM or if the VM cannot be migrated to the destination node in a predefined time window, migration is re-attempted by excluding the corresponding destination. Otherwise, the VM is migrated to the destination vehicular node. A migration attempt is marked as unsuccessful if a certain number of migration attempts fail. An unsuccessful migration requires intervention of the RSU. Thus, the VM is directed to the RSU if the migration cannot be completed.

The authors have proposed two approaches against random selection of the destination vehicular node. The first approach is called the Vehicular Virtual Machine Migration with Least Workload (VVMM-LW) whereas the second approach is called Vehicular Virtual Machine Migration with Mobility-Awareness (VVMM-MA). The former ranks the v nearest vehicles with respect to their current workload and selects the one(s) with the lightest workload. The latter uses the vehicles' trajectories and estimates the future location of all vehicles in the vehicular cloud and excludes the ones that are forecasted to go off the grid. For those who are forecasted to remain in the grid, the algorithm runs the VVMM-LW to select the destination vehicular node to migrate the virtual machine. The authors have shown that VVMM-MA can improve the performance of random selection policy by up to 60% under highly congested traffic conditions whereas the improvement is still above 35% under lightly congested traffic scenarios.

It is worthwhile noting that VM migration is typically a bandwidth intensive task taking many minutes. Hence, handling frequent migration among moving vehicles is a grand challenge of vehicular clouds. Furthermore, although encryption of data in transit is critical, it is resource and time intensive. These issues still remain open for the researchers in this field.

5.4 CONTEXT-AWARENESS

Context-awareness in a vehicular cloud is emergent for smart city applications for various reasons. In Ref. [81], the authors propose a behavior pattern recognition methodology to detect anomalies in driver behaviors, and inform the other drivers in the vicinity for their safety. On the other hand, Santa and Gmez-Skarmeta propose a context-aware information-provisioning scheme for V2V and V2I communications for road safety [82]. The vehicles are identified by the RFID technology, and by keeping track of vehicles, current traffic and road condition information is obtained as provided to all drivers in the vehicular network. Wan et al. have extended these ideas to propose a cloud-assisted context-aware architecture [65]. A multilayer architecture is proposed for context-aware vehicular cloud implementation. The three layers are summarized subsequently:

- *Vehicular computational layer* is the upmost layer where context-aware driver behavior detection system is implemented. The behavior detection module communicates with other vehicular nodes to share context-aware road and safety information. Furthermore, the behavior detection module also shares this information with mobile users who wish to access these services via smart phones.

- *Location computational layer* is below the vehicular computational layer, and it consists of the RSUs deployed at specific locations to exchange information with onboard equipment units. Thus, whenever a vehicle is outside the range of a vehicular network, it can still access the road-side information via RSUs. The location computational layer is connected to the Internet and receives service from the cloud computational layer.
- *Cloud computational layer* provides context-aware cloud services through interconnected clouds of automotive multimedia content cloud, traffic authority cloud, location-based service cloud, automotive manufacturer cloud, and other application clouds. This layer provides context-aware cloud services to vehicular drivers, traffic authorities or vehicular social networks (VSNs).

As mentioned previously, vehicular social networks and context-aware vehicular security are two key components of this architecture. Vehicular social networks are envisioned to be an inseparable part of vehicular clouds they will primarily serve for traffic data mining and mobile crowd sensing [65]. Context-awareness in vehicular security is necessary to reconfigure the security policies based on the changes in the user's context. The authors have proposed a context-aware vehicular security framework that consists of data collection, policy management, anomaly detection and trust management modules. Data collection module collects data such as time, road conditions, velocity, that would reveal context information. The context information is passed to the policy management, anomaly detection and trust management modules. When an anomaly is detected, the trust management unit assesses the trustworthiness of the vehicular node that has been detected to have misbehaved. The authors showed that if context-aware vehicular cloud framework is adopted as an alternative to the traditional traffic routing, travel times can be reduced by around 50% as the distance traveled increases.

6 OPEN ISSUES AND FUTURE DIRECTIONS IN VEHICULAR SMART CITY SYSTEMS

Vehicular networks still experience several challenges that have to be addressed before they are widely adopted by smart cities. As mentioned previously, VANETs operate on a mature communication infrastructure however VANETs can be enhanced by incorporating cloud-inspired operational model as the data collected is huge and needs to be analyzed, interpreted and communicated. Therefore vehicular clouds in smart cities need novel and effective solutions for virtual machine management, vehicular node security, vehicular driver's privacy and context-aware services via mobile crowdsensing over vehicular social networks.

Vehicular VM management calls for novel solutions that fulfill service quality requirements. Moreover, migration efficiency is still an open issue, thus new algorithms to ensure minimum VM migration latency and minimum service disruption. Furthermore virtualization-based vulnerabilities have to be addressed in vehicular VM management and migration.

As security and privacy are the grand challenges in any cloud system, vehicular clouds have to incorporate robust solutions to avoid unauthorized access to the vehicular resources. Continuous authorization techniques that incorporate detection of anomalous patterns into the existing authorization schemes will improve robustness of vehicular clouds. Indeed, anomaly detection will require analysis of massive amount of unstructured data. Therefore, cloud-based big data analytics solutions will have to be integrated into the cloud computational layer in a multilayered vehicular cloud architecture.

While behavior analysis and anomaly detection will improve security, due to computing overhead, degradation in service quality should be expected. Hence, the researchers working in this field should also address the trade-off between security-privacy and service quality.

Vehicular social networks form another emerging field to accelerate the performance of vehicular clouds. VSNs can help vehicular clouds make use of crowdsensed data. Having said that, mobile cloud-based crowdsensing systems experience several challenges. As reported in Ref. [83], injection of specious information into crowdsensed data may introduce public safety vulnerabilities in case of an emergency [84]. Trustworthy crowdsensing schemes [83] can be enhanced by new trust derivation models [85]. As mentioned previously, Sybil attacks threaten the lifetime and the reliability of the vehicular cloud; hence VSNs call for behavior recognition-based Sybil detection techniques to improve the robustness of the vehicular cloud.

7 SUMMARY

Connected vehicles have various application areas in future smart cities. Having an established communication infrastructure and standards, VANETs can be adopted in these applications. However, due to continuously increasing demand for computation, storage and communications, a cloud inspired model is required for vehicular networks. This chapter has provided a survey of vehicular networks for smart city applications, and presented an overview of the studies that pave the way toward implementation of vehicular clouds in smart cities. To this end, the chapter has presented the VANET architecture and the vehicular cloud architectures utilizing the VANET infrastructure. Then the challenges and solutions in VANETs and vehicular clouds for smart cities have been presented in detail. The chapter has also dedicated one section to discussing future directions and open issues in this topic.

REFERENCES

[1] Yovanof GS, Hazapis GN. An architectural framework and enabling wireless technologies for digital cities & intelligent urban environments. Wireless Pers Commun 2009;49(3):445–63.

[2] Nam T, Pardo TA. Conceptualizing smart city with dimensions of technology, people, and institutions. In: Proceedings of the 12th annual international digital government research conference: digital government innovation in challenging times. ACM, New York, NY, USA, 2011. p. 282–91.

[3] Giffinger R, Gudrun H. Smart cities ranking: an effective instrument for the positioning of the cities? Archit City Environ 2010;4(12):7–26.

[4] Washburn D, Sindhu U. Helping CIOs understand "smart city" initiatives. Defining the smart city, its drivers, and the role of the CIO. *Forrester*, Research Report, 11 February, 2010. <https://www.forrester.com/report/Helping+CIOs+Understand+Smart+City+Initiatives/-/E-RES55590>; [Online].

[5] Gerla M, Lee E-K, Pau G, Lee U. Internet of vehicles: from intelligent grid to autonomous cars and vehicular clouds," in Internet of Things (WF-IoT), 2014 IEEE World Forum on, March 2014. p. 241–46.

[6] Lee E, Lee EK, Gerla M, Oh SY. Vehicular cloud networking: architecture and design principles. IEEE Commun Mag 2014;52(2):148–55.

[7] Hu X, Wang L, Sheng Z, TalebiFard P, Zhou L, Liu J, Leung VCM. Towards a service centric contextualized vehicular cloud. In: Proceedings of the fourth ACM international symposium on development and analysis of intelligent vehicular networks and applications. ACM, New York, NY, USA, 2014, DIVANet 14. p. 73–80.

[8] Kantarci B. Cyber-physical alternate route recommendation system for paramedics in an urban area. In: IEEE wireless communications and networking conference (WCNC), March 2015. p. 2276–81.

[9] Research and Innovative Technology Administration (RITA) U.S. Department of Transportation (US DOT), About ITS. <http://www.its.dot.gov/faqs.htm>; 2015. [Online].

[10] Taleb T, Sakhaee E, Jamalipour A, Hashimoto K, Kato N, Nemoto Y. A stable routing protocol to support its services in vanet networks. IEEE Trans Vehicular Technol 2007;56(6):3337–47.

[11] Losilla F, Garcia-Sanchez AJ, Garcia-Sanchez J, Garcia-Haro J. On the role of wireless sensor networks in intelligent transportation systems. In: 2012 14th international conference on transparent optical networks (ICTON); July 2012. p. 1–4.

[12] Yang Y, Bagrodia R. Evaluation of vanet-based advanced intelligent transportation systems. In: Proceedings of the sixth ACM international workshop on vehicular internetworking; 2009. p. 3–12.

[13] Hartenstein H, Laberteaux KP. VANET: vehicular applications and inter-networking technologies. United Kingdom: John Wiley & Sons Ltd; 2010.

[14] Lee U, Gerla M. A survey of urban vehicular sensing platforms. Comput. Netw. 2010;54(4):527–44. Advances in wireless and mobile networks.

[15] Gerla M, Kleinrock L. Vehicular networks and the future of the mobile internet. Comput. Netw. 2011;55(2):457–69. Wireless for the future internet.

[16] Wisitpongphan N, Bai F, Mudalige P, Sadekar V, Tonguz O. Routing in sparse vehicular ad hoc wireless networks. IEEE J. Selected Areas Commun 2007;25(8):1538–56.

[17] IEEE Standard for Message Sets for Vehicle/Roadside Communications. IEEE Standard 1455–1999. p. 1–134; September 1999.

[18] IEEE 802.11p, Amendment to Standard for Information Technology-Telecommunications and Information Exchange Between Systems-Local and Metropolitan Area Networks-Specific requirements Part 11 Wireless LAN Medium Access Control (MAC) and Physical Layer (PHY) Specifications-Amendment 7 Wireless Access in Vehicular Environment. IEEE Standard IEEE 802.11p, version 2010; 2010.

[19] IEEE Standard 1609.0-2013. IEEE Guide for Wireless Access in Vehicular Environments (WAVE) - Architecture; 2013.

[20] IEEE Standard 1609.2-2016. Standard for Wireless Access in Vehicular Environments (WAVE) Security Services for Applications and Management Messages; 2016.

[21] IEEE Standard 1609.3-2016. Approved Draft Standard for Wireless Access in Vehicular Environments (WAVE) Networking Services; 2016.

[22] IEEE Standard 1609.4-2016. Standard for Wireless Access in Vehicular Environments (WAVE) Multi-Channel Operations; 2016.

[23] IEEE 1609.6. Remote Management Service (Under Development).

[24] IEEE Standard 1609.11-2010. Over the-Air Data Exchange Protocol for Intelligent Transportation Systems (ITS); 2010.

[25] IEEE Standard 1609.12-2016. Identifier Allocations; 2016.

[26] Research and Innovative Technology Administration (RITA) U.S. Department of Transportation (US DOT). Fact Sheet on IEEE 1609—Family of Standards for Wireless Access in Vehicular Environments (WAVE). <http://www.standards.its.dot.gov/Factsheets/PrintFactsheet/80>; [Online].

[27] FCC 47 CFR 90 Telecommunications, Private land mobile radio services, 371 – 377: Regulations governing the licensing and use of frequencies in the 5850–5925 MHz band for dedicated short-range communications service (DSRCS).

[28] ETSI EN 302 571 V1.2.1 (2013-09). Intelligent Transport Systems (ITS); Radio communications equipment operating in the 5855 MHz to 5925 MHz frequency band; September 2013.

[29] ETSI Standard (Draft). Intelligent transport systems (ITS); access layer specification for intelligent transport systems operating in the 5 Ghz frequency band. ETSI EN 302 663 V1.2.0 (2012-11); 2012.

[30] Status of ITS Communication Standards Document HTG3-1 Version: 2012-11-12. EU-US ITS Task Force Standards Harmonization Working Group Harmonization Task Group 3; November 12, 2012.

[31] Willke TL, Tientrakool P, Maxemchuk NF. A survey of inter-vehicle communication protocols and their applications. IEEE Commun Survey Tutorials 2009;11(2):3–20.

[32] Karagiannis G, Altintas O, Ekici E, Heijenk G, Jarupan B, Lin K, Weil T. Vehicular networking: a survey and tutorial on requirements, architectures, challenges, standards and solutions. IEEE Commun Surveys Tutorials 2011;13(4):584–616.

[33] Sommer C, Dressler F. Vehicular networking. United Kingdom: Cambridge University Press; 2014.

[34] Arif S, Olariu S, Wang J, Yan G, Yang W, Khalil I. Datacenter at the airport: reasoning about time-dependent parking lot occupancy. IEEE Trans Parallel Distrib Syst 2012;23:2067–80.

[35] Dinh H, Lee C, Niyato D, Wang P. A survey of mobile cloud computing: architecture, applications and approaches. Wireless Commun Mobile Compu 2013;13:1587–611.

[36] Mousannif H, Khalil I, al Moatassime H. Cooperation as a service in VANETs. J Universal Comput Sci 2011;17:1202–18.

[37] Refaat TK, Kantarci B, Mouftah HT. Dynamic virtual machine migration in a vehicular cloud. IEEE Symp Comput Commun (ISCC) 2014. p. 1–6.

[38] Olariu S, Hristov T, Yan G. The next paradigm shift: from vehicular networks to vehicular clouds. In: Basagni S, Conti M, Giordano S, Stojmenovic I, editors. Chapter in *Mobile Ad hoc Networking: Cutting Edge Directions*. Wiley-IEEE Press; 2013. p. 645–700.

[39] Chung T-Y, Chen Y-M, Hsu C-H. Adaptive momentum-based motion detection approach and its application on handoff in wireless networks. Sensors 2009;9(7):5715.

[40] Yang X, Liu J, Vaidya NF, Zhao F. A vehicle-to-vehicle communication protocol for cooperative collision warning. MOBIQUITOUS 2004. In: The first annual international conference on mobile and ubiquitous systems: networking and services; August 2004. p. 114–23.

[41] Whaiduzzaman Md, Sookhak M, Gani A, Buyya R. A survey on vehicular cloud computing. J Netw Comput Appl 2014;40:325–44.

[42] Bordley L, Cherry CR, Stephens D, Zimmer R, Petrolino J. Commercial motor vehicle wireless roadside inspection pilot test, part B: stakeholder perceptions. Transportation Research Board 91st Annual Meeting; 2012.

[43] Hussain R, Son J, Eun H, Kim S, Oh H. Rethinking vehicular communications: merging VANET with cloud computing. In: IEEE 4th international conference on cloud computing technology and science; December 2012. p. 606–9.

[44] Boutaba R, Zhang Q, Zhani MF. Virtual machine migration in cloud computing environments: benefits, challenges, and approaches. In: Mouftah HT, Kantarci B, editors. Communication Infrastructures for Cloud Computing. IGI Global, Hershey, PA; 2013. p. 383–408.

[45] Bifulco F, Amitrano CC, Tregua M. Driving smartization through intelligent transport. Chinese Bus Rev 2014;13:243–59.

[46] Kley F, Lerch C, Dallinger D. New business models for electric cars: a holistic approach. Energ. Policy 2011;39(6):3392–403.

[47] Rodier CJ, Shaheen SA. Transit-based smart parking: an evaluation of the San Francisco Bay Area Field test. Transport Res Part C 2010;18(2):225–33.

[48] Senart A, Kurth S, Roux GL. Assessment framework of plug-in electric vehicles strategies. In: 2010 First IEEE international conference on smart grid communications (SmartGridComm); 2010. p. 155–60.

[49] Berthon B, Guittat P. Rise of the intelligent city. <http://www.cas-uk.com/SiteCollectionDocuments/PDF/Accenture-Outlook-Rise-of-the-Intelligent-City-Sustainability.pdf>; 2011 [Online].

[50] Greengard S. Smart transportation networks drive gains. Commun ACM 2014;58(1):25–7.

[51] Open Automotive Alliance. <http://openautoalliance.net> [Online].

[52] Parker Application. <http://theparkerapp.com/> [Online].

[53] Aparcar Application. <http://aparcar.com/> [Online].

[54] Waze Application. <https://www.waze.com/> [Online].

[55] Barba CT, Aguiar LU, Igartua MA, Parra-Arnau J, Monedero DR, Forn J, Pallars E. A collaborative protocol for anonymous reporting in vehicular ad hoc networks. Comput Standard Interf 2013;36(1):188–97.

[56] Benjumea V, Lopez J, Troya JM. Specification of a framework for the anonymous use of privileges. Telemat Inform 2006;23(3):179–95.

[57] Arkian HR, Atani RE, Pourkhalili A, Kamali S. Cluster-based traffic information generalization in vehicular ad-hoc networks. Vehicular Commun 2014;1(4):197–07.

[58] Gerla M, Tsai J T-C. Multicluster, mobile, multimedia radio network. Wireless Netw 1995;1(3):255–65.

[59] Venkataraman H, Delcelier R, Muntean G-M. A moving cluster architecture and an intelligent resource reuse protocol for vehicular networks. Wireless Netw 2013;19(8):1881–900.

[60] Arkian HR, Atani RE, Pourkhalili A, Kamali S. A stable clustering scheme based on adaptive multiple metric in vehicular ad-hoc networks. J Inform Sci Eng 2015;31(2).

[61] Tal I, Muntean G-M. User-oriented fuzzy logic-based clustering scheme for vehicular ad-hoc networks. In: 2013 IEEE 77th vehicular technology conference (VTC Spring); June 2013. p. 1–5.

[62] Mershad K, Artail H. Finding a star in a vehicular cloud. IEEE Intelligent Transport Syst Mag 2013;5(2):55–68.

[63] Alazawi Z, Altowaijri S, Mehmood R, Abdljabar MB. Intelligent disaster management system based on cloud-enabled vehicular networks. In: 11th International conference on ITS telecommunications (ITST); August 2011. p. 361–8.

[64] Amici R, Bonola M, Bracciale L, Rabuffi A, Loreti P, Bianchi G. Performance assessment of an epidemic protocol in VANET using real traces. Procedia Comput Sci 2014;40(0):92–9. In: 4th International conference on selected topics in mobile & wireless networking (MoWNet2014).

[65] Wan J, Lao S, Zhang D, Yang LT, Lloret J. Context-aware vehicular cyber-physical systems with cloud support: architecture, challenges, and solutions. IEEE Commun Mag 2014;52:106–13.

[66] Yan G, Rawat DB, Bista BB. Towards secure vehicular clouds. In: Proceedings of complex, intelligent and software intensive systems (CISIS); 2012. p. 370–5.

[67] Aloqaily M, Kantarci B, Mouftah HT. Provisioning delay effect of partaking a trusted third party in a vehicular cloud. Global Information Infrastructure Symposium (GIIS); September 2014. p. 1–3.

[68] Yan G, Wen D, Olariu S, Weigle MC. Security challenges in vehicular cloud computing. IEEE Trans Intelligent Transport Syst 2013;14/1:284–94.

[69] Pearson S, Benameur A. Privacy, security and trust issues arising from cloud computing. In: IEEE second international conference on cloud computing technology and science (CloudCom); 2010. p. 693–702.

[70] Younis YM, Kifaya K. Secure cloud computing for critical infrastructure: a survey. Liverpool John Moores University, United Kingdom, Technical Report; 2013.

[71] Ristenpart T, Tromer E, Shacham E, Savage S. Hey, you, get off of my cloud: exploring information leakage in third-party compute clouds. In: Proceedings of ACM conference on computer and communications security (CSS); November 2009. p. 199–212.

[72] Chang S, Qi Y, Zhu H, Zhao J, Shen X. Footprint: detecting Sybil attacks in urban vehicular networks. IEEE Trans Parallel Distrib Syst 2012;23/5:1103–14.

[73] Raya M, Hubaux JP. The security of vehicular ad hoc networks. 3rd ACM workshop on security of ad hoc and sensor networks; 2005. p. 1–11.

[74] Raya M, Huabux JP. Securing vehicular ad hoc networks. Comput Security 2007;15/1:39–68.

[75] Almutairi AA, Ghafoor A. Risk-aware virtual resource management for multitenant cloud datacenters. IEEE Cloud Comput 2014;1/3:34–44.

[76] Hsin-Yi T, Siebenhaar M, Miede A, Yu-Lun H, Steinmetz R. Threat as a service? virtualization's impact on cloud security. IT Professional 2012;14/1:32–7.

[77] Gerla M. Vehicular cloud computing. 11th Annual Mediterranean ad hoc networking workshop (Med-Hoc-Net); June 2012. p. 152–5.

[78] Whaidduzzaman M, Sookhak M, Ghani A, Buyya R. A survey on vehicular cloud computing. J Netw Comput Appl 2014;40:325–44.

[79] Anastasopolou K, Tryfonas T, Kokolakis S. Strategic interaction analysis of privacy-sensitive end-users of cloud-based mobile apps. HCI2014 international-human aspects of information security, privacy, and trust; June 2014.

[80] Aloqaily M, Kantarci B, Mouftah HT. On the impact of quality of experience (QoE) in a vehicular cloud with various providers. In: Proceedings of HONET-PFe; December 2014. p. 1–5.

[81] Al-Sultan S, Al-Bayatti A, Zedan H. Context-aware driver behaviour detection system in intelligent transportation systems. IEEE Trans Vehicular Technol 2013;62/9:4264–75.

[82] Santa J, Gmez-Skarmeta A. Sharing context-aware road and safety information. IEEE Pervasive Comput 2009;8/3:58–65.

[83] Kantarci B, Mouftah HT. Trustworthy sensing for public safety in cloud-centric Internet of Things. IEEE Internet Things J 2014;1/4:360–8.

[84] Besaleva LI, Weaver AC. Applications of social networks and crowdsourcing for disaster management improvement. In: International conference on social computing (SocialCom); September 2013. p. 213–9.

[85] Duan J, Gao D, Yang D, Foh CH, Chen H. An energy-aware trust derivation scheme with game theoretic approach in wireless sensor networks for IoT applications. IEEE Internet of Things J 2014;1(1):58–69.

SMART HOME CYBERSECURITY CONSIDERING THE INTEGRATION OF RENEWABLE ENERGY

Y. Liu*, S. Hu*, J. Wu†, Y. Shi†, Y. Jin**, Y. Hu‡, X. Li‡

*Department of Electrical and Computer Science, Michigan Technological University, Houghton, MI, USA;
†Department of Computer Science and Engineering, Notre Dame University, Notre Dame, IN, USA;
**Department of Electrical Engineering and Computer Science, University of Central Florida, Orlando, FL, USA;
‡Institute of Computing Technology, Chinese Academy of Science, Beijing, China

1 INTRODUCTION

The concept of smart home has gained significant popularity in the recent past. It enables the automatic control of household appliances scheduling, leading to the improved energy usage efficiency and reduced economical cost. Refer to Fig. 9.1 for the smart home system. Under the popular advanced metering infrastructure (AMI), the smart meter at a home system automatically receives the periodically updated utility pricing information sent from utility. Subsequently, the smart controller which implements the smart home scheduling algorithms will be activated to automatically schedule various home appliances with the target of reducing electricity bill, that is, shifting the heavy energy load off the peak pricing hours [8]. When there are multiple customers in a local community, game theoretic scheduling could be applied to achieve a community wide optimal solutions for smart home scheduling [9]. The aforementioned process involves two pricing schemes which are popularly used together. The first one is called real-time pricing which is to bill the customers on the basis of the energy usage during the past time window, whereas the second one is called guideline pricing, which is to guide the customers on when would be expensive such that the customers should avoid [7].

On the other hand, renewable energy has become indispensable components in a modern electricity grid. There are multiple levels of renewable energy sources such as system level, community level, and home level. The integration of renewable energy reduces the need of conventional fuel energy, which mitigates the pollution and emission of greenhouse gas. In addition to the benefits, renewable introduces more challenges such as uncertainty. Thus, the dispatch of renewable energy has been studied in many existing works to ensure the reliability and security in the power system. Although the system level and community level renewable energy are usually generated by the large scale wind farm and solar farms, the home level renewable energy is usually a PV penal installed on the roof of a house, which can be controlled by the smart home controller. It plays an important role in the smart home system. It

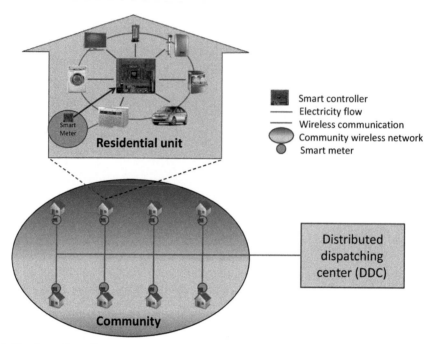

FIGURE 9.1 The Smart Home System

supplies energy to customers directly using the local resource, mitigating the burden of power generation and transmission.

There is an ongoing deployment of a novel technology called netmetering into the smart grid. Such a technology allows the customer to sell the excessively generated renewable energy back to grid for being rewarded. In fact, net metering has already been implemented in 27 states in United States [2,3] and the reward depends on the electricity price as well as the electricity market regulations that vary in different states.

Despite various advances in smart home technology, such a technology is vulnerable to cyberattacks. A particularly important example is pricing cyberattack where hackers manipulate the guideline electricity pricing seen at the smart meters such that the electricity bills of hackers can be reduced and the local power system balance can be disturbed. This cybersecurity problem aggravates with the integration of renewable energy. The main reason is that renewable energy sources are associated with the inherent uncertainties. Since the total energy demand is equal to those requested from customers, this in turn induces the uncertainties to the energy demand from the utility. The introduction of net metering technique, which allows the customers to sell the excessively generated renewable energy back to the grid, further increases variations in energy supply. There are some technical advances in developing detection and defense frameworks against pricing cyberattacks such as the single event detection technique proposed in Ref. [7] and the partially observable Markov decision process (POMDP) based long-term detection technique proposed in Ref. [6]. However, none of them considers the renewable energy integration or impact of net metering.

In this book chapter, we will describe the impact of the integration of renewable energy and net metering technology on the smart home pricing cyberattack detection. Renewable energy and net metering change the grid energy demand, which is considered by the utility when designing the guideline price. If the cyberattack detection technique does not consider this, the detection performance would be degraded. This motivates us to develop a new smart home pricing cyberattack detection framework which handles the aforementioned impact into the detection through developing a net metering aware energy load prediction which is further integrated into the cyberattack detection. In summary, this book chapter describes a renewable energy and net-metering-aware energy-load-prediction technique which is developed on the basis of the cross entropy optimization. On the basis of this energy-load-prediction technique, a smart home pricing cyberattack-detection technique considering the net-metering impact is then presented, which is the first such work in the problem context. This technique significantly improved the POMDP-based smart home cyberattack detection framework developed in Ref. [6]. The simulation results demonstrate that our new framework can significantly improve the detection accuracy from 65.95 to 95.14% compared to Ref. [6]. Although most parts of this book chapter are adopted from our previous work [1], improved presentations and more simulation results are provided to further increase the clarity of the techniques.

2 PRELIMINARIES

Consider a community consisting of N customers. Refer to Fig. 9.2. Each customer is supplied by energy from both the grid and the home level PV panel. Smart home technique is deployed such that each customer $n \in N$ has a set of home appliances A_n to be scheduled by a smart controller. The customers keep receiving the guideline price from the utility, and each customer schedules the energy consumption in the next 24 h which is divided into H time slots.

2.1 ENERGY CONSUMPTION

Each customer uses smart home scheduling technique to schedule the home appliances. For home appliance $m \in A_n$, denoted by X_m the set of power levels. At each time slot, the customer n chooses a power level $x_m^h \in X_m$ for the home appliance m subject to the following constraints. (1) The total energy consumption of home appliance m over the time horizon is equal to the required energy consumption of the specified task E_m such that $\sum_{h=1}^{H} x_m^h e_m^h = E_m$, where e_m^h is the actual execution time period of home

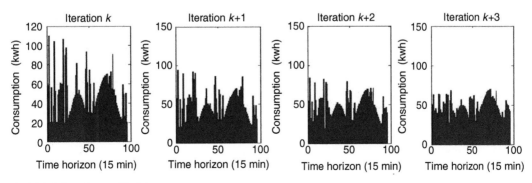

FIGURE 9.2 The System Model

appliance m at time slot h. (2) The home appliance m starts to work no earlier than the required start time α_m and completes the task no later than the deadline β_m such that $x_m^h = 0, \forall h < \alpha_m$ or $h > \beta_m$. Denote by l_n^h the energy consumption of customer n at time slot h. Thus, the community energy load L_h is calculated as $L_h = \sum_{h=1}^{H} l_n^h$. At time slot h, the total energy consumption of customer n is equal to the total energy consumption of all the home appliances such that $\sum_{m \in A_n} x_m^h e_m^h = l_n^h$.

2.2 NET METERING

Each customer is installed with a home-level PV panel and a rechargeable battery. The PV panel serves as a DG unit to the customer. The battery can store the residual energy for future use. Denote by θ_n^h the renewable energy generated by the PV panel of the customer n at time slot h, which is assumed to be approximately known in advance through prediction. The renewable energy generated in the whole community is calculated as $\Theta_h = \sum_{n=1}^{N} \theta_n^h$. The energy storage of customer n at time slot h is denoted by b_n^h, which is upper bounded by B_n such that $0 \le b_n^h \le B_n$.

Net metering says that each customer can sell the energy generated by the PV panel and stored in the battery back to the grid. Thus, at each time slot, the customer could purchase energy from the grid or sell energy to the grid. Denote by y_n^h the energy trading amount of customer n at time slot h. The customer purchases energy from the grid if $y_n^h > 0$ and sells energy to the grid if $y_n^h < 0$. The energy sold by a customer could be consumed by some neighbors in the same community. Thus, the total amount of energy purchased from the utility is $\sum_{n=1}^{N} y_n^h$. For each customer, the battery storage at each time slot is constrained by

$$b_n^{h+1} = b_n^h + \theta_n^h + y_n^h - l_n^h \qquad (9.1)$$

2.3 MONETARY COST

The popular quadratic pricing model is used to compute the monetary cost for purchasing energy from the utility. Thus, the total monetary cost of the community at time slot h is $p_h \left(\sum_{n=1}^{N} y_n^h \right)^2$ [9], where p_h is the guideline price. Each customer is paid with a partial price when selling energy to the grid, which is denote by $\frac{p_h}{W}$, where $W > 1$ is a constant. Thus, the total monetary cost of customer n at time slot h is given as

$$C_n^h = \begin{cases} p_h \left(\sum_{i=1}^{N} y_i^h \right) y_n^y, & \text{if } y_n^h \ge 0 \\ -\frac{p_h}{W} \left(\sum_{i=1}^{N} y_i^h \right) y_n^h, & \text{if } y_n^h < 0 \end{cases} \qquad (9.2)$$

Note that the utility pays the customer with the rate $\frac{p_h}{W}$ for selling energy back to the grid and sells it to other customers with price p_h. The difference between those two prices is cost of the utility due to supporting net metering. Plugging Eq. (9.1) into Eq. (9.2), the monetary cost of customer n at time slot h can be rewritten in terms of energy consumption, renewable energy generation and battery storage as

$$C_n^h = \begin{cases} p_h \left[\left(\sum_{i=1,i\neq n}^{N} y_i^h \right) + b_n^{h+1} - b_n^h - \theta_n^h + l_n^h \right] \\ \left(b_n^{h+1} - b_n^h - \theta_n^h + l_n^h \right), \text{if } b_n^{h+1} - b_n^h - \theta_n^h + l_n^h \ge 0 \\ -\frac{p_h}{W} \left[\left(\sum_{i=1,i\neq n}^{N} y_i^h \right) + b_n^{h+1} - b_n^h - \theta_n^h + l_n^h \right] \\ \left(b_n^{h+1} - b_n^h - \theta_n^h + l_n^h \right), \text{if } b_n^{h+1} - b_n^h - \theta_n^h + l_n^h < 0 \end{cases} \qquad (9.3)$$

3 ENERGY LOAD PREDICTION CONSIDERING NET METERING

3.1 GAME FORMULATION CONSIDERING NET METERING

Given the aforementioned model of smart home technique, each customer n aims to minimize his/her monetary cost within the next 24 h, which depends on the energy scheduling and net metering/energy b_n^h trading. Each customer aims to assign the power level of each home appliance x_m^h and determines the battery storage b_n^h. This naturally leads to a game among customers as follows.

Net metering aware energy consumption scheduling game
- Players: customers $\{1,2,3, \ldots, N\}$
- Shared information: y_n^h
- Optimization problem:

$$\min \sum_{h=1}^{H} C_n^h$$

$$s.t. \sum_{h=1}^{H} x_m^h e_m^h = E_m$$

$$\sum_{m \in A_n} x_m^h e_m^h = l_n^h$$

$$b_n^{h+1} = b_n^h + \theta_n^h + y_n^h - l_n^h$$

$$x_m^h = 0, \forall h < \alpha_m \; or > \beta_m$$

$$C_n^h \begin{cases} p_h\left[\left(\sum_{i=1,i\neq n}^{N} y_i^h\right) + b_n^{h+1} - b_n^h - \theta_n^h + l_n^h\right] \\ \left(b_n^{h+1} - b_n^h - \theta_n^h + l_n^h\right), if \; b_n^{h+1} - b_n^h - \theta_n^h + l_n^h \geq 0 \\ -\dfrac{p_h}{W}\left[\left(\sum_{i=1,i\neq n}^{N} y_i^h\right)\right] + b_n^{h+1} - b_n^h - \theta_n^h + l_n^h \\ \left(b_n^{h+1} - b_n^h - \theta_n^h + l_n^h\right), if \; b_n^{h+1} - b_n^h - \theta_n^h + l_n^h < 0 \end{cases}$$

Decision variables: x_m^h and b_n^h

Each customer tends to follow the aforementioned game to minimize the monetary cost, which means that solving it can predict the energy load in the future given the guideline price.

3.2 PROBLEM SOLVING

The iterative approach is a standard way to solve the energy consumption scheduling game in which each customer solves Problem P1 assuming the total energy trading of other customers is fixed in each iteration. After the new solution is obtained, each customer updates the energy trading to solve Problem P1. This is repeated until convergence. The complete procedure is described in Algorithm 1. In line 4, the customer determines x_m^h while assuming it is fixed using the dynamic-programming-based method proposed in Ref. [8]. In line 5, the customer determines the optimal battery storage using the stochastic optimization algorithm presented. Problem P1 is nonconvex in terms of the battery storage. To circumvent this difficulty, we present a stochastic optimization algorithm based on cross entropy optimization to compute the battery storage that minimizes the monetary cost.

Algorithm 1 Net Metering Aware Energy Load Prediction Algorithm

---

Require: E_m, θ_n^h, α_m, β_m, $\mathcal{X}_m$ and p_h

1: **while** Not converge **do**
2:     **for** Each customer n **do**
3:         **while** Not converge **do**
4:             Solve Problem **P1** using dynamic programming based method to compute x_m^h assuming b_n^h is fixed.
5:             Solve Problem **P1** using cross entropy optimization based method to compute b_n^h assuming x_m^h is fixed.
6:         **end while**
7:     **end for**
8: **end while**
9: **return** x_m^h and b_n^h

Cross entropy optimization method is a stochastic optimization technique based on importance sampling [4]. For completeness, some theoretic foundation of the cross-entropy optimization method is included as follows. Consider the following optimization problem as a generalization of **P1**.

$$\min f\left(b_n\right)$$
$$s.t. b_n \in B$$
$$b_n = \left\{b_n^1, b_n^2, \ldots, b_n^H\right\} \tag{9.4}$$

where $b_n = \left\{b_n^1, b_n^2, \ldots, b_n^H\right\}$ is the battery storage over the time horizon and B is the feasible set of b_n. In the cross-entropy optimization method, a probability density function (PDF) in b is employed, which is denoted by $\rho(b, p)$ while p characterizes the PDF. The cross-entropy optimization method aims to find the maximum value of ϵ such that $P\left[f\left(b_n\right) \le \epsilon\right] \to 0$. Since $P\left[f\left(b_n\right) \le \epsilon\right]$ cannot be known analytically, the cross-entropy optimization method evaluates it using Monte-Carlo simulations. Denote by

$$\delta(\epsilon) = P\left[f\left(b_n\right) \le \epsilon\right] \tag{9.5}$$

an indicator function and $\left[B_1, B_2, \ldots, B_K\right]$ a set of samples generated from the PDF $\rho(b)$. Thus, the indicator function is evaluated by

$$\delta(\epsilon) = \sum_{k=1}^{K} P\left[f\left(B_k\right) \le \epsilon\right] \tag{9.6}$$

However, $P\left[f\left(b_n\right) \le \epsilon\right]$ is a rare event since it approaches zero eventually. Thus, a large amount of samples are needed to compute the maximum value of ϵ. To circumvent this difficulty, the cross-entropy optimization method utilizes the idea of importance sampling. It updates the PDF $\rho(b, p)$ to improve the generated samples such that they will locate in an area with better objective values. Denote by $\theta(b, p)$ the optimal PDF. The estimation of $\delta(\epsilon)$ can be presented by

$$\hat{\delta}(\epsilon) = \frac{1}{K}\sum_{k=1}^{K}P\Big[f(B_k)\le\epsilon\Big]\frac{\rho(B_k)}{\theta(B_k)} \tag{9.7}$$

Define by $\theta^*(b) = \dfrac{P\Big[f(B_k)\le\epsilon\Big]\rho(b,p)}{\delta(\epsilon)}$ the optimal PDF. Thus, $\hat{\delta}(\epsilon)$ can approach the optimal value of $\delta(\in)$. The cross-entropy method finds the optimal PDF by minimizing the Kullback–Leibler distance, which is equivalent to solving the optimization problem

$$\max{}_p \int \theta^*(b)\ln\rho(b,p)db, \tag{9.8}$$

Thus, the optimal PDF in $\rho(b, p)$, p^* can be estimated as

$$\hat{p}^* = \arg{}_p \max\frac{1}{K}\sum_{k=1}^{K}P\Big[f(B_k)\le\epsilon\Big]\frac{\rho(B_k,u)}{\theta(B_k,w)}\ln\rho(B_k,p). \tag{9.9}$$

In the optimization of battery storage, we repeatedly generate samples using the PDF $\rho(b, p)$ and update it through solving Eq. (9.8) until convergence.

4 IMPACT OF NET METERING TO PRICING CYBERATTACK DETECTION

The smart home system is vulnerable to cyberattacks. The malicious hacker can attack the smart meter and manipulate the received guideline price. This can mislead the smart home scheduling of the customers and impact the energy load. As demonstrated by Ref. [7], the cyberattacks can significantly increase the peak-to-average ratio (PAR) of the energy load, which can impact the stability of the power grid. In terms of detection, Ref. [6] proposes to predict the future guideline price from the historical data using support vector regression (SVR) and compares it with the received guideline price. Since this technique works only for single event detection, subsequently a partially observable Markov decision process–based detection is developed for long-term monitoring and detection. However, net metering changes the grid energy demand which also changes the guideline pricing. Since the detection framework in Ref. [6] involves the guideline pricing prediction, it could be compromised if the net metering impact is not considered. It motivates us to integrate the aforementioned cross-entropy optimization–based grid energy prediction (and thus the guideline price prediction) into the pricing cyberattack detection to improve detection accuracy.

4.1 SVR-BASED SINGLE EVENT DETECTION

At each single time slot, the cyberattack is detected on the basis of the comparison between the received guideline price and historical data. Since the electricity price tends to be similar in short term, SVR is deployed to predict the guideline price using only the historical data [10]. However, such a prediction is not accurate since the energy demand prediction also needs to consider the net metering impact.

As is known, the energy demand depends on the current guideline price as well as the renewable energy generation. Thus, we consider both the impact of the historical guideline price and the renewable energy in our SVR model. Denote by $p = \{p_1, p_2, p_3,...,p_t\}$ the vector of guideline price from time slot 1 to T. Denote by $V = \{\Theta_1, \Theta_2, \Theta_3,...\Theta_t\}$ the vector of renewable energy generation from time slot 1 to T. Denote by $D = \{L_1, L_2, L_3,...L_t\}$ the energy demand from time slot 1 to T. We define a function $G(p, V, D)$, which models the predicted guideline price using the difference between the total energy demand and renewable energy. Thus, the single event defense technique originally proposed in Ref. [6] is modified as follows.

- Predict the guideline price P_v using SVR from the time series $G(p, V, D)$.
- The customers conduct smart home scheduling simulation with predicted and received guideline prices, respectively.
- Compare the PAR with predicted and received guideline prices, defined by P_p and P_r, respectively.
- Cyberattack is reported if $P_p - P_r > \delta_p$, where δ_p is a predefined threshold.

4.2 POMDP-BASED LONG-TERM DETECTION

The single event detection technique is based on the PAR increase in a single time slot, which cannot address the cumulative impact [6]. Furthermore, the transient variation of the electricity price can reduce the detection accuracy. Thus, a partially observable Markov decision process (POMDP) based long-term detection technique is proposed in Ref. [6]. For completeness, some details of this technique are included as follows.

The POMDP technique [5] takes the real world state $s \in S$ as the input and generates actions $a \in A$ as the output [5]. Since the real-world state cannot always be perfectly known, the decision maker needs to estimate the state from the observation $o \in O$. Thus, state, observation, and action are the three key components of POMDP. A general POMDP problem is denoted by S, O, A, T, Ω, R. In this problem, the state is defined as the number of hacked smart meters such that $S = \{s_0, s_1, s_2,...s_N\}$, where s_i means that there are totally i smart meters hacked. Similarly, $O = \{o_0, o_1, o_2,...o_N\}$, where o_i means that there are i smart meters hacked according to the observation. In our long-term detection technique, the SVR-based single event detection technique is used to obtain the observation. Given the current state and observation, the decision maker has two available actions in the set $A = \{a_0, a_1\}$. a_0 means ignoring the cyberattack and continuing monitoring the system. a_1 means checking and fixing the hacked smart meters.

POMDP models the mappings between the states when an action is taken. When taking action a, the state transits from s to s' with probability $T(s', a, s)$, which is called transition probability. Meanwhile, the decision maker receives a reward $R(s', a, s)$. The mapping between states and observations is defined as $O(s', a, s) = P(o \mid a, s)$, which is the probability of observation o conditioned on state s and action a. In this problem, the state transition probability $T(s', a, s)$ and observation function $O(s', a, s) = P(o \mid a, s)$ are trained on the basis of the historical data. The reward function $R(s', a, s)$ is defined on the basis of the losses taken by each hacked smart meter and the labor cost for checking and repairing the hacked smart meters. Given the model, the POMDP technique aims to optimize the discounted expected reward through picking the optimal action, which is formulated as a Bellman equation. Refer to Refs. [5,6] for more details of POMDP technique.

Our detection technique is illustrated in Fig. 9.3. Given the predicted and received guideline prices, the net-metering-aware energy-load prediction technique is involved in computing the PAR increase.

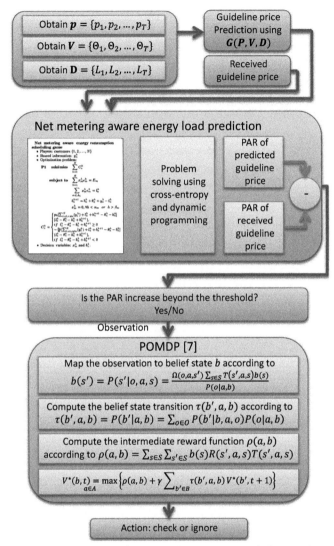

FIGURE 9.3 Algorithmic Flow for the Net Metering Aware Energy Load Prediction and Smart Home Pricing Cyberattack Detection Technique

5 SIMULATION RESULTS

In this section, simulations are conducted to analyze the impact of cyberattacks and the impact of net metering to our defense techniques. In the simulation, we consider different sizes of communities consisting of 100, 200, 300, 400, and 500 customers, respectively. The setup of the energy consumptions of the customers is similar to the previous works [6,7].

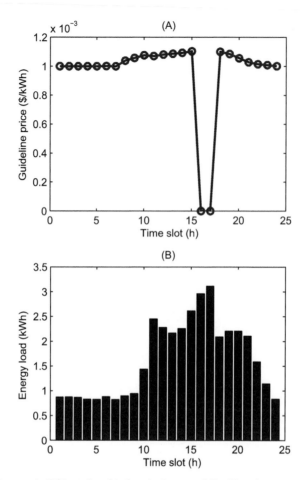

FIGURE 9.4 Without Cyberattack, Without Considering the Impact of Net Metering

(A) Received guideline price and predicted guideline price. (B) Predicted energy load according to the predicted guideline price in (a), PAR = 1.4700.

The impacts of cyberattacks and net metering on energy load are analyzed and the results are shown in Figs 9.4–9.6 From those figures, we make the following observations.

- Refer to Fig. 9.4 for the prediction technique without considering the impact of net metering even if it is actually deployed. Shown in Fig. 9.4A is the received guideline price without cyberattack and the predicted guideline price using the SVR-based method in Ref. [7], respectively. The predicted guideline price does not match the received guideline price well such that it forms a peak from 12:00 to 14:00 while it is a gap in the received guideline price. Shown in Fig. 9.4B is the predicted energy load using this predicted guideline price. The PAR is 1.4700.
- Refer to Fig. 9.5 for the prediction technique considering net metering. Shown in Fig. 9.5A is the received guideline price without cyberattack and the predicted guideline price considering

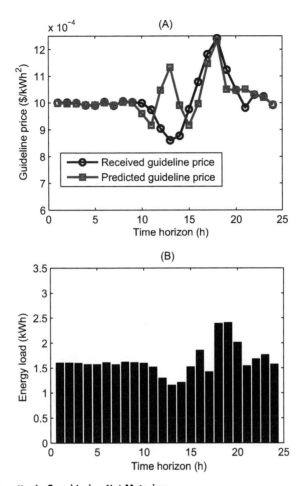

FIGURE 9.5 Without Cyberattack, Considering Net Metering

(A) Received guideline price without cyberattack and predicted guideline price. (B) Predicted energy load according to the predicted guideline price in (A), PAR = 1.3986.

net metering, respectively. The predicted guideline price matches the received one better than the SVR-based method in Ref. [7]. Shown in Fig. 9.5B is the predicted energy load using the predicted guideline price in Fig. 9.5A. The PAR is 1.3986. Comparing with this result, the PAR corresponding to the predicted energy load in Fig. 9.4B is $\dfrac{1.4700-1.3986}{1.3986} = 5.11\%$ higher.

- Refer to Fig. 9.6 for the impact of cyberattack. Shown in Fig. 9.6A is the manipulated guideline price. The price is manipulated to be zero between 16:00 and 17:00. Shown in Fig. 9.6B is the energy load under cyberattack. Corresponding to the manipulated guideline price, the energy load reaches a peak at 16:00 and 17:00. The PAR is 1.9037. This is $\dfrac{1.9037-1.4700}{1.4700} = 29.50\%$

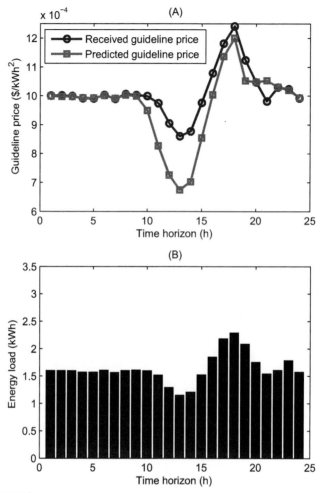

FIGURE 9.6 With Cyberattack

(A) The manipulated received guideline price. (B) Energy load corresponding to the manipulated received guideline price, PAR = 1.9037.

higher than the predicted energy load in Fig. 9.4B and $\dfrac{1.9037-1.3986}{1.3986} = 36.11\%$ higher than the predicted energy load in Fig. 9.5B.

In order to demonstrate the performance of the detection technique, we randomly generate cyberattacks and conduct the simulations for the time horizon of 48 h. The detection technique with and without considering net metering are compared in terms of detection accuracy, PAR and labor cost for detecting hacks and fixing hacked smart meters. The detection accuracy is defined as $\theta = 1 - \dfrac{|i-j|}{j}$ if j smart meters are actually hacked while i smart meters are reported to be hacked. To demonstrate the robustness of our technique, we conduct the simulation in the testcase with 100, 200, 300, 400, and 500 customers, respectively. The detection accuracies are shown in Figs. 9.7–9.11, respectively. The PAR and labor cost are shown in Table 9.1. From those figures and table, we make the following observations.

- The results for the 100-customer testcase are shown in Fig. 9.7 and Table 9.1, respectively. Refer to Fig. 9.7. The detection technique considering net metering has a detection accuracy of 95.35% on average while it is 67.45% when the impact of net metering is not considered. The detection techniques with and without considering net metering are also compared in terms of the corresponding labor cost and PAR of the energy load. Refer to Table 9.1. Without any detection technique, the PAR of the energy load is 1.6860. Using the detection technique without considering net metering, the PAR is decreased to 1.5435. Using the defense technique considering net metering, the PAR is further reduced to 1.3943. The detection technique considering net metering can reduce the PAR by $\dfrac{1.5435-1.3943}{1.5435}=9.67\%$ at the cost of increasing the labor cost by $\dfrac{1.0048-1}{1}=0.48\%$ compared to that without considering net metering.

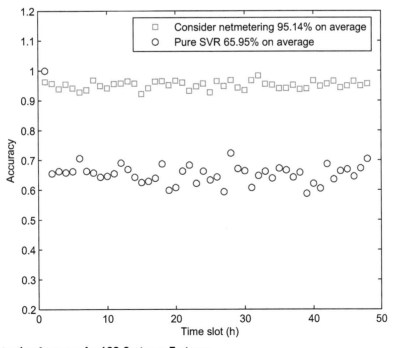

FIGURE 9.7 Detection Accuracy for 100-Customer Testcase

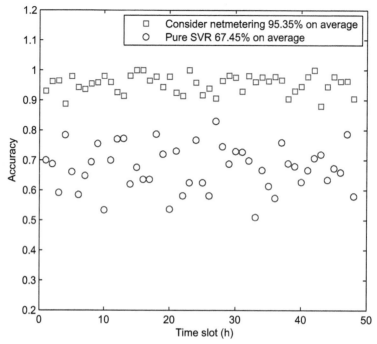

FIGURE 9.8 Detection Accuracy for 200-Customer Testcase

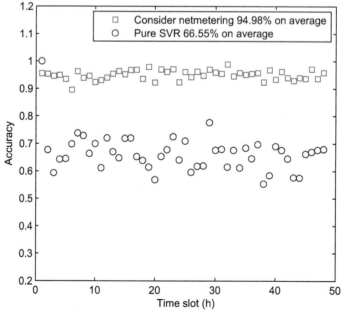

FIGURE 9.9 Detection Accuracy for 300-Customer Testcase

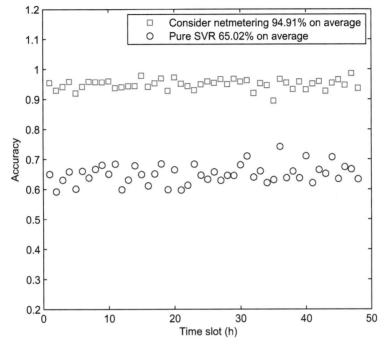

FIGURE 9.10 Detection Accuracy for 400-Customer Testcase

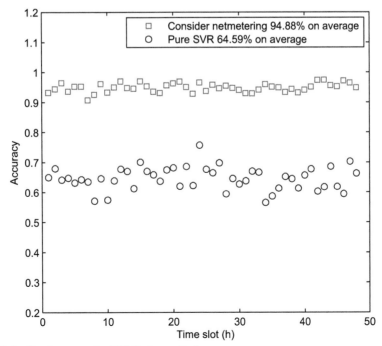

FIGURE 9.11 Detection Accuracy for 500-Customer Testcase

Table 9.1 Simulation Results of Detection Techniques

Testcase	Parameter	No Detection	Detection Without Considering Net Metering	Detection Considering Net Metering
100-Customer	PAR	1.6860	1.5435	1.3943
	Normalized labor cost	—	1	1.0048
200-Customer	PAR	1.6856	1.5417	1.4035
	Normalized labor cost	—	1	1.0037
300-Customer	PAR	1.6875	1.5464	1.4015
	Normalized labor cost	—	1	1.0033
400-Customer	PAR	1.6872	1.5452	1.4116
	Normalized labor cost	—	1	1.0124
500-Customer	PAR	1.6509	1.5422	1.4112
	Normalized labor cost	—	1	1.0067

- The results for the 500-customer testcase are shown in Fig. 9.11 and Table 9.1, respectively. Refer to Fig. 9.11. The detection technique considering net metering has a detection accuracy of 95.14% on average, whereas it is 65.95% when the impact of net metering is not considered. The detection techniques with and without considering net metering are also compared in terms of the corresponding labor cost and PAR of the energy load. Refer to Table 9.1. Without any detection technique, the PAR of the energy load is 1.6509. Using the detection technique without considering net metering, the PAR is decreased to 1.5422. Using the defense technique considering net metering, the PAR is further reduced to 1.4112. The detection technique considering net metering can reduce the PAR by $\frac{1.5422-1.4112}{1.5422}=8.49\%$ at the cost of increasing the labor cost by $\frac{1.0067-1}{1}=0.67\%$ compared to that without considering net metering. The results of other testcases are similar to the 100-customer and 500-customer testcases.

The net metering helps reduce the peak demand from the grid rather than increasing it. Thus, the PAR could be significantly reduced. On the other hand, the detection detection of cyberattacks depends on PAR. The predicted guideline price considering net metering leads to a lower PAR than that without considering net metering. The cyberattack induced PAR increase is smaller than the actual increase when net metering is not considered. This is why the detection technique without considering net metering cannot detect around 30% of the cyberattacks as demonstrated by our simulation results.

6 CONCLUSIONS

In this book chapter, the impact of the integration of renewable energy and net metering technology on the smart home pricing cyberattack detection is studied. Renewable energy and net metering change the grid energy demand, which is considered by the utility when designing the guideline price. If the cyberattack detection technique does not explore this, the detection performance would be degraded. This motivates us to develop a new smart home pricing cyberattack detection framework which handles the aforementioned impact into the detection. The simulation results demonstrate that our new framework can significantly improve the detection accuracy for about 30% compared to the state-of-the-art technique.

REFERENCES

[1] Liu Y, Hu S, Wu J, Shi Y, Jin Y, Hu Y, Li X. Impact assessment of net metering on smart home cyberattack detection. Design Automation Conference (DAC), 2015 52nd ACM/EDAC/IEEE, San Francisco, CA, 2015, pp. 1–6.

[2] http://www.solarcity.com/learn/understanding-netmetering.aspx [Online].

[3] Distributed generation and renewable energy current programs for businesses. http://docs.cpuc.ca.gov/published/newsrelease/7408.htm [Online].

[4] Botev ZI, Kroese DP, Rubinstein RY, et al. The cross entropy method for optimization. In: Govindaraju V, Rao CR, editors. Machine Learning: Theory and Applications, vol. 31. Chennai: Elsevier BV; 2013. p. 35–59.

[5] Kaelbling LP, Littman ML, Cassandra A. Planning and acting in partially observable stochastic domains. Artif Intell 1998;101(1):99–134.

[6] Liu Y, Hu S, Ho T-Y. Leveraging strategic defense techniques for smart home pricing cyberattacks. IEEE Trans Depend Secure Comput 2016;13(2):220–5.

[7] Liu Y, Hu S, Ho T-Y. Vulnerability assessment and defense technology for smart home cybersecurity considering pricing cyberattacks. In: Proceedings of IEEE/ACM international conference on computer-aided design; 2014. p. 183–190.

[8] Liu L, Zhou Y, Liu Y, Hu S. Dynamic programming based game theoretic algorithm for economical multi-user smart home scheduling. In: Proceedings of IEEE midwest symposium on circuits and systems; 2014. p. 362–365.

[9] Mohsenian-Rad A, Wong V, Jatskevich J, Schober R, Leon-Garcia A. Autonomous demand-side management based on game-theoretic energy consumption scheduling for the future smart grid. IEEE Trans Smart Grid 2010;1(3):320–31.

[10] Tuomas K, Rossi F, Lendasse A. Ls-svm functional network for time series prediction. In: Proceedings of European symposium on artificial neural networks; 2006.

SMART HOME SCHEDULING AND CYBERSECURITY: FUNDAMENTALS

10

Y. Liu, S. Hu

Department of Electrical and Computer Science, Michigan Technological University, Houghton, MI, USA

1 INTRODUCTION

The smart home technique facilitates the end usages of electricity energy through the automatic control of home appliances of the customers. In the smart home infrastructure, all the home appliances of the customer are connected to the smart controller installed in the home of each customer, which receives the electricity-pricing information from the utility and optimizes the energy consumption scheduling. According to the dynamic-pricing curve provided by the utility, the customer can shift most of the energy consumption off the peaking pricing hours to the non-peak ones, thus reducing the electricity bill. Generally speaking, there are two popular pricing schemes which are usually used together in the prevailing US electricity market, which are real-time pricing and guideline pricing, respectively. Although the customers are charged by real-time pricing on the basis of the energy consumption in the past time window, guideline pricing predicts the future electricity price and provides it as a reference. Thus, the customers can schedule their energy consumptions accordingly. From the utility's point of view, a smartly designed guideline price can help balance the energy load and reduces the peak energy usage. This mitigates the pressure of the energy generation, transmission and distribution.

The smart home technique is supported by the advanced metering infrastructure (AMI), which enables the two-way communications between the utility and customers. The AMI is the communication system of smart grid, which consists of backbone network and local area networks [5]. Taking advantage of this infrastructure, the utility transmits the pricing information to the substation through the backbone network. Subsequently, the substation transmits the pricing information to the target customers through the local area network [4]. There are various wired and wireless communication protocol available such as WiFi, 802.11, ZigBee, WiMAX, IEEE 802.16e, broadband PLC, and long term evolution (LTE) [6]. In the real implementation of the AMI, different combinations of the protocols can be chosen depending on the availability. Refer to Fig. 10.1 for a simplified AMI system. There are some existing smart home scheduling works utilizing this infrastructure such as [3,7]. In Ref. [7], a linear-programming-based smart home scheduling technique is proposed to schedule the energy consumption of the home appliances for electricity bill reduction. In Ref. [3], game theory is utilized to solve the smart home scheduling problem among multiple customers. Even

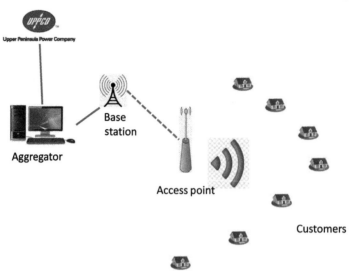

FIGURE 10.1 System Model

more smart home scheduling techniques have been investigated to handle the smart home scheduling problems with more complex constraints.

Despite the benefit it offers, the smart home system is vulnerable to cyberattacks. A hacker can launch cyberattacks on the basis of the connection and devices. Especially, the hacker can easily tamper a smart meter and manipulate the pricing information received by it, thus misleading the smart home scheduling of the customers. As is known, the modern smart meters are usually microcontroller based and installed with advanced embedded operation systems, which is vulnerable to virus and cyberattacks from the internet. For example, the smart meter of Texas Instruments is based on MSP430, which supports two-way communication such that the system can be remotely upgraded [8]. Remote upgrading makes the smart meter vulnerable to cyberattacks. In fact, multiple public media reports witness that various smart devices such as security camera, smart TV, smart doorlock, smart refrigerators, smart bulbs, and powered outlets became the targets of hackers [9–11].

In this book chapter, the impact of pricing cyberattacks to the smart home and the local community is analyzed. As is shown in our previous work [1], manipulating the guideline price can reduce the expense of the cyberattacker and increase the peak energy usage in the local power system. The detection techniques are then proposed to mitigate the impact of the cyberattacks and protect the smart home system. In order to identify the cyberattack, we predict the future guideline price from the historical data using support vector regression (SVR) and compare it with the one received by the smart meter. Since the electricity-pricing curve tends to be similar in short term, manipulation can be detected if the difference between the received guideline price and the predicted one are significantly different. Since the purpose of the detection technique is to limit the impact to local community and customers, the impact differences are defined to quantify the bill and the peak-to-average ratio (PAR) increase comparing the predicted guideline pricing curve and the received guideline pricing curve. When the impact differences are larger than some prespecified values, the cyberattack is reported to the utility to request for further check which is associated with human interaction.

This framework is used to detect the cyberattacks at each single time slot, which is defined as single event detection technique. The performance of the aforementioned single event detection technique is limited since it does not have a long-term view. The single event detection technique reports an alert if the impact differences are beyond the threshold. However, in many cases the impact differences are minor at each time slot while they can accumulate to be significant in long term. For example, suppose the threshold is set as 2%, which means that the cyberattack is reported if the received guideline electricity price can increase the average bill by 2%. However, if the hacker manipulates the electricity pricing to increase the bill by 1.9% each time slot, no cyberattack will be detected. However, the cumulative bill increase during the long term would be significant. The cumulative impact can be mitigated by lowering the thresholds. However, this could increase the labor cost for checking the smart meters significantly due to the fluctuation of the electricity-pricing curve. To tackle this problem, a long-term detection technique using partially observable Markov decision process (POMDP) is proposed in Ref. [1], which has the ingredients such as probabilistic state transition, reward expectation, and policy transfer graph to account for the cumulative cyberattack impact and its potential future impact. The general attacks targeting electricity pricing in smart grid have been studied in existing literature. The work in Ref. [12] proposes a jamming attack such that the attacker jams the communication network to block the updated electricity price. The work in Ref. [13] introduces false data injection attack to manipulate the real-time monitoring data to interfere the state estimation, which damages the power grid. Generally speaking, the cyberattacks in the smart home context includes the cyberattack on both input and output of smart meter as well as smart meter itself. Although the works in Refs. [12] and [13] present the cyberattacks targeting the input, the works in Refs. [14,15] studies energy theft, which is the cyberattack targeting the output such that the hackers attack their own smart meters to decrease the measurements of energy usage, thus saving the electricity bill of themselves. These works do not address smart home pricing cyberattacks since they do not consider the impact due to smart home scheduling. To mitigate the impacts of those attacks, defense techniques have been designed in literature. The work in Ref. [16] proposes a likelihood ratio-test-based algorithm to detect the malicious data attack on smart grid state estimation. The work in Ref. [17], proposes detection technique for jamming attack in time critical networks. The works in Refs. [18,19] study the countermeasure techniques against false data injection attack on the basis of sparse optimization and Kalman filter, respectively. Again, these works do not address the pricing cyberattacks in the smart home context.

In this chapter, the vulnerability of the smart home infrastructure will be assessed and the detection technologies against pricing cyberattacks will be discussed, which are based on machine learning techniques. In addition, the advanced control theoretical partially observable Markov decision process will be discussed. The salience of the algorithms will be demonstrated through simulation results.

2 SMART HOME SYSTEM PRELIMINARIES
2.1 SMART HOME SYSTEM MODEL

A set of $\mathcal{N} = \{1, 2, \ldots, N\}$ customers supplied by a utility are considered in the system. Refer to Fig. 10.1. Each customer schedules the energy consumption during the next 24 h from the current moment, which is divided into H time slots and let $\mathcal{H} = \{1, 2, \ldots, H\}$. Each customer is equipped with a smart meter, which receives the electricity price from the utility and sends the measurement of real-time energy consumption to the aggregator. In the following, we will give an overview of the game model and solution of smart home scheduling, which have been developed in our previous work [20].

Each customer $n \in N$ has a set of home appliances denoted by A_n. At each time slot h, the home appliance $m \in A_n$ works under power level $x_m^h \in X_m$, where X_m is the set of available power levels for home appliance m. In general, there are two categories of home appliances, namely, manually controlled home appliances and automatically controlled ones. The category of manually controlled home appliances consists of TV set, computer, refrigerator etc. The category of automatically controlled home appliances consists of washing machine, cloth dryer, dish washer, electric vehicle (EV) etc. It is worth noting that heating, ventilation, and air conditioning (HVAC) system could be classified into both categories. For example, the customer can adjust the working power level of the air conditioner manually according to the temperature in the room. However, there also exists situation that the customer needs the temperature to reach a certain level at a certain time point, in which the air conditioner works in the automatic mode. For each customer n, denote by P_n the set of manually controlled home appliances and denote by Q_n the set of automatically controlled home appliances. Thus, $A_n = P_n \cup Q_n$. For customer n, the total energy consumption of the manually controlled home appliances at time slot h is denoted by $l_{n,h}^p$, one has

$$l_{n,h}^p = \sum_{m \in P_n} x_{m,h} t_{m,h} \tag{10.1}$$

where $t_{m,h}$ is the actual execution time of home appliance m at time slot h. At time slot h, the total energy consumption of automatically controlled home appliances is denoted by $l_{n,h}^q$, one has

$$l_{n,h}^q = \sum_{m \in Q_n} x_{m,h} t_{m,h} . \tag{10.2}$$

For each automatically controlled home appliance $m \in Q_n$, the power level is chosen subject to the following constraints.

(1) The total energy consumption of the home appliance m is equal to the required total energy consumption E_m. That is,

$$E_m = \sum_{h \in H} x_{m,h} t_{m,h} . \tag{10.3}$$

(2) For a specified task, the home appliance m needs to be executed after the earliest start time α_m and before the deadline β_m such that

$$x_{m,h} = 0, \forall h < \alpha_m \ or \ h > \beta_m . \tag{10.4}$$

At each time slot h, the total energy load of the community is denoted by L_h, and thus

$$L_h = \sum_{n \in N} \left(l_{n,h}^p + l_{n,h}^q \right). \tag{10.5}$$

As is mentioned before, there are two types of electricity prices in the smart home system, which are guideline electricity price and real-time electricity price, respectively. The guideline electricity price is provided to the customers to facilitate smart home scheduling, whereas the real-time electricity price is used in computing the bill.

In real-time pricing, at each time slot the monetary cost of energy consumption depends on the total energy load of the grid. In this paper, the quadratic cost function is used to compute the total monetary cost of all the customers, which is a popular pricing model used in literature [3,23]. The total monetary cost at time slot C_h is given as

$$C_h = a_h L_n^2 \qquad (10.6)$$

where a_h is the pricing parameter which models the relationship between energy consumption and monetary cost. At each time slot, the monetary cost is distributed to each customer according to the energy usage of the customer. For customer n, the monetary cost is denoted by $C_{n,h}$ where

$$C_{n,h} = \left(l_{n,h}^p + l_{n,h}^q \right) \frac{C_h}{L_h} = a_h \left(l_{n,h}^p + l_{n,h}^q \right) L_h. \qquad (10.7)$$

2.2 SMART HOME SCHEDULING

Given the aforementioned constraints, one can formulate a game where each customer n aims to minimize the individual monetary cost. Since the monetary cost of each customer n depends on the total energy load of the community including all the other customers, the scheduling of one customer has an impact on others. This naturally leads to a game. The monetary cost of each customer n can be divided into two parts. One has

$$C_{n,h} = a_h \left(l_{n,h}^p + l_{n,h}^q \right) \left(l_{n,h}^p + l_{n,h}^q \right) + a_h \left(l_{n,h}^p + l_{n,h}^q \right) l_{-n,h}, \qquad (10.8)$$

where $l_{-n,h}$ is the community wide energy load excluding the energy consumption of customer n at time slot h and

$$l_{-n,h} = \sum_{i \in N, i \neq n} \left(l_{i,h}^p + l_{i,h}^q \right). \qquad (10.9)$$

The game can be then formulated as follows [20].
Game model:

- *Players*: All the customers in the system.
- *Payoff function*: $P\left(l_{n,h}^p \mid l_{-n,h} \right) - C_{n,h} = -a_h \left(l_{n,h}^p + l_{n,h}^q \right) \left(l_{n,h}^p + l_{n,h}^q \right) - a_h \left(l_{n,h}^p + l_{n,h}^q \right) l_{-n,h}.$
- *Shared information*: $l_{-n,h}$.
- *Problem formulation*:

$$\text{P1} \quad \min_{m \in Q_n, \forall h \in \mathcal{H}, x_m^h \in X_m} a_h \left(l_{n,h}^p + l_{n,h}^q \right) L_h.$$

$$\text{subject to } l_{n,h}^p = \sum_{m \in P_n} x_{m,h} t_{m,h}.$$

$$l_{n,h}^p = \sum_{m \in Q_n} x_{m,h} t_{m,h}$$

$$E_m = \sum_{h \in H} x_{m,h} t_{m,h}$$

$$x_{m,h} = 0, \forall h < \alpha_m \text{ or } h > \beta_m$$

$$L_h = \sum_{n \in N} \left(l_{n,h}^p + l_{n,h}^q \right)$$

In the game, each customer aims to minimize the total payment through scheduling the energy consumption of the automatically controlled home appliances. Nash equilibrium is achieved when no one can further reduce his/her own monetary cost without changing that of any other customer [24]. To solve this game, a decentralized technique proposed in our previous works [20,22] is used. This is an iterative algorithm. In each iteration, each customer solves the Problem P1 using the dynamic-programming-based algorithm while assuming the energy consumption of all the others, that is, $l_{-n,h}$, is fixed [20]. After each iteration, each customer obtains the new energy consumption information from all others according to their updated scheduling solutions. With the updated energy consumption of each customer, $l_{-n,h}$ is updated and each customer uses dynamic programming to schedule their home appliances with the updated $l_{-n,h}$. This process is iterated until convergence. For the scheduling of each customer, the home appliances are scheduled one after another. The method for the scheduling of single home appliance is described as follows [21].

The energy consumption of a home appliance is scheduled while assuming those of all others are fixed. Thus, if the power level of home appliance i at time slot $x_{i,h}$ is fixed, the total energy load can be directly calculated and the electricity bill can be derived. The dynamic-programming-based smart home scheduling algorithm aims to choose the power levels of a home appliance at each time slot to minimize the electricity bill. It proceeds one time slot after another from time slot 1. For time slot h, the accumulative energy consumption and corresponding electricity bill are computed for combination of power levels from time slot 1 to h. Among the combinations with the same accumulative energy consumption, only the one with the lowest monetary cost is maintained because it is the only one that could be a part of the final solution. An example is included as follows.

Consider a home appliance with three power levels and an operation period of four time slots. The electricity bill corresponding to each time slot and power level is given in Table 10.1. From time slot 1 to h, the combination of power levels is recorded as $\left(x_{i,1}, x_{i,2}, \ldots, x_{i,h}\right)$. Starting from time slot 1, all power levels (1), (2), (3) are maintained because each of them has a unique accumulative energy consumption. Proceeding to time slot 2, the combination of power levels (1,2) and (2,1) have the same accumulative energy consumption, which is 3. However, the electricity bills are 3.5 and 3.2, respectively. Thus, only (2,1) is maintained since it has a lower electricity bill. Such a technique is applied to all other combinations of power levels. Thus, (1,1), (2,1), (3,1), (3,2), (3,3) are maintained from the original 9 combinations in the first two time slots. Combining them with all the available power levels at the third time slot, totally 15 combinations are generated for the total three time slots. Subsequently, (1,1,1), (2,1,1), (3,1,1), (3,1,2), (3,1,3), (3,2,3), and (3,3,3) are maintained. Proceeding to time slot 4 in the same way as before, totally 21 combinations are generated and (1,1,1,1), (2,1,1,1), (3,1,1,1), (3,1,1,2), (3,1,1,3), (2,1,2,3), (3,1,3,3), (2,3,3,3), and (3,3,3,3) are maintained. Suppose the given task requires an energy consumption of 5. The combination (3,1,1,1) is chosen since it has the same amount of total energy consumption with the requirement.

Table 10.1 Electricity Bill Corresponding to Each Power Level				
Monetary Cost ($)	**Time Slot 1**	**Time Slot 2**	**Time Slot 3**	**Time Slot 4**
Power level 1	1	1.2	1.5	1.1
Power level 2	2	2.5	2.8	2.4
Power level 3	3	3.8	4.1	3.6

```
1:  for Each customer do
2:      for Each home appliance do
3:          Initialize set A = ∅ as the record of combinations of power levels
4:          Initialize set Cmin = ∅ as the record of monetary cost
5:          Initialize set Cen = ∅ as the record of accumulative consumption
6:          for h = ts : te do
7:              Calculate the customer's monetary cost of each power level from Ei according to Equation 1.7.
8:          end for
9:          for h = ts : te do
10:             for A1 ∈ A do
11:                 for E1 ∈ Ei do
12:                     Current = {A1, E1}
13:                     Calculate the corresponding monetary cost C and accumulative energy consumption Ce of Current
14:                     if C < Cmin(Ce) and Ce ∈ Cen then
15:                         A(Ce) = Current and Cmin(Ce) = C
16:                     else if Ce ∉ Cen then
17:                         Cen = Cen ∪ Ce
18:                         A(Ce) = Current and A = A ∪ A(Ce)
19:                         Cmin = Cmin(Ce)
20:                     end if
21:                 end for
22:                 Select the combination from A whose energy consumption is the same as required
23:             end for
24:         end for
25:     end for
26: end for
```

ALGORITHM 10.1 The Scheduling Algorithm for Single Home Appliance

The complete description of the smart home scheduling algorithm for multiple customers is presented in Algorithm 10.1. Three sets A, C_{en}, and C_{min} are used to store the combinations of power levels, accumulative consumptions and electricity bills, respectively. The array *current* is a temporary variable to store the temporarily generated combination of power levels. In line 3, 4, and 5, the sets A, C_{en}, and C_{min} are initialized as empty sets. In line 7, the electricity bills corresponding to each power level at each time slots are computed, which is similar to those in Table 10.2. In line 12, each power level is combined with the existing combination of power levels. The corresponding electricity bill C and

Table 10.2 Daily Energy Consumption and Regular Execution Duration of Automatically Controlled Home Appliances [20]

Home Appliance	Daily Consumption	Execution Duration
Washing machine	1.2–2 kWh	1–3 h
Dish washer	1.2–2 kWh	1–3 h
Cloth dryer	1.5–3 kWh	1–3 h
EV	9–12 kWh	4–8 h
Air conditioner	2–3 kWh	2–3 h
Heater	2–3 kWh	2–3 h

accumulative energy consumption are calculated in line 13. From line 14 to line 20, the sets A, C_{en}, and C_{min} are updated. The updating rule is as follows. If there is a combination of power levels with accumulative energy consumption C_e in A denoted by $A(C_e)$, it is replaced by *current* if its corresponding monetary cost $C_{min}(C_e)$ is greater than $C.C_{min}(C_e)$ is also updated as C. If there is no combination of power levels with accumulative energy consumption C_e, Current is included in A as $A(C_e)$. *Ce* and C are also included in Cen and *Cmin* as *Cen(Ce)* and $C_{min}(C_e)$, respectively. When all time slots have been proceeded, the combination of power levels with required accumulative energy consumption is returned in line 22.

3 PRICING CYBERATTACKS

Consider the communication infrastructure of the AMI in Fig. 10.1. First, the utility transmits the pricing information to a central computer in the local community (substation) through Internet. Second, such pricing information is broadcast to smart meters through Internet or WiFi network or a combination of them, depending on the communication infrastructure of the local community. For example, in a hierarchical infrastructure where a community consists of multiple subcommunities, pricing information is forwarded to each subcommunity through internet and is then broadcast inside the subcommunity through the WiFi network. In a WiFi network, there are some access points, which serve as the agents to receive the pricing information from the subcommunity and forward it to the smart meters inside the subcommunity.

The aforementioned popular infrastructure is vulnerable to at least three attacking strategies. First, one can directly hack the computer in the substation and modify the pricing information there. Subsequently, the pricing information forwarded to the whole community could be a faking one. Second, one can block an access point in the WiFi network using the jamming attack (ie, sending excessive requests to the access point), create a fake access point, and send the faking pricing information to the smart meters covered by the fake access point. Third, one can hack the smart meter and modify the pricing it receives. As is indicated by Ref. [25], "commercial smart devices including smart meters are often designed and manufactured utilizing off-the-shelf components and/or solutions. Therefore, security protection methods can only be applied at the application/network level and can hardly cover the hardware infrastructure. As a result, attackers can easily bypass firmware verification and install malicious OS kernel in the device to remotely control the smart device." For example, an attacker could remotely manipulate the guideline pricing received at a smart meter without being realized. With the different difficulty levels in implementation, the cyberattacker can choose which one to use in practice.

3.1 CYBERATTACK FOR BILL REDUCTION

The first possible cyberattack is to fake the guideline pricing curve such that the utility bill of the cyberattacker can be reduced at the cost of bill increase of others in the community. Consider the following scenario. The guideline electricity price in the early morning such as 1:00 am to 8:00 am are usually not high due to limited amount of human activities. However, if a cyberattacker schedules a large load during this period, it could still be expensive. Therefore, if the cyberattacker fakes the guideline-pricing curve such that the electricity price during 1:00 am and 8:00 am is very high, then almost no customer in the community will schedule energy during this period. Subsequently, the cyberattacker can schedule his/her own large load there, resulting in the significant reduction of his/her own bill. Of course, such a reduction comes from the increase of the bill of other customers. The procedure for the cyberattack for bill reduction using pricing manipulation is as follows.

1. Determine the starting time ts and ending time te for the hacker to schedule his/her own energy load.
2. Manipulate the guideline electricity prices received at the target smart meters such that after manipulation the guideline prices are high from ts to te and low at other time slots.
3. Schedule his/her own energy load from ts to te. When the guideline electricity price is high from ts to te, the customers tend not to schedule the energy consumption there according to the smart home scheduling. This reduces the energy load during these time slots, and results in the decrease of the real-time electricity price there. Subsequently, the cyberattack could schedule the energy consumption from ts to te, and makes profit through saving his/her own bill at the cost of increasing the bill of other customers.

3.2 CYBERATTACK FOR FORMING THE PEAK ENERGY LOAD

The second possible attack is to fake the guideline pricing curve such that a peak energy usage can be formed. Consider the following scenario. The guideline electricity price at 8:00 pm is usually expensive since the utility discourages the excessive energy usage during this period which is typically occupied with various human activities (eg, watching TV). If a cyberattacker creates a fake guideline pricing curve with very low price at this slot, significant amount of energy (eg, laundry load) will be accumulated there. This will form a peak in the energy usage, which could significantly impact the power system stability. The procedure for the cyberattack for forming a peak energy load using pricing manipulation is as follows.

1. Determine the starting time ts and ending time te of peak energy usage hours.
2. Manipulate the guideline electricity prices received at the target smart meters such that after manipulation the guideline prices are very low from ts to te.
3. A peak energy load will be formed from ts to te. If the guideline electricity price is very low from ts to te, the customers tend to schedule large energy load there due to smart home scheduling. This increases the energy load during this time period which could potentially form a peak in energy consumption.

4 SINGLE EVENT DETECTION TECHNOLOGY

The aforementioned two pricing cyberattacks need to significantly perturb the guideline pricing curves. The key to design the countermeasure is to identify the guideline pricing manipulation. Machine learning and statistical data analysis techniques would be natural choices since the typical guideline pricing curves should be similar to each other in a short term. In this work, we choose SVR since it tends to produce robust results [26].

After computing the predicted guideline price curve, one can compare it with the guideline pricing curve received at the smart meter. For comparison, one could compute the maximum difference between the two curves and signal an alert when it is larger than a threshold. To explore this idea, we define a set of thresholds, called the maximum tolerable impact differences. An example maximum tolerable impact differences could be up to 2% increase in PAR and up to 5% increase in bill. Given the predicted guideline pricing curve and the received guideline pricing curve, one can perform smart home scheduling simulations to compute the average bill and PAR for each of the two guideline pricing curves, and then compute the differences between them, called the actual impact differences. When the actual impact differences are larger than the maximum tolerable impact differences, a potential pricing attack is spotted and an alert signal will be sent to the utility to request for further check which could need some amount of human interaction. This work will develop such a framework.

Denote by vectors a_p and a the predicted guideline electricity price and the received guideline electricity price, respectively. Denote by Δ_B and Δ_P the actual impact differences in bill and PAR, respectively. We also define two thresholds δ_B and δ_P which are the maximum tolerable impact differences in bill and PAR, respectively. If $\Delta_B > \delta_B$ or $\Delta_P > \delta_P$, the smart meter treats the guideline electricity price as being manipulated due to cyberattacks. After computing the predicted guideline pricing curve a_p, one can perform smart home scheduling simulations to compute the bill B_p and the PAR P_p. Similarly, one can compute the bill B and the PAR P through using the received guideline pricing curve. Subsequently, the bill increase rate is computed as $\Delta_B = \dfrac{B - B_p}{B_p}$ and the PAR increase is computed as $\Delta_P = \dfrac{P - P_p}{P_p}$. If $\Delta_B > \delta_B$ or $\Delta_P > \delta_P$, an alert will be signaled.

5 LONG-TERM DETECTION TECHNIQUE
5.1 MOTIVATION

As mentioned in Section 1, the single event detection would have low detection rate when the thresholds in impact differences are above those used by hackers. A straightforward solution is to lower the thresholds. However, this would result in significant false alarm due to the fluctuation of the electricity pricing. To tackle this limitation, we propose to integrate the single event detection with POMDP [27] such that the cyberattacks can still be identified while the false alarm is mitigated. The rational behind this idea is that due to the inherent long-term view provided by POMDP, even if one uses low thresholds in single event detection, the false alarm can be ameliorated through various POMDP ingredients such as the probabilistic state transition and reward expectation. Thus, the long-term detection technique can identify most of the possible cyberattacks without increasing the cost due to false alarms.

POMDP is an advanced control theoretic technique, which takes as input the real world states (eg, those corresponding to different cyberattack impacts) and generate actions (eg, fixing the hacked smart meters) as output. Since in reality one cannot directly obtain the state, POMDP approximately computes the states on the basis of the observations which are possibly with some uncertainties. For completeness, an overview of the POMDP technique proposed in Ref. [27] is presented as follows.

A typical POMDP problem is described as a tuple $\langle S, O, A, T, \Omega, R \rangle$ [27,28]. The finite set S denotes the state space of the system. The finite set A denotes the action space containing all the available actions for the decision maker (eg, the local community in our problem). The transition from the state s to s' while taking action a is defined using the corresponding transition probability $P(s' | a, s)$, namely, $T(s', a, s) = P(s' | a, s)$. After taking action α at sate s, the decision maker receives a reward $R(s', a, s)$ if the next state is s'. The observation space of the system state is described as the finite set O. The mapping between states and observations is defined as $O(s', a, s) = P(o | a, s)$, which is the conditional probability that the observation is o while action and state are a and s, respectively. The decision maker estimates the system state on the basis of observation and takes action based on it. Thus, state, observation and action are the three key components of POMDP.

The state space is given by the set $S = \{s_0, s_1, s_2, \ldots, s_N\}$, where each state $s_i \in S$ denotes that i smart meters are hacked. Thus, there are totally $N+1$ states for a community with N customers. The observation space is given by the set $O = \{o_0, o_1, o_2, \ldots, o_N\}$, in which $o_i \in O$ denotes that i smart meters are observed to be hacked. During the long run, the smart meters are continuously monitored and the utility or local community has the updated the observation of the system during each time slot.

Since the exact current state is not known, the decision maker estimates the number of hacked smart meters from the observation and judges if the cyberattack can introduce significant expected system loss. If it is the case, the decision maker needs to check and fix the hacked smart meters. Otherwise, it chooses to ignore the cyberattack since checking and repairing the smart meters are associated with labor cost. Thus, the decision maker has two available actions.

- action a_0, monitor the system and ignore the cyberattacks;
- action a_1, apply single event detection technique on each single smart meter to detect the hacked smart meters and fix them.

Note that the POMDP-based long-term detection technique contains observation as a key component, which is obtained by the single event detection technique. Each customer uses SVR to predict the guideline price and conduct smart home scheduling simulation with it. If the PAR or electricity bill corresponding to the received guideline price is sufficiently higher than that corresponding to the predicted guideline price, an alert signal is generated. The decision maker counts the total number of alerts and sets the observation according to it. For example, if there are n alerts, the observation is obtained as on.

One might wonder why integrating single event detection to POMDP would improve the detection rate while mitigating false alarm. This is essentially due to the inherent long-term view provided by POMDP. Even if low thresholds in single event detection induce false alarms, they can be ameliorated through various POMDP ingredients such as the probabilistic state transition and reward expectation, while the detection rate is significantly improved. The estimation of the state is defined as belief state b. The properties described in Ref. [27] for belief state are summarized as follows.

- The belief state $b = \{b(s_0), b(s_1), \ldots, b(s_N)\}$ denotes the confidence level associated with each state. For example, the belief state for our problem can be $\{0.8, 0.2, 0, \ldots, 0\}$, which means that the current system state is s_0 with probability 0.8, s_1 with probability 0.2 and other states with probability 0 according to the estimation. Each time when the new observation is obtained, the belief state is also updated according to the Bayesian rule as follows [27].

$$b(s') = P(s' \mid o, a, s) = \frac{\Omega(o, a, s') \sum_{s \in S} T(s', a, s) b(s)}{P(o \mid a, b)}, \tag{10.10}$$

where $P(o \mid a, b)$ is a normalizing factor. As is seen from Eq. (10.10), the new belief state depends on the last belief state, current observation as well as the last action. It also depends on state transition function and observation function, which are the underlying regulations of the POMDP.

- The state transition function T(s′, a, s) represents the transition of real world states, which are not completely observable. Thus, the belief state transition is modeled by [27]

$$\tau(b',a,b) = P(b'\,|\,a,b) = \sum_{o \in O} P(b'\,|\,b,a,o) P(o\,|\,a,b) \qquad (10.11)$$

In Eq. (10.11), $P(b'|a,b)$ is the same normalizing factor as in Eq. (10.10). When b, a and o are all given, b' is deterministic and can be directly obtained from Eq. (10.10). Thus, $P(b'\,|\,b,a,o)$ is given as [27]

$$P(b'\,|\,b,a,o) = \begin{cases} 1, \text{ if } b,\, a,\, o \to b' \\ o, \text{ otherwise} \end{cases} \qquad (10.12)$$

- The reward function _ is defined on the basis of the belief state [27].

$$\rho(a,b) = \sum_{s \in S} \sum_{s' \in S} b(s) R(s',a,s) T(s',a,s) \qquad (10.13)$$

which is the reward when the decision maker takes action a at belief state b.
- The decision maker will optimize the long-term expected reward. However, the future event imposes more uncertainty than the current. Thus, a discount factor is introduced to reduce the importance of future event in the optimization target and the expected reward is modified as discounted expected reward defined as $E\left[\sum_{t=0}^{\infty} \gamma^t r_t\right]$. r_t is the reward in step t. The optimal value of the discounted expected reward given the current belief state b is denoted as $V^*(b)$ such that [27]

P3 $V^*(b) = \max_{a \in A}\left\{\rho(a,b) + \gamma \sum_{b' \in B} \tau(b',a,b) V^*(b')\right\}$. Each time the decision maker intends to find the optimal action. He or she needs to consider both the reward in the present step and the discounted expected reward in the long future. During the long run, the decision maker receives the update of the observation each time slot, from which he/she estimates the belief state of the system. Given the belief state of the current system state, the decision maker obtains the optimal action by solving the optimization problem P3. The procedure for solving a POMDP problem is presented in the example given later.

5.2 OUR POMDP-BASED DETECTION

In our problem context, there are three types of transition functions, namely, state transition function, belief state transition function and observation transition function. We already show the belief state transition function in Eq. (10.11). In the formulation of the POMDP, state transition probability indicates the mapping between the system states at two adjacent time slots. Each action leads to the update of the system state. Action a0 does not reduce the number of hacked smart meters under the influence of the decision maker. The hacker can continue the cyberattack if it is not eliminated.

For example, it can hack more smart meters to increase the impact to the power system. Thus, the state transition function for action a_0 is given as

$$T\left(s_i, a_0, s_j\right) = P_a\left(s_i\,|\,s_j\right), \qquad (10.14)$$

where $P_a\left(s_i \mid s_j\right)$ is the probability that the hacker will hack i smart meters at next time slot if j smart meters are hacked at this moment.

In contrast to other works which use artificial model for state transition, this work uses model training to compute $P_a\left(s_i \mid s_j\right)$ from the historical data. Although the real system state is not known in history, since action a_0 does not change the system state, we can develop the state transition from the observation. In order to compute the state transition probability from the historical data, define an observation transition function as $T'\left(o', a_0, o\right)$, which is the probability that the observation transits from o to o'. Thus, the state transition can be derived from

$$T\left(s', a_0, s\right) = \sum_{o \in O} \sum_{o' \in O} T'\left(o', a_0, o\right) P\left(s \mid a_0, o\right) P\left(s' \mid a_0, o'\right), \tag{10.15}$$

where $P\left(s \mid a_0, o\right)$ is the conditional probability that observation o is obtained after taking action a_0 while the real state is s. $P\left(s \mid a_0, o\right)$.

$$P\left(s \mid a_0, o\right) = \frac{P\left(s \mid a_0, o\right) P(s)}{\sum_{s' \in S} P\left(s' \mid a_0, o\right) P\left(s'\right)}. \tag{10.16}$$

In Eq. (10.16), $P\left(s \mid a_0, o\right)$ is the observation function defined in Eq. (10.18). Since the probability for each system state P(s) is not known, it is approximated by $P(o)$. $T'\left(o', a_0, o\right)$ is obtained from the historical data. For example, before action a_0 is taken, o_0 appears 10 times. After action a0, there are 5 times o_0 transiting to o_0, 3 times to o_1 and 2 times to o_2. Thus we have $T'\left(o_0, a_0, o_0\right) = 0.5$, $T'\left(o_1, a_0, o_0\right) = 0.3$ and $T'\left(o_2, a_0, o_0\right) = 0.2$. Derived from the historical data, the state transition function takes into consideration the long-term effect of cyberattacks. Taking action a_1, the utility or local community applies the single event detection technique to each individual smart meter to determine whether it is hacked and the hacked smart meters will be fixed. The transition function for a_1 is given as follows.

$$T\left(s_i, a_0, s_j\right) = \begin{cases} 1, & \text{if } s_i = s_0 \\ 0, & \text{otherwise} \end{cases}. \tag{10.17}$$

The observation after taking a_0 is given as

$$O\left(o_i, a_0, s_j\right) = P\left(o_i \mid a_0, s_j\right), \tag{10.18}$$

in which $P\left(o_i \mid a_0, s_j\right)$ is the probability that there are j smart meters hacked while i ones are observed to be hacked. If action a_1 is taken, all hacked smart meters are checked and fixed by the single event detection technique, which is also the belief of the decision maker. Therefore, we have

$$O\left(o_i, a_1, s_j\right) = \begin{cases} 1, & \text{if } o_i = o_0 \\ 0, & \text{otherwise} \end{cases}. \tag{10.19}$$

The cyberattacks induces loss to the power system. This is modeled by the reward functions. For the ease of optimization, the system loss is formulated as follows. Give two parameters S_1^* and S_2^*, where $S_1^* \leq i \leq S_2^*$. If the total number of hacked smart meters $S_1^* \leq i \leq S_2^*$, the system loss is C_1^L. If $i > S_2^*$, the system loss is $C_1^L + C_2^L$. The action a_1 needs human interaction and is associated with a labor cost. Without loss of generality, the labor cost can be separated into two parts. The on-site inspection cost C^I and smart meter recovery cost C^R. Once action a_1 is taken, the on-site inspection cost is C^I is no matter smart meter is fixed or not. Fixing each smart meter costs C^R. Thus, the reward for a_0 and a_1 are given as follows.

$$R\left(s_i,a_0,s_j\right)= \begin{cases} C_1^L, \text{ if } S_1^* \le i \le S_2^* \\ C_1^L + C_2^L, \text{ if } i > S_2^* \\ 0, \text{ otherwise} \end{cases} \tag{10.20}$$

$$R\left(s_i,a_0,s_j\right) = -C^I - \left(j-i\right)C^R. \tag{10.21}$$

At each time slot, the long-term detection technique consists of two phases. (1) Monitor the system state and obtain the observation. (2) Solve the POMDP problem to compute the optimal policy and take corresponding action. According to Problem **P3**, the optimal action is based on the belief state, which depends on the current observation, last belief state as well as the action taken last step. Summarizing all the possible combinations of observations and corresponding policies, a policy transfer graph can be achieved, which is the output of POMDP. It shows the optimal action the decision maker can take given the current situation. The complete procedure of the proposed long-term detection technique is summarized in Fig. 10.2.

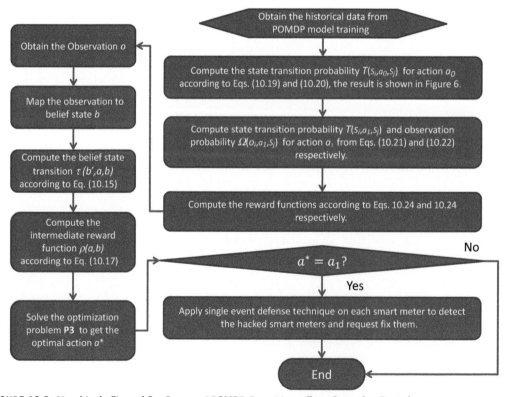

FIGURE 10.2 Algorithmic Flow of Our Proposed POMDP-Based Long-Term Detection Technique

6 CASE STUDY FOR LONG-TERM DETECTION TECHNIQUE

Refer to Fig. 10.3. Consider a mini-community with two customers. A basic idea to define the states is that

- s_0: no smart meter hacked
- s_1: smart meter 1 is hacked
- s_2: smart meter 2 is hacked
- s_3: both smart meters are hacked.

There are totally two actions which are a_0: ignore the cyberattacks and continue to monitor the smart meters and a_1: check and repair the hacked smart meters. Each action can lead to the changing of the current state and corresponding to the new state, a new observation is obtained by the decision maker. In order to solve the POMDP problem, we need to consider the future transition of the states. This can be obtained by

- learning from historical data,
- calibrating the mapping from observation to state,
- applying Bayesian rule, according to Eqs. (10.15) and (10.16).

Using this method, the state transfer graph depending on each action is obtained as shown in Fig. 10.4. For example, starting from state s_0, the transition probabilities are 0.5, 0.2, 0.2, and 0.1 to s_0, s_1, s_2, and s_3, respectively conditioned on a_0.

Since the detection method cannot be 100% accurate, the exact state is unknown to the decision maker. Thus, the decision maker depends on the estimation of the states to solve the POMDP problem. The estimation is defined as belief state b. For example, if there is no hacked smart meter, we have $(s_0)=1$ $b(s_1)=0$, and $b(s_2) = 0$. It is updated according to the Bayesian rule if a new observation is obtained according to Eq. (10.10). Note that checking and repairing the smart meters lead to labor

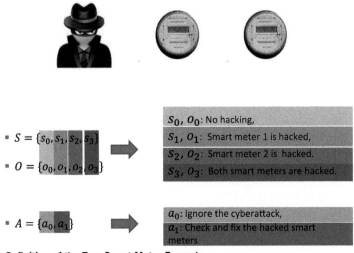

FIGURE 10.3 State Definition of the Two-Smart Meter Example

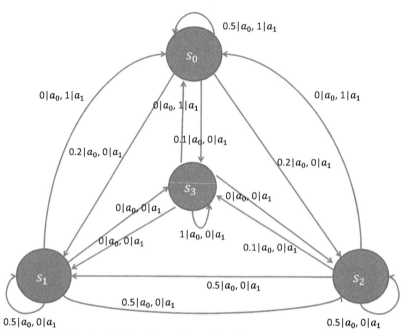

FIGURE 10.4 State Transfer Graph of the Two-Smart Meter Example

costs. Similarly, a hacked smart meter can cause economical losses to the system. Thus, each action also corresponds to a reward. In order to solve the POMDP problem, the decision maker needs to consider the future expected reward and choose the action with the maximum one. Since the estimation of future event is less accurate than the current one, the reward of a future action is multiplied by a discount factor. Suppose the current belief state is $b(s_0) = 0.8, b(s_1) = 0$, and $b(s_2) = 0$, the procedure to compute the optimal action is depicted in Fig. 10.5. The maximum discounted expected rewards of each combination of belief state and action are denoted by $R(s_0, a_0), R(s_0, a_1), R(s_1, a_0)$, and $R(s_1, a_1)$, respectively. The optimal action is a_0 if $0.8R(s_0, a_0) + 0.2R(s_1, a_0) > 0.8R(s_0, a_1) + 0.2R(s_1, a_1)$. Otherwise, the optimal action is a_1.

As is discussed previously, the decision maker aims to choose the action with the maximum discounted expected reward. Assume the discount factor is 0.9 and we look three steps forward. At the current time slot t, there are two available actions. For each action taken at time slot t, there are two future available actions at time slot $t+1$. Thus, there are four available actions at time slot $t+1$ in total. Similarly, there are 8 available actions at time slot $t+2$. Computing the optimal expected discounted reward is actually computing the optimal path from the root to a leave of the tree. It can be solved using dynamic programming, which is a standard algorithm to solve this problem. $R(s_0, a_0), R(s_0, a_1), R(s_1, a_0)$, and $R(s_1, a_1)$ are all calculated in this approach.

Using the standard solver of POMDP [29], a policy transfer graph can be computed as shown in Fig. 10.6. Initially, there is no hacked smart meter. Thus, we start from node e_0 and take the corresponding action a_0. If the obtained observation is o_0, we remain taking action a_0. If the obtained observation is o_1, o_2 or o_3, we transfer to node e_1 and take the corresponding action a_1. After taking a_1, o_0 is the only possible observation. Thus, the system returns to node e_0.

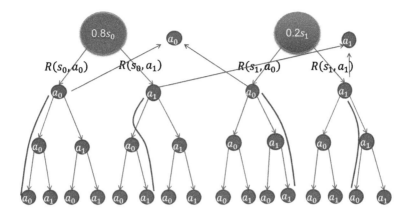

$$V^*(b) = \max\{0.8R(s_0, a_0) + 0.2R(s_1, a_0), 0.2R(s_1, a_1) + 0.8R(s_0, a_1)\}$$

$$a = \left(0.8R(s_0, a_0) + 0.2R(s_1, a_0) > 0.2R(s_1, a_1) + 0.8R(s_0, a_1)\right)? a_0 : a_1$$

FIGURE 10.5 Computing Optimal Policy for the Two-Smart Meter Examples

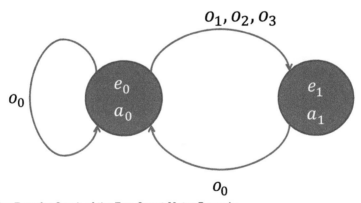

FIGURE 10.6 Policy Transfer Graph of the Two-Smart Meter Example

7 SIMULATION

The proposed algorithms are implemented using MATLAB and C programming language and simulations are conducted to analyze the impacts of smart home pricing cyberattacks and the performance of detection techniques. In the simulation setup, a community consisting of 500 customers is considered, and each customer is equipped with a smart meter connected with the AMI. For each customer, there are both manually controlled home appliances and automatically controlled home appliances. In our testcases, the average energy load due to manually controlled home appliances is shown in Fig. 10.7. The range for the daily energy consumption and regular execution duration of the automatically

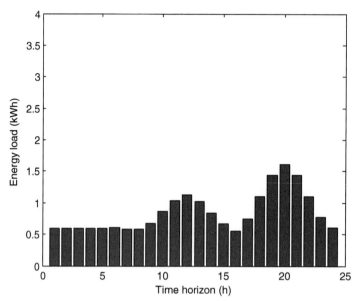

FIGURE 10.7 Average Energy Load by Manually Controlled Home Appliances (Background Energy Load)

controlled home appliances are shown in Table 10.2, which is similar to our previous works [2,20]. The simulation duration is set to the next 24 h divided into hourly time slots. The quadratic pricing model is used. Therefore, y-axes in Fig. 10.8A, Fig. 10.9A, Fig. 10.10A, and Fig. 10.11A, B show the quadratic coefficients in pricing and one needs to multiple them by the energy load in the corresponding time slots to obtain unit price.

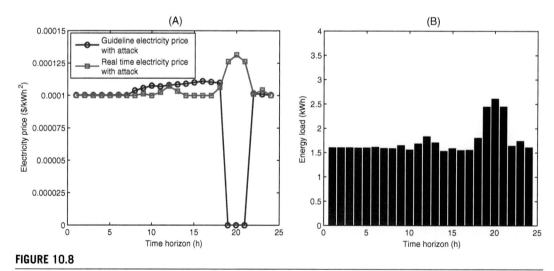

FIGURE 10.8

(A) Guideline electricity price and real-time electricity price without cyberattack. (B) Average energy load (PAR = 1.107) without cyberattack.

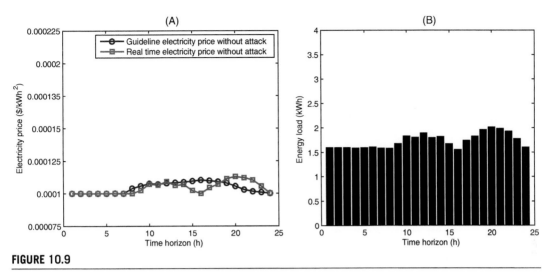

FIGURE 10.9

(A) Guideline electricity price and real-time electricity price with pricing cyberattack for bill reduction. (B) Average energy load (PAR = 1.358) with pricing cyberattack for bill reduction.

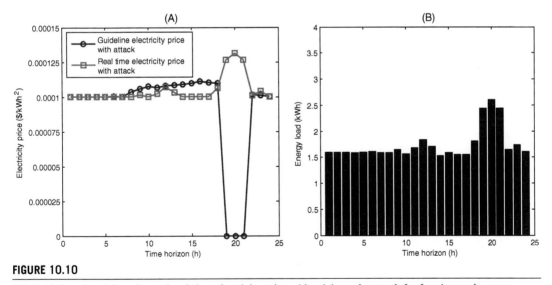

FIGURE 10.10

(A) Guideline electricity price and real-time electricity price with pricing cyberattack for forming peak energy.
(B) Average energy load (PAR = 1.502) with pricing cyberattack for forming peak energy.

7.1 CYBERATTACK FOR BILL REDUCTION

In this simulation, the smart home scheduling results with the unattacked guideline electricity price and the attacked guideline electricity price are compared. For the cyberattack, the hacker manipulates the guideline price and create a peak price from 1:00 am to 6:00 am and makes the rest flat. Refer to

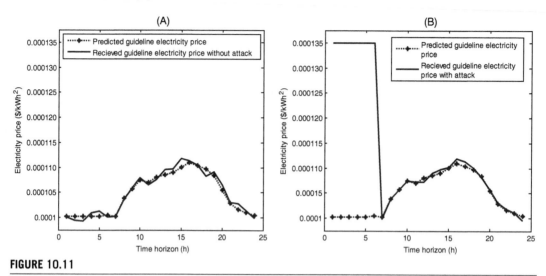

FIGURE 10.11

(A) Without cyberattack: for received guideline pricing, bill = $3.83, PAR = 1.170; and for predicted guideline pricing, bill = $3.82, PAR = 1.153, $\Delta_B = -0.26\%$ and $\Delta_P = -1.45\%$. (B) With cyberattack: for received guideline pricing, bill = $3.83, PAR = 1.170; and for predicted guideline pricing, bill = $4.09, PAR = 1.203, $\Delta_B = 6.79\%$, and $\Delta_P = 2.82\%$.

Fig. 10.8 for the results without attack and Fig. 10.9 for the results with attack, respectively. We make the following observations.

- From Fig. 10.8A, one sees that the guideline price well matches the real-time price, which is as expected.
- From Fig. 10.8B, one sees that the energy load is well distributed over the time horizon.
- In contrast, Fig. 10.9A shows that the guideline price and real-time price are significantly different. The reason is that when a smart home scheduler sees a high guideline price from 1:00 am to 6:00 am, it tends not to schedule the load there, which can be clearly seen from Fig. 10.9B. Note that there are still scheduled appliances during that time period due to (1) the background energy such as refrigerator and (2) the appliances which are required to be scheduled there due to starting time and ending time constraints. Due to the reduced energy usage from 1:00 am to 6:00 am, the real-time price at these time slots are lower than what it should be. Since the quadratic pricing model is used, the unit price is computed as the multiplication of quadratic coefficient (y axis) and the energy load. From 1:00 am to 6:00 am, the unit price is $0.0812 per kWh without cyberattack and is $0.0528 per kWh with cyberattack on average, which is a 34.3% reduction. Thus, if the hacker schedules his/her own load during this time period, a significant reduction in his/her own bill can be achieved.

This bill reduction of the hacker comes from the bill increases of other customers. Using the unattacked guideline electricity price, each customer pays $3.82 on average. However, using the attacked guideline price, the average bill increases to $4.12, which is 7.9% higher.

- As a byproduct, the cyberattack will also impact the energy load balancing. Comparing Figs. 10.8B and 10.9B, the PAR of the energy load from unattacked guideline electricity price is 1.107 while it becomes 1.358 with the attacked guideline electricity price.

7.2 CYBERATTACK FOR FORMING A PEAK ENERGY LOAD

In this simulation, the smart home scheduling results with the unattacked guideline electricity price and the attacked guideline electricity price are compared. For this cyberattack, the hacker manipulates the guideline price and creates a dip from 7:00 pm to 9:00 pm. Refer to Fig. 10.8 for the results without attack and Fig. 10.10 for the results with attack, respectively. We make the following observations.

- Fig. 10.10A shows that the guideline price and real-time price are significantly different. The reason is that when a smart home scheduler sees a low guideline price from 7:00 pm to 9:00 pm, it tends to schedule a large amount of load there, which can be clearly seen from Fig. 10.10B. The PAR of the energy load from unattacked guideline electricity price is 1.107 while it becomes 1.502 with the attacked guideline electricity price which is increased by 35.7%. This means that the cyberattacks can significantly unbalance the local energy load.
- Due to the peak energy usage from 7:00 pm to 9:00 pm, the real-time price at these time slots are higher than what it should be. From 7:00 pm to 9:00 pm, after converting quadratic pricing to unit price, one can obtain that the unit price without cyberattack is $0.160 per kWh and with cyberattack is $0.111 per kWh on average, which is a 43.9% increase. The average daily bill of each customer is $4.02, which is 5.24% more.

7.3 SINGLE EVENT DETECTION TECHNIQUE

In this simulation, the performance of our proposed single event detection technique is evaluated. Given a set of the historical guideline electricity prices of last 7 days, a predicted guideline electricity price is computed using the SVR on these data. We have tested on various pricing cyberattacks. To present our simulation results, we assume that the hacker chooses to use a bill reduction cyberattack through increasing the guideline electricity price during some time slots. Refer to Fig. 10.11 for the simulation results. The maximum tolerable impact differences are set as $\delta_B = 5\%$ for average bill increase and $\delta_p = 2\%$ for PAR increase, respectively. We make the following observations.

- When there is no pricing cyberattack, Fig. 10.11A shows the received guideline electricity price and the predicted guideline electricity price. Comparing with the predicted electricity price, the average bill pay increase is -0.26%, smaller than δ_B and the PAR increase is -1.45%, smaller than δ_p. Thus, the received guideline electricity is regarded as normal.
- When there is pricing cyberattack, Fig. 10.11B shows the received guideline electricity price and the predicted guideline electricity price. Comparing with the predicted electricity price, the average bill increase is 6.79%, larger than δ_B and the PAR increase is 2.82%, larger than δ_p. In this case, the electricity pricing manipulation is detected.

7.4 LONG-TERM DETECTION TECHNIQUE

In the simulation, we consider five different communities consisting of 100, 200, 300, 400, and 500 customers, respectively, to evaluate the performance of the long-term detection technique. We analyze the performance of the proposed long-term detection technique using the policy transfer graph obtained by solving the POMDP problem. In order to show the advantage of our proposed method, the results are compared with those obtained from three other methods. In the first method, we repeatedly use the single event detection technique and recover the smart meters reported to be hacked, which is defined as the heuristic method. In the second method, a Bollinger-bands-based detection technique is deployed, which is a standard statistical data analysis method that is widely used in financial research, stock trading and time series analysis [30]. It computes the moving average and standard deviation of the historical guideline price during the last seven days. Subsequently, it computes an upper band and a lower band, which are K times the deviation above and below the moving average, respectively. They are known as Bollinger bands. Cyberattack is reported if the received guideline price is outside the bands. In the last method, no defense technique is deployed. The cyberattacks are randomly generated and the simulations are conducted for the time horizon of 48 h. The three methods are evaluated using observation accuracy, PAR increase, bill increase and labor cost for detecting hacks and fixing hacked smart meters. At each time slot, the observation accuracy is defined using the difference between the number of actually hacked smart meters and the number of smart meters reported to be hacked by each method. Precisely, the observation accuracy is defined as $\theta = 1 - \dfrac{|i-j|}{j}$ if the observation is o_i and the system state is s_j (ie, j smart meters are actually hacked while i smart meters are reported to be hacked). The observation accuracies of the proposed method and the heuristic method for different testcases are shown in Figs. 10.12 and 10.13A, B, and Fig. 10.14, respectively. The results on bill increase, PAR increase and labor cost are shown in Table 10.3. For the 500-customer testcase, we make the following observations.

- Comparing with the results of no defense technique, the bill increase of and PAR increase are reduced by $1 - \dfrac{0.118}{1} = 88.2\%$ and $1 - \dfrac{3.42\%}{31.3\%} = 89.1\%$, respectively.

- The Bollinger-bands-based detection technique has an accuracy of only 36.52%. Comparing with the Bollinger-bands-based detection technique, our proposed long-term technique can reduce the PAR increase by $\dfrac{4.57\% - 3.42\%}{4.57\%} = 35.15\%$ and the normalized bill increase by $\dfrac{0.127 - 0.118}{0.127} = 7.09$, while still saving the normalized labor cost by $\dfrac{1.690 - 1.093}{1.093} = 54.62\%$.

  This is because the Bollinger-bands-based detection technique is a general statistical data analysis method without considering the specific problem nature of the smart home pricing cyberattacks.

- Comparing with the heuristic method, our POMDP method has much better detection accuracy. It has the accuracy of 97.73% while heuristic method has the accuracy of 51.15%. In addition, our POMDP method can reduce the PAR and bill increase by $\dfrac{1 - 3.42\%}{8.40\%} = 59.3\%$ and $\dfrac{1 - 0.118}{0.313} = 62.3\%$, respectively, at the expense of increasing the labor cost by only $\dfrac{1.0816 - 1}{1} = 8.16\%$ comparing with heuristic method since more smart meters need to be fixed.

Again, the large improvements in detection accuracy, PAR and bill outweigh the small expense increase for our POMDP method. It demonstrates that our proposed long-term detection technique can more effectively detect the cyberattacks considering the cumulative effect, which cannot be handled by the single event detection technique. The results of the other testcases are shown in Fig. 10.12A, Fig. 10.12B, Fig. 10.13A, and Fig. 10.13B, respectively, which are similar with those 500-customer testcase.

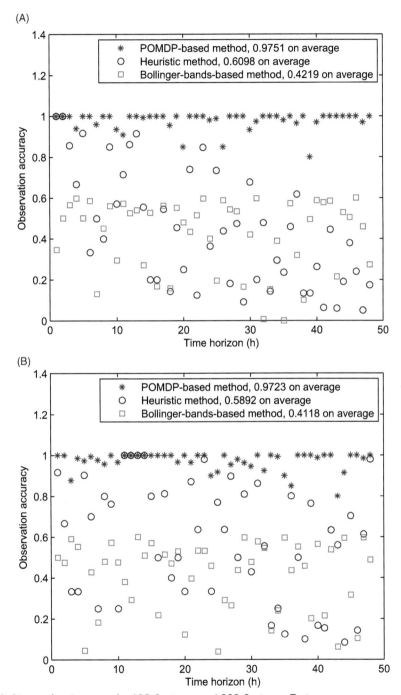

FIGURE 10.12 Observation Accuracy for 100-Customer and 200-Customer Testcases

(A) Observation accuracy for 100-customer testcase. (B) Observation accuracy for 200-customer testcase.

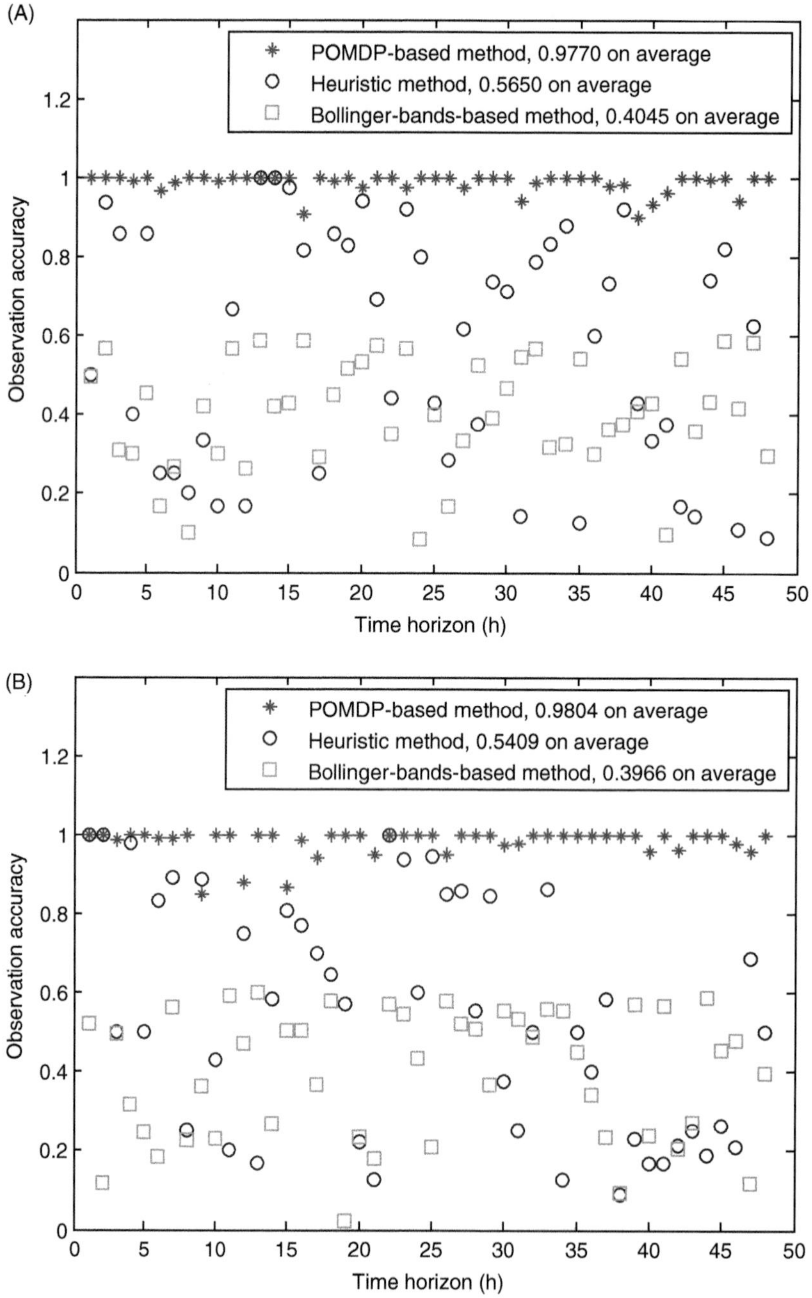

FIGURE 10.13 Observation Accuracy for 300-Customer and 400-Customer Testcases

(A) Observation accuracy for 300-customer testcase. (B) Observation accuracy for 400-customer testcase.

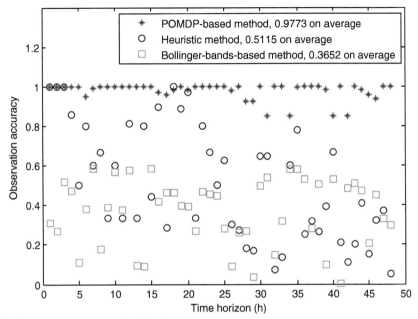

FIGURE 10.14 **Observation Accuracy for 500-Customer Testcase**

Testcase	Parameter	No Detection	Bollinger Bands Method	Heuristic Method	POMDP-Based Method

Table 10.3 Simulation Results of Detection Techniques

Testcase	Parameter	No Detection	Bollinger Bands Method	Heuristic Method	POMDP-Based Method
100-Customer	PAR increase	32.1%	4.03%	6.85%	2.87%
	Normalized bill increase	1	0.115	0.284	0.098
	Normalized labor cost	—	1.690	1	1.093
200-Customer	PAR increase	30.4%	4.29%	7.33%	3.26%
	Normalized bill increase	1	0.118	0.289	0.102
	Normalized labor cost	—	1.601	1	1.074
300-Customer	PAR increase	31.8%	4.47%	7.82%	3.24%
	Normalized bill increase	1	0.121	0.296	0.107
	Normalized labor cost	—	1.635	1	1.058
400-Customer	PAR increase	30.6%	4.52%	8.29%	3.31%
	Normalized bill increase	1	0.125	0.309	0.112
	Normalized labor cost	—	1.669	1	1.087
500-Customer	PAR increase	31.3%	4.57%	8.40%	3.42%
	Normalized bill increase	1	0.127	0.313	0.118
	Normalized labor cost	—	1.731	1	1.0816

8 CONCLUSIONS

In this chapter, the smart home technique is presented, which enables the customers to schedule the energy consumption, thus avoiding using electricity energy during the peak pricing hours. This results in the reduction of electricity bill from the customer's perspective and improvement of the energy-load balance from the utility's perspective. Despite its effectiveness, this exposes the smart home system to malicious attacks. The impact of pricing cyberattacks is studied in the smart home context.

Two attacking strategies are developed, which can increase the electricity bill of the customers and the peak energy usage of the energy load, respectively. The detection techniques have been developed using partially observable Markov decision process to mitigate the impact of the cyberattacks. The simulation results demonstrate that the proposed long-term detection technique has the detection accuracy of more than 97% with the significant reduction in PAR and bill compared to a natural heuristic algorithm which repeatedly uses the single event detection technique.

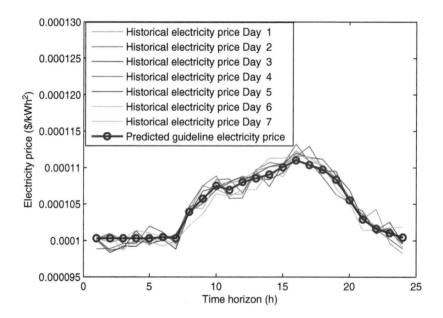

REFERENCES

[1] Liu Y, Hu S, Ho T-Y. Leveraging strategic defense techniques for smart home pricing cyberattacks. Accepted to IEEE transactions on dependable and secure computing.

[2] Liu Y, Hu S, Ho T-Y. Vulnerability assessment and defense technology for smart home cybersecurity considering pricing cyberattacks. In: Proceedings of IEEE/ACM international conference on computer-aided design (ICCAD); 2014. p. 183–90.

[3] Mohsenian-Rad A, Wong V, Jatskevich J, Schober R, Leon-Garcia A. Autonomous demand-side management based on game-theoretic energy consumption scheduling for the future smart grid. IEEE Trans Smart Grid. 2010;1(3):320–31.

[4] Khan R, Khan J. A heterogeneous wimax-wlan network for ami communications in the smart grid. In: Proceedings of IEEE international conference on smart grid communications; 2012. p. 710–15.

[5] Wang W, Zhuo L. Cyber security in the Smart Grid: Survey and challenges. Comput Netw 2013;57(5): 1344–71.

[6] Mohassel RR, Fung A, Mohammadi F, Raahemifar K. A survey on advanced metering infrastructure. Int J Electr Power Energy Syst 2014;63:473–84.

[7] Chen X, Wei T, Hu S. Uncertainty-aware household appliance scheduling considering dynamic electricity pricing in smart home. IEEE Trans Smart Grid 2013;4(2):932–41.

[8] <http://www.ti.com/solution/docs/appsolution.tsp?appId=407> [Online].

[9] <http://www.cnn.com/2013/08/02/tech/innovation/hackable-homes/> [Online].

[10] <http://www.nbcnews.com/tech/internet/smart-refrigerators-hacked-send-out-spam-report-n11946> [Online].

[11] <http://money.cnn.com/2014/09/24/technology/security/bash-bug/> [Online].

[12] Li H, Han Z. Manipulating the electricity power market via jamming the price signaling in smart grid. In: Proceedings of IEEE GLOBECOM workshops; 2011. p. 1168–72.

[13] Xie L, Mo Y, Sinopoli B. False data injection attacks in electricity markets. In: Proceedings of IEEE international conference on smart grid communications; 2010. p. 226–31.

[14] Liao C, Ten C-W, Hu S. Strategic FRTU deployment considering cybersecurity in secondary distribution network. IEEE Trans Smart Grid 2013;4(3):1264–74.

[15] Zhou Y, Chen X, Zomaya A, Hu S. A dynamic programming algorithm for leveraging probabilistic detection of energy theft in smart home," in manuscript.

[16] Kosut O, Jia L, Thomas R, Tong L. Malicious data attacks on smart grid state estimation: attack strategies and countermeasures. In: Proceedings of IEEE international conference on smart grid communications; 2010. p. 220–25.

[17] Lu Z, Wang W, Wang C. From jammer to gambler modeling and detection of jamming attacks against time-critical traffic. In: Proceedings of IEEE INFOCOM; 2011. p. 1871–79.

[18] Liu L, Esmalifalak M, Ding Q, Emesih V, Han Z. Detecting false data injection attacks on power grid by sparse optimization. IEEE Trans Smart Grid 2014;5(2):612–21.

[19] Manandhar K, Cao X, Hu F, Liu Y. Detection of faults and attacks including false data injection attack in smart grid using kalman filter. IEEE Trans Control Netw Syst 2014;1(4):370–79.

[20] Liu L, Zhou Y, Liu Y, Hu S. Dynamic programming based game theoretic algorithm for economical multi-user smart home scheduling. In: Proceedings of IEEE midwest symposium on circuits and systems (MWSCAS); 2014. p. 362–65.

[21] Liu L, Yang X, Huang H, Hu S. Smart home scheduling for cost reduction and its implementation on FPGA. J Circuit Syst Comput 2015;24(4).

[22] Liu L, Liu Y, Wang L, Zomaya A, Hu S. Economical and balanced energy usage in the smart home infrastructure: a tutorial and new results. IEEE Trans Emerg Topics Comput 2015;3(4):556–70.

[23] Liu Y, Hu S, Huang H, Ranjan R, Zomaya A, Wang L. Game theoretic market driven smart home scheduling considering energy balancing. Accepted to IEEE Systems Journal.

[24] Rosen J. Existence and uniqueness of equilibrium points for concave *n*-person games. Econometrica 1965;33(3):520–34.

[25] Jin Y. Personal communication.

[26] Chang C-C, Lin C-J. LIBSVM: A library for support vector machines. ACM Trans Intell Syst Technol 2011;2. 27:1–27:27.

[27] Kaelbling LP, Littman ML, Cassandra A. Planning and acting in partially observable stochastic domains. Artif Intell 1998;101(1):99–134.

[28] <http://cs.brown.edu/research/ai/pomdp/index.html> [Online].

[29] <http://www.pomdp.org/code/index.shtml> [Online].

[30] Lento C, Gradojevic N, Wright C. Investment information content in bollinger bands? Appl Financial Econ Lett 2007;3(4):263–7.

ADVANCED OPTICAL NETWORK ARCHITECTURE FOR THE NEXT GENERATION INTERNET ACCESS

11

C.A. Kyriakopoulos*, G.I. Papadimitriou*, P. Nicopolitidis*, E. Varvarigos[†]

*Department of Informatics, Aristotle University, Thessaloniki, Greece;
[†]Department of Computer Engineering and Informatics, Patras University, Patras, Greece

1 INTRODUCTION

1.1 CONVERGENCE TO THE DIGITAL COMMUNICATION AGE

Heading to the modern age, city authorities have to adjust the local structure of communications to achieve speed, quality, low energy consumption, and consistency. Traditional structures cannot satisfy the late hunger for instant communication and transfer of a huge amount of data, for example, video streaming.

Smart cities of the future [1] will be based upon digital technologies to reduce costs and resource consumption. The performance of data transfers between interconnected parts of the city can increase when it is based on a next generation optical network that offers high data rates and low error rate. The efficiency of communications can boost productivity and at the same time, communications will avoid the hurdle of long and different geographical distances that are impeding quality of service (QoS). It is important for a network architecture to be consistent and deployed throughout the city, so issues about interoperability and compatibility are avoided.

Maintaining and extending old communication structures cannot be a goal to confront the foreseeable future. The cost of maintenance increases as equipment gets old and not being produced abundantly. Also, experienced personnel lacks knowledge of legacy technologies that are being replaced constantly by manufacturing plants. Customer needs for higher throughput and more efficiency leaves no alternative choice than the adoption of a network architecture that is scalable, efficient, provides instant access and fulfills the promise of the 2020 European agenda.

A new network architecture [2] will be described in this chapter that is suitable for providing internet access and intercommunication to the end users, overcoming interoperability issues and offering high available bandwidth to fulfill connectivity needs for the years to come. Also, methods that aggregate and predict traffic that can coexist with the new architecture will be described in detail. These methods offer better resource utilization, leading to higher network throughput or lower energy consumption [3], according to current customer needs.

Smart Cities and Homes. http://dx.doi.org/10.1016/B978-0-12-803454-5.00011-0

1.2 INTRODUCING THE PANDA ARCHITECTURE

A passive optical network (PON) is a telecommunications network that uses point-to-multipoint logical connections in which passive optical components such as splitters are used to enable a single optical fiber to serve multiple end points. A PON consists of an optical line terminal (OLT) which is located at the service provider's central office (CO) and a number of optical network units (ONUs), shared by end users. A PON reduces the amount of fiber, CO equipment and cost in general which is required compared to point-to-point architectures. A PON is a type of a fiber-optic access network. In most cases, downstream signals are broadcast to all end users sharing multiple fibers. Encryption is used to prevent eavesdropping. Upstream signals are transferred using a multiple access protocol, based on time division multiple access (TDMA).

The main part of the proposed PANDA architecture [2] that will be described in this chapter consists of a PON. This type of network was chosen due to the high bandwidth it is able to offer by using wavelength division multiplexing (WDM) [4] in the optical domain and at the same time, low error rate and consistency. Up to 10 Gbps are available to downstream traffic (data that end-users receive depending on the configuration and originate from the backbone network topology) and 1 or 2 Gbps (also depending on the configuration) are available to clusters of end users (since fiber to the curb architecture—FTTC is present) for upstreaming data. Upon this network topology, techniques for aggregating and predicting traffic data [5] are applied to increase performance, availability, and all other important metrics.

The last mile from a home to the curb consists of conventional DSL connectivity, so there can exist compatibility with preinstalled network equipment. It is still far away from the era with full optical connectivity being available. At the point of aggregation of end user data that begin from a city block, resides an ONU. This is the component where signal conversion from electrical to optical takes place. When the signal enters the optical domain, it gets transferred to the OLT, that is, part of CO and then to the optical backbone network. In case the last-mile architecture was replaced by an all-optical network, the cost of deployment and maintenance would increase to a point where the performance benefits would be shadowed and this would not be a viable candidate solution.

The main idea of the PANDA architecture (Fig. 11.1) is based on a modern optical network for transferring data from city blocks to the backbone and vice versa. Due to preexisting (preinstalled) access equipment and the high cost of replacing it with similar optical equipment, the last mile that connects end-users to the optical network is based on conventional DSL copper equipment. Since the optical network does not impose any type of bottleneck to the end users, they are able to achieve high bandwidth utilization according to their configuration. Also, this architecture of providing access to the end users is compatible with the directions of the European 2020 Agenda.

1.3 INCREASING PERFORMANCE WITH TRAFFIC AGGREGATION AND PREDICTION

The PANDA architecture can undergo performance improvement when it is combined with advanced traffic aggregation and prediction strategies [6]. Modern network architectures must be able to create space for performance improvement when a higher layer of computation logic can take advantage and exploit the physical layer possibilities in an advanced way. Since the physical layer of PANDA is considered an advanced platform [2] and is able to serve user- needs for the years to come, it is candidate for being used by higher layer algorithms.

Traffic prediction is another important aspect that paves the way for further performance improvement, either at the backbone or access network. Backbone networks benefit from efficient routing,

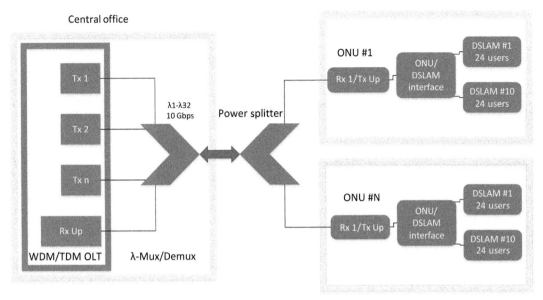

FIGURE 11.1 PANDA Architecture

better resource utilization or low energy consumption, among others. Access networks undergo improvements during dynamic bandwidth allocation (DBA) [7] from the CO structure to its connected ONUs. The importance is obvious when a part of the network that can accurately predict traffic for a large time span, to be able to efficiently preallocate resources aiming at decreasing performance metrics like the average data-serve delay or power consumption.

A node is able to accumulate traffic that originates from different sources and heads to different destinations. Traffic aggregation serves the purpose of grouping data that share a common trait such as a QoS parameter. Network performance improves, for example, better bandwidth utilization, when packets are transferred as groups. When it comes to access networks like PANDA, an ONU accumulates data that originate from different DSL users and forwards them to the backbone network through the CO. This procedure improves when advanced aggregation techniques apply.

Four categories of data-aggregating methods [8] will be analyzed later in this chapter. These methods relate to parameters such as the size of the burst or the time window of the aggregation. Depending on a specific aim of performance improvement, the algorithms consider an aforementioned parameter as the main focus and develop accordingly. Also, the basic models of traffic prediction that fit into the PANDA architecture will be introduced.

2 PANDA ARCHITECTURE
2.1 VDSL ENDUSER ACCESS

The main standard that connects end-users to their respective ONU is VDSL2, and from cabinet to the OLT there is a PON with 10 Gbps downstream and shared upstream of 1 or 2 Gbps (depending on the configuration). The point of transition between the two types of networks consists of 48-port VDSL2

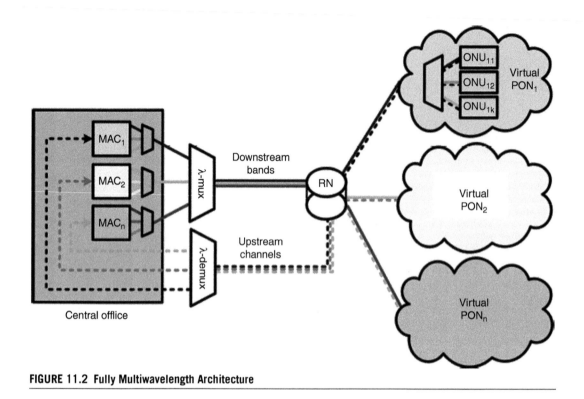

FIGURE 11.2 Fully Multiwavelength Architecture

DSLAMs and the target deployment is FTTC. There are also Ethernet aggregators of 1 or 2 Gbps that aggregate upstream traffic. Usually, these aggregators act as a broadband access switch inside the cabinet which multiplexes traffic from up to 16 DSLAMs through single or double GbE links. The aggregator consists of a 10 GbE multiplexed uplink to the optical MAC unit (ie, ONU) which interfaces the cabinet with the main PON [2].

2.2 MULTIWAVELENGTH PASSIVE OPTICAL NETWORK

The PON physical layer standard is the XG-PON [9] that demonstrates data rates up to 10 Gbps. The CO uses 16–32 wavelengths, each capable of carrying the aforementioned rates. The optical network (Fig. 11.2) terminates on the ONU, an optical subsystem that interconnects the PON with the Ethernet aggregator at the VDSL side of the network architecture. Wavelengths are passively routed over the optical infrastructure. Two deployment scenarios are utilized according to the distance between the CO and the ONUs: (1) under urban environment and when distances are limited, a power splitter is used at remote nodes (RNs). There is multicasting stemming from the OLT—with every ONU receiving all wavelengths- of data. When a frame is transferred to the destination, the respective ONU filters out all the wavelengths targeting different ONUs and keeps the data of its own. The side effect of this procedure is that the power is split according to the number of ONUs, creating extensive losses. So, limiting the distance is important for having small propagation losses. (2) When long distances are introduced, a demultiplexer is used instead of a splitter. That way, every ONU only receives its own wavelength and the reach is extended significantly. No additional filtering is required at every destination.

On the other hand, all ONUs share a common wavelength for upstreaming data. This is the case because if multiple wavelengths were allowed, tunable or multiwavelength sources would have been required at every ONU, increasing the network cost and complexity. Several upstream wavelengths render bandwidth allocation a challenging task. The CO dynamically allocates bandwidth (ie, DBA) to avoid collisions between ONU transmissions over the shared segment of the network. Typical DBA algorithms are designed to work efficiently when a single wavelength is shared among the ONUs [10].

2.3 THE MAC LAYER

There are many similarities of the MAC layer (Fig. 11.3) to XG-PON [9]. A single wavelength is utilized for both downstream and upstream traffic. In XG-PON, a time slot lasts 125 μs and includes a number of frames. The nature of the logical topology depends upon passive equipment, so every downstream frame is broadcast and chosen from the destination ONU according to its embedded XGEM identifier [9]. In other words, every ONU inspects all broadcast frames and keeps the one that targets itself as destination.

Among the similarities (frame duration, XGEM structure, etc) there are also subtle differentiations, since the downstream procedure is not truly broadcast. There are two implementation options in the physical layer (passive splitter or WDM) that actually offer a point-to-point downstream connection between the CO and ONUs and if the XGEM identifiers are absent, the CO includes an additional subsystem that assigns frames to ONUs on the basis of their wavelength, as shown in Fig. 11.3. From a practical perspective, the frame to wavelength mapping can be extracted from the (XGEM identifier, ONU, wavelength) triplet at the CO. Moreover, if the XGEM frame structure is maintained, the XG-PON ONUs are fully compliant with PANDA in the downstream direction.

The fact of using a single wavelength to upstream data, makes PANDA compatible with XG-PON in this data direction. So, the ONUs structure their upstream data in bursts and the CO is responsible for allowing to transmit them according to a schedule (also called Dynamic Bandwidth Assignment). During this procedure, the CO ensures that there is not overlapping of upstream frames at the single wavelength connection between itself and the main splitter (or WDM according to each implementation). There are two ways for the CO to be aware of ONU bandwidth requirements: (1) explicitly through received reports and (2) by inspecting the amount of empty upstream frames. According to these actions, it is able to apply a sophisticated algorithm for achieving better resource utilization of the network.

2.4 LINES RATES AND POPLATION COVERAGE

The actual line rates PANDA achieves during operation (Table 11.1), depend on the number of end-users and DSLAMs. The maximum number of DSLAMs per ONU is 48. End-users connect to their respective DSLAM with speed of 21, 42, or 83 Mbps. Traffic from each ONU to the OLT is carried with a predefined wavelength and its rate equals to 1 or 2 Gbps. On the other hand, downstream traffic is carried with a wavelength of 10 Gbps rate. The optical segment of the network doesn't impose any type of bottleneck to any DSL user, so the latter is able to exploit the full specifications of its service contract.

The upstream user rate depends on the number of ONUs and the number of users per ONU. The number of wavelengths equals to 16 or 32 and the number of ONU users reaches 120, 240, or 480,

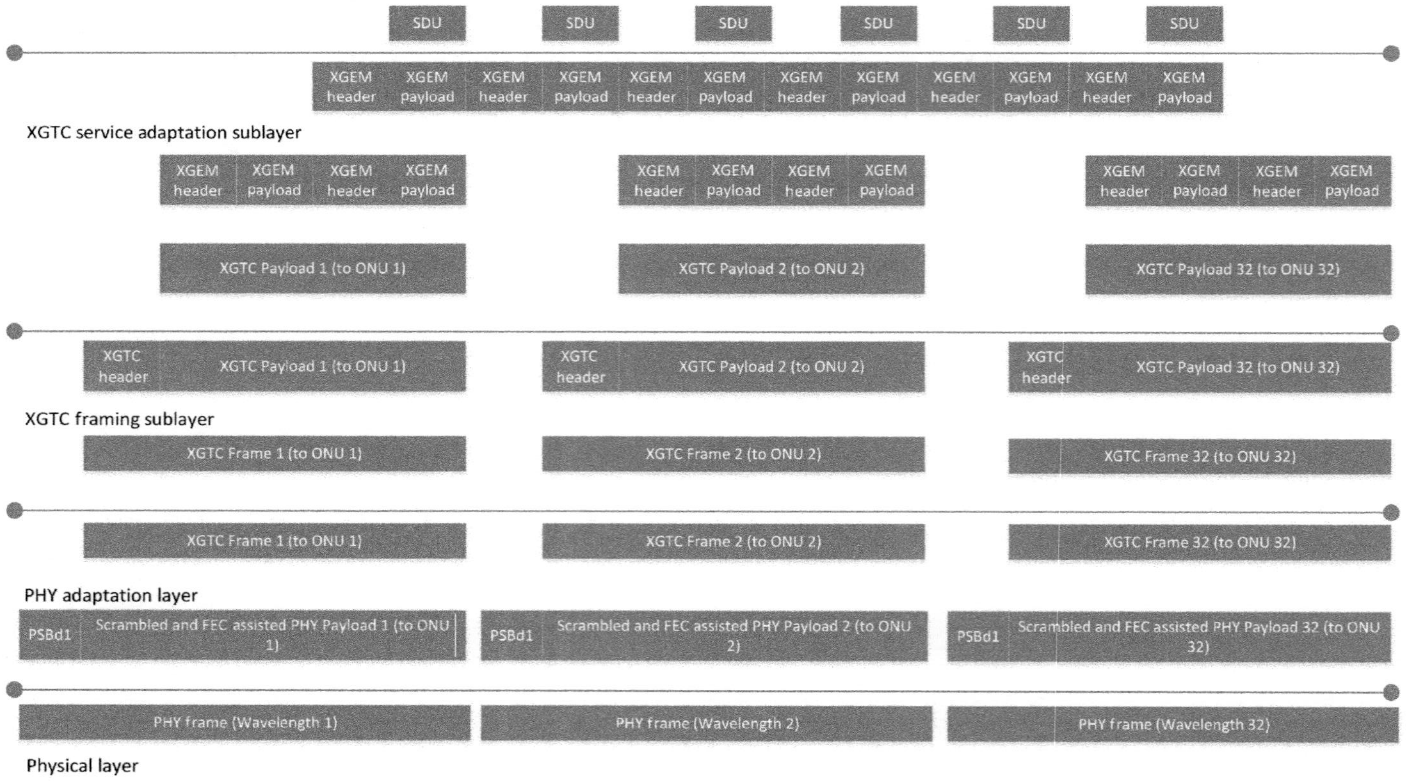

FIGURE 11.3 The MAC Layer

Table 11.1 Attainable Rates and Total Users

D/S per ONU (Gbps)	Wave-lengths	Users per DSLAM	U/S per ONU (Gbps)	DSLAM Uplink (Gbps)	DSLAMs per ONU	Users	D/S per User (Mbps)	U/S per User (Mbps)
10	16	24	10	1	10	3840	42	2.60
10	16	48	10	1	10	7680	21	1.30
10	16	24	10	2	5	1920	83	5.21
10	16	48	10	2	5	3840	42	2.60
10	32	24	10	1	10	7680	42	1.30
10	32	48	10	1	10	15360	21	0.65
10	32	24	10	2	5	3840	83	2.60
10	32	48	10	2	5	7680	42	1.30

depending on the DSLAM configuration. So according to these data, user speeds range from 0.65 to 5.15 Mbps which is typical for an asymmetric protocol that promotes downstream rates. In case only one user shows activity, the peak can reach 50 Mbps.

The total number of users that PANDA architecture can serve is determined by the number of ONUs and the number of users per ONU. The inference that can be drawn from the two tables is that the number of users, ranges from 1920 to 15360, all connected to a single CO.

2.5 EXTENSIONS

The actual numbers of Table 11.2 show that PANDA is capable of serving thousands of DSL users, each one of them communicating with speed range of 21–100 Mbps. This is a viable situation concerning short to midterm connectivity needs. Possible extensions of the proposed architecture are useful for rendering it viable for long term user coverage, extending its scale and scope.

Taking advantage of VDSL bonding-vectoring that can extend access to several hundreds of Mbps or point to point optical ethernet, PANDA is able to offer end-user access up to 1 Gbps. Since there is not any type of bottleneck and users can reach the nominal speeds, the limits of the proposed architecture are depicted in Table 11.2. These numbers show that 1600 PANDA users in a CO can benefit of 200 Mbps speed, but when the speed reaches 1 Gbps, their number declines to 320.

Table 11.2 shows the clear trade-off between the minimum guaranteed rates and the number of users in PANDA, considering the upper bandwidth wavelength limit being 10 Gbps. Under these circumstances, extending the number of users requires more wavelengths and more than one ONU per cabinet. This leads to a linear growth of the user-number, that is, 128 wavelengths can feed 1280 users under 1 Gbps speed. This situation breaks compatibility with the XG-PON standard since advanced DBA methods have to be utilized, that are specifically designed for this type of optical network.

Another option is to employ more than one COs, allowing many layers of PANDA to operate simultaneously. Included in pros is that this method attains the same number of end users, offers better upstream data rates and extensively reuses current optical fibers. On the other hand, extensive

Table 11.2 Line Rate and Population Coverage

Wavelengths	Users per ONU	Users	D/S per User (Mbps)	U/S per User (Mbps)
16	50	800	200	12.5
16	25	400	400	25.0
16	17	267	600	37.5
16	13	200	800	50.0
16	10	160	1000	62.5
32	50	1600	200	6.3
32	25	800	400	12.5
32	17	533	600	18.8
32	13	400	800	25.0
32	10	320	1000	31.3

wavelength planning is required and a combination of waveband demultiplexers, WDMs, and passive optical splitters.

2.6 SCALABILITY STUDY

There are three main scenarios that have been studied under the scalability perspective:

- Provision of at least 30 Mbps to 50% of the connected end-users and at least 10 Mbps to the rest. This case assumes the penetration of 52 Mbps VDSL access to half of active connections.
- Provision of at least 30 Mbps to all users. This case assumes that there is full VDSL penetration to access users.
- Provision of 100 Mbps to 50% of the users and 30 Mbps to the rest. This case assumes the introduction of VDSL profile 17a with cross-talk cancelations.

The first case is a short-term solution where the aggregated rate at the ONU site is less than 5 Gbps, that is, at least two ONUs can be grouped together under the same wavelength channel. A grouping of eight ONUs is possible using the same wavelength. There is decrease to four ONUs for the case of 30 Mbps connectivity and to two for 100 Mbps. These data are valid for urban access which is the most widespread access scenario.

A medium-term scalability scenario assumes penetration rate of 75% which is the current status quo in most European areas. The grouping factor of ONUs per wavelength is half in comparison to the short-term scenario, that is, 4, 2, or 1 ONU per wavelength is required respectively for the cases of 30 Mbps to 50% of users, 30 Mbps to all users and 100 Mbps to half of users. The last case is realized with one wavelength per ONU. In practice though, it is observed that the aggregated rate ranges from 6.8 to 7.6 Gbps which leaves spare bandwidth to the 10 Gbps upper limit per wavelength. This allows further extension of the user number.

The long term scalability scenario is the predicted by European 2020 agenda and assumes full penetration. The grouping factor remains the same as in the previous case, but single wavelengths are

Table 11.3 Geographical Coverage for Different Data Rates

Nominal Rate (Mbps)	Convergence Distance (m)	Coverage for Different Areas				
		Dense Urban (%)	Urban (%)	Suburban (%)	Rural (%)	Total Coverage (%)
100	150	98.4	86.0	19.1	37.7	80.20
>80	500	100	100	96.1	85.1	99.30
>50	900	100	100	100	87.9	99.88
>30	1200	100	100	100	88.9	99.89

utilized 100% (all 10 Gbps each one offers). The aggregated traffic of this scenario ranges from 8.9 to 10.1 Gbps, considering 100 Mbps access rate per user.

2.7 GEOGRAPHICAL STUDY

The distribution of users in four different areas is considered as basis for a geographical study, according to the characteristics of a copper-based distribution network. These areas are (1) dense urban, (2) urban, (3) suburban, and (4) rural. This study (Table 11.3) denotes that—using profile 17a—downstream and upstream rates of 100 and 50 Mbps respectively can be achieved.

A large set of representative areas was used to produce the average values. Real customer rates will be reduced due to crosstalk effects, impairments of line, quality and number of connections. All customers are able to achieve nominal rates in dense urban areas and 80 Mbps in urban areas, though all of them can peak at 100 Mbps. In suburban and rural areas, connectivity drops at 19.1% and 37.7% respectively. This is due to the distance from the RN most users undergo. DSL attenuation is tightly correlated to the distance increase. The main inference is that the majority of users can achieve nominal rates. There is also a small percentage of 11.1% in rural areas that is not capable to achieve rates higher than 30 Mbps.

The geographical coverage results can be found in Table 11.3. By observing the right column (percentages of coverage in respect to the total number of subscribers), almost all users can achieve nominal rates of 80 Mbps and 80% of users are served with the peak rate of 100 Mbps. So, the PANDA architecture in combination with the advanced VDSL2 (profile 17a) standard, is fully scalable and extends beyond the standards set by European Union for the Agenda 2020.

3 AGGREGATION AND PREDICTION OF TRAFFIC

3.1 INTRODUCTION

One of the main goals of Traffic Prediction as a concept is the ability to predict the short-term bandwidth requirements [11] while the network is operating. DBA benefits the most from prediction since the upstream channel is shared fairly (according to applied policies) among ONUs taking into account their current bandwidth requirements, avoiding that way empty time windows that increase the average delay time in their data queues. This is the case because the OLT that is responsible for the bandwidth allocation, is able to predict future requirements on behalf of ONUs and distribute it more efficiently and fairly. This type of prediction can be applied at the PON side, that is, after the backbone network topology and before the DSL access architecture. Clusters of users that belong to specific ONUs, benefit from the decrease of their latency metrics.

The prediction of future traffic fluctuations with increased accuracy [12], is an important task for future networks. Accuracy is related to a specified time window of prediction and usually, as the window size increases the accuracy decreases. So, it is important for a traffic prediction method [13] to be able to function properly with as much accuracy as possible, on a large window size. This large window size helps in taking better decisions at the operator's side. On the other hand, minimizing the error of prediction assists in taking the right decisions that boost network's performance. In practice, there is no linear correlation between accuracy and time window size—accuracy decreases disproportionally as prediction's window size increases.

Another aspect is the amount of network resources that are reserved for minimizing the uncertainty of prediction. The windows size is also correlated to the ability of network's administration and operation. Traffic characteristics that are captured from measurements are fed to the prediction model and drive its performance. The way these characteristics accumulate has also an explicit impact to accuracy [12].

Efficient traffic aggregation [14] is important at the CO premises. Aggregated data from ONU upstream traffic will eventually be routed to the backbone optical network, so prediction is important to ease the process, for a traffic aggregation scheme such as optical burst switching (OBS) [15]. This scheme, for example, is based on aggregation of data at backbone network nodes that will be send to a destination node, remaining entirely in the optical domain during the transfer. A control packet will be initially dispatched to the destination, just to create the array of lightpaths that will carry the burst along the path. So, the basic traffic aggregation schemes that can co-exist with PANDA will be presented as well.

The next section of this chapter will elaborate on the main principles of traffic aggregation and prediction schemes.

3.2 TRAFFIC ANALYSIS AND PREDICTION

The definition of traffic prediction [16] is based on a continuous time stochastic process $Y(t) = X(t) + \mu$, where μ is the mean rate and $X(t)$ is a random process having continuous integrated spectrum and zero mean. This is the input signal to the predictor and the output is $\hat{Y}(t+\tau)$ which uses τ as the prediction interval. The error cannot be higher than 20% and is provided by the normalized τ-step predictor.

$$\overline{err}(\tau) \equiv \frac{\hat{Y}(t+\tau) - \hat{Y}(t+\tau)}{\hat{Y}(t+\tau)} \tag{11.1}$$

The maximum prediction time window is provided by

$$MPI \equiv \max\left\{\tau \mid \overline{Perr}(\tau, \varepsilon) \leq P_e\right\} \tag{11.2}$$

Where ε is the percentage of the error, with probability P_ε and specified prediction confidence interval $(P_\varepsilon, \varepsilon)$.

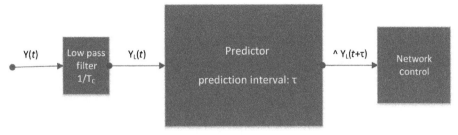

FIGURE 11.4 Traffic Prediction

A high MPI value with the error being inside the confidence interval, means that the prediction in this case is preferable due to the ability to ease the network operating procedure $\overline{Perr(\tau,\varepsilon)=P_r[err(\tau)>\varepsilon]}$ (Fig. 11.4).

Gaussian processes are usually utilized to approach network traffic characteristics, especially in backbone networks where the concentration of a high number of data sources explicitly affect their performance. The ARMA prediction model is utilized for detecting the statistical properties that define predictability and also to analyze traffic aggregation. According to the Wold decomposition theorem, a static Gaussian process $Y(t)=X(t)+\mu$ can uniquely be represented [16] by a moving average of static Gaussian white noise $n(t)$.

$$Y(t) = \int_0^\infty h_\mu \eta(t-\mu)du + \mu \qquad (11.3)$$

Where h_u is a real number function of u with a given real process $Y(t)$. According to Kolmogorov, the τ-step predictor is represented by a static Gaussian property.

$$\hat{X}(t+\tau) = \int_0^\infty h_{\mu+r} \eta(t-\mu)du \qquad (11.4)$$

The variance is expressed by the following formula.

$$\hat{\sigma}_\tau^2 = \int_0^\tau h_\mu^2 du = \sigma^2 - \int_\tau^\infty h_\mu^2 du \qquad (11.5)$$

The MPI, when $C = \sigma/\mu$, is provided by

$$MPI \equiv \max\left\{\tau \mid \frac{\hat{\sigma}_\tau^2}{\sigma_\tau^2} \leq O(P_e,\varepsilon,C)\right\} \qquad (11.6)$$

Where

$$O(P_e,\varepsilon,C) = \left[\frac{1}{C^2\varphi^2(1-P_e)} - 1\right]\frac{\varepsilon^2}{1-\varepsilon^2} \qquad (11.7)$$

The circular Markov-modulated poisson process (CMPP) [16] is used for confronting real network traffic traces. This model not only captures second grade statistical data as ARMA, but also first grade which also cannot be distributed under Gaussian. This model performs well in queue analysis.

The traffic rate in CMPP is $\vec{r} = [r_1, ..., r_i, ... r_n]$ and there is a circular transition two-dimension array Q. According to Markov properties, the most efficient predictor is

$$\hat{Y}(t+\tau) \equiv E[Y(t+\tau) | Y(t) = r_i]$$
$$= \sum_{K=1}^{N} r_K [e^{Q_\tau}]_{lk}$$
$$= \frac{1}{N} \sum_{k=1}^{N} r_k \sum_{l=1}^{N} e^{\lambda_{n\tau}} W^{l(i-k)} \tag{11.8}$$

Also, MPI equals to $MPI \equiv \max \left\{ \tau \mid P_{\overline{err}}(\tau, \varepsilon) < P_{\varepsilon} \right\}$, with

$$P_{\overline{err}}(\tau, \varepsilon) = \frac{1}{N} \sum_{i=1}^{N} P_{\overline{err_i}}(\tau, \varepsilon) = \frac{1}{N} \sum_{i=1}^{N} \sum_{(j)_i^*} [e^{Q_\tau}]_{ij} \tag{11.9}$$

There are also adaptive traffic prediction models that are based on training and are categorized in two sets. The first set is used to detect the parameters of the prediction model. The second set is dedicated to the evaluation of data and is used to compare the prediction data with actual ones, aiming at the extraction of conclusions about predictor's performance.

The prediction model works for every input at timestamp $t - 1$, calculating the corresponding $\hat{y}(t)$. The input is fed with previously recorded values like $y(t-i)$ for every variable that is calculated at $t-1$ timestamp.

During training, which is the phase of model parameters' detection, the set of data is entered as input into the algorithm. The set of training data consists of input and output values. The algorithm estimates the model parameters which provide the minimum error between outputs and real traffic values.

At the prediction phase, the input is only fed with estimation data to predict the value of $\hat{y}(t)$. The main criterion that is used for evaluating the validity of prediction values is the root mean square error (RMSE).

$$RMSE = \sqrt{\frac{\sum_{t=1}^{n} [y(t) - \hat{y}(t)]^2}{n}} \tag{11.10}$$

The real output is $y(t)$ and the estimated is $\hat{y}(t)$. The number of input data is n. Then, RMSE calculates the error between the real values that have been recorded and the estimated ones [5] (Fig. 11.5).

3.3 TRAFFIC AGGREGATION STRATEGIES

Data can be transferred as bursts inside a backbone optical network. That way, performance increases as was shown in related research [15]. A lot of algorithms have been proposed to aggregate data at the nodes of the network and most of them preserve local queues for the accumulation of data that originate from different sources, according to their destination or the requirements of QoS. The algorithms select a number of packets from their queues and send them as bursts to the destination node. Two parameters that are essential for the aggregation is size and time. According to these values, the collection completes and the transmission procedure initiates. These algorithms are classified into four categories [8] (Fig. 11.6).

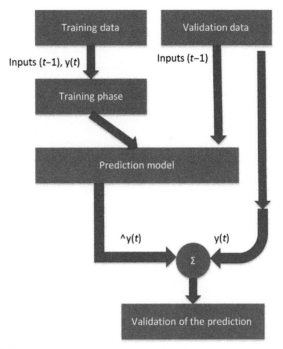

FIGURE 11.5 Training Model

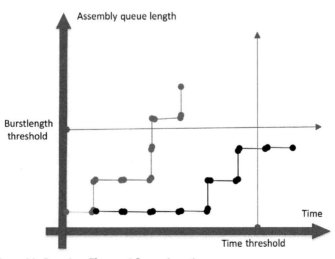

FIGURE 11.6 Burst Assembly Based on Time and Queue Length

The first category includes algorithms that are based on time with a threshold T being the completion criterion of the aggregating procedure. At the same time, there is a lowest limit of size B_{min} (padding can be used if necessary). The second category includes algorithms that are based on the size of the burst and when this reaches a predefined value, the transmission of the burst initiates. There are pros and cons in the first two categories. An algorithm that is based on the size, does not have time limit, so it is possible when the network load is low to delay the transmission, resulting in degradation of performance. If it is based on time and the network has high load, the average delay increases with negative impact to network's performance.

It is essential for an algorithm to be able to be executed with acceptable results under all conditions. This is the basic idea for a third category which combines the first two. It uses three thresholds: time, size and the minimum burst size. The fourth category is based on the dynamic adjustment of these thresholds, either when it correlated to time or size. These algorithms are adaptive to the current network traffic which is recorded from measurements during its operation. The side-effect of these methods that belong to the last category is the increase of complexity during the prediction and estimation of thresholds.

Given the fact that algorithms that are based on time and the size of the burst, aggregate packets they receive inside a predefined time window, the use of an electronic buffer is a prerequisite. This procedure changes the characteristics of traffic from packet to the burst level. For the multidimensional understanding of an OBS network, the aggregation procedure must be analyzed. In case this procedure follows a Poisson distribution which has exponentially distributed arrival times and the size of packets is constant, the distribution of burst size can be calculated as follows.

$$P_r\{Burstlength = l\} = P_r\{n_p P = l\} = \frac{\left(\dfrac{T}{\mu}\right)^{\frac{l}{p}}}{\left(\dfrac{l}{p}-1\right)!} e^{\frac{-T}{\mu}} \tag{11.11}$$

That way there is approach of a Gamma distribution and when T/μ is large, of a Gaussian with average $PE[n_p]$ and variation $P^2E[n_p]$. The number of included packets in a burst can be constant when the packet size P and the threshold B of burst's length are also constants, and equals to $n_p = E[n_p] = B/P$. Arrival times between packets are random variables, so the times between bursts are the summary of them, following a Gaussian distribution with average $\mu E[n_p]$ and variation $\mu^2 E[n_p]$. The characteristics of aggregated traffic in a hybrid algorithm (size and time) are correlated to a Gaussian distribution. Using algorithms with dynamic aggregation, the time between arrivals of bursts and also the size, depend on the arrival rate.

Multiple rate aggregation [17] is defined as the data aggregation of multiple optical sources which provide data with different rate and different destinations. In a backbone optical network there are two types of nodes that aggregate data, those that aggregate end-user data from the access network and the rest that aggregate IP flows in a light tree. The main goal is the creation of a light tree set with the maximum aggregation. When an access network receives content from the light trees, it delivers it to the end-users. There is the possibility of multiple-transponder utilization in a backbone router that aggregates data. The transponders at the heads and queues, correspond to roots and leafs of light trees. The transfer rate in a transponder is usually higher that the required of a multiple IP flow, just to avoid any kind of bottleneck.

The requirements in a type of network such as this are: (1) the use of the shortest path for the formation of the light tree, (2) the relation of an IP flow with multiple receivers and the light trees is one-to-many, (3) a multiple IP flow cannot be split into many light trees, (4) transponders in a light tree must be of the same type and (5) intermediate OXC elements lack grooming capabilities.

Finally, the field of energy efficiency [18] in optical networks can benefit from advanced aggregation techniques and statistics that relate to aggregated data. Applying energy-efficient algorithms, nodes are put into sleep mode [19] that have no high impact on traffic aggregation, so energy consumption decreases. Given the fact these idle devices can have impact later during network's operation, it is important for traffic prediction algorithms to be available for execution.

4 DBA SIMULATION AND RESULTS
4.1 PANDA DBA STRUCTURE

DBA is the process of allocating upstream transmission opportunities by the OLT to its connected ONUs. This process is assisted by activity status reporting from the ONUs, or implicitly by the observation of idle XGEM frames at the OLT side. DBA is performed effectively when there is adaptive reaction to upstream traffic, or through short and long term traffic prediction. The benefits that stem from the elaborate and effective bandwidth allocation, allow more subscribers to access the network, consistency in their provided services and at the same time, the transfer of high-quality media data that are encoded with variable bit rate methods.

In conventional TDM-style EPONs, the scheduling procedure [20] relates to the upstream transmissions on the single wavelength channel. In WDM EPONs the DBA procedure is extended to scheduling the upstream transmissions of multiple upstream wavelengths, when supported by the ONUs. In other words, in WDM EPONs the scheduling is related to both the length of the transmission and the wavelength channel it will be fulfilled at. Next, a comparison of two broad paradigms for dynamically allocating grants for upstream transmissions on the different upstream wavelengths will take place, that is, online and offline DBA scheduling.

In an online scheduler, the OLT performs the scheduling of every ONU at the time it receives its corresponding report message. This action indicates that it does not need to carry global knowledge of bandwidth requirements of the other ONUs. A basic online scheduling policy for the WDM PON is the OLT to schedule the upstream transmission for an ONU on the first available wavelength channel, the latter supports. The amount of the bandwidth (ie, the byte length of the granted time window) that is allocated to an ONU can be determined according to any of the existing DBA mechanisms for single-channel PONs.

In an offline scheduler, the ONUs are scheduled for transmission once the OLT has received the report messages from all ONUs, so it can have a global view of all bandwidth requirements the ONUs have and perform their scheduling accordingly. Since there is a single timestamp an offline scheduler makes scheduling decisions for all ONUs at, all of the reports, which are appended to the end of the data payload of a grant, from the previous DBA cycle should be received. This requires that the scheduling algorithm be executed after the OLT receives the end of the last ONU's gated transmission window. After this procedure, a gap between scheduling grant cycles is introduced, which is dubbed as the interscheduling cycle gap (ISCG). The length of the ISCG on a wavelength channel is equal to the computation time of the grant cycle, the transmission time for the grant and the round trip time (RTT) to the first ONU scheduled on the wavelength in the next round.

Dynamic bandwidth assignment in PANDA is the process by which the OLT allocates upstream transmission opportunities to the traffic-bearing entities within ONUs, based on dynamic indication of their activity and their configured traffic contracts. An ONU can include many different traffic entities that are identified uniquely and fulfill different subscriber services. The activity status indication can be either explicit through buffer status reporting, or implicit through transmission of idle XGEM[a] frames during the upstream transmission opportunities. In comparison to static bandwidth assignment, the DBA mechanism in PANDA improves the typical XG-PON upstream bandwidth utilization by reacting adaptively to the ONUs' burst traffic patterns. There are also practical benefits of DBA. Firstly, more subscribers can be added to the access network due to the efficient bandwidth use. Secondly, subscribers can benefit from enhanced services such as those requiring variable rate with peaks and bursty traffic that extends beyond the levels that can reasonably be allocated statically.

In XG-PON (similarly to PANDA), the recipients of bandwidth units are uniquely identified by their Allocation IDs (Alloc-IDs). The semantic behind this scheme is that an ONU can carry more than one Alloc-IDs devoted to different paid services and the OLT treats them according to its enforced policy. Regardless of the number of Alloc-IDs assigned to each ONU, the number of XGEM ports multiplexed onto each Alloc-ID, and the actual physical and logical queuing structure implemented by the ONU, the OLT models the traffic aggregate associated with each Alloc-ID as a single logical buffer and, for the purpose of bandwidth assignment, considers all Alloc-IDs specified for the given PON to be independent peer entities on the same level of logical hierarchy. Every downstream frame contains a BWMap, that is, a data structure which specifies the timing and size of upstream transmissions by the ONUs. The DBA function then provides input to the OLT upstream scheduler, which is responsible for generating the bandwidth maps. Every ONU extracts these data that correspond to its own Alloc-IDs and schedules for transmission data that reside in its corresponding queues.

Dynamic bandwidth assignment in XG-PON encompasses the following functional requirements. These functions apply on the level of individual Alloc-IDs and their provisioned bandwidth component parameters. The XG-PON OLT is required to support DBA. (1) Inference of the logical upstream transmit buffer occupancy status. (2) Update of the instantaneously assigned bandwidth according to the inferred buffer occupancy status within the provisioned bandwidth component parameters. (3) Issue of allocations according to the updated instantaneous bandwidth. (4) Management of the DBA operations.

Depending on the ONU buffer occupancy inference mechanism, two DBA methods can be distinguished: status reporting (SR) DBA is based on explicit buffer occupancy reports that are solicited by the OLT and submitted by the ONUs in response; traffic monitoring (TM) DBA is based on the OLT's observation of the idle XGEM frame pattern and its comparison with the corresponding bandwidth maps.

The PANDA OLT should support a combination of both TM and SR DBA methods and be capable of performing the aforementioned DBA functions in an efficient and fair manner. The specific efficiency and fairness criteria can be based on overall PON utilization, the individual ONU's performance, tested against the corresponding objectives, and comparative performance tested for multiple ONUs.

A PANDA ONU should support DBA status reporting, and should transmit upstream DBA reports as instructed by the OLT. The status reporting DBA method involves in-band signaling between the OLT and the ONUs, which is an inherent part of the XGTC[b] specification. The algorithmic details of how the OLT applies the reported or inferred status information, the entire specification of the traffic

---

[a]Sublayer data partitioning unit.
[b]Part of the XG-PON standard.

monitoring DBA method, as well as the details of the OLT upstream scheduler, which is responsible for the BWMap generation, are outside the XGTC layer scope, and their implementation is left to the OLT vendor.

When QoS is taken into account from the OLT, there is predefined notation from the standard that assists in the DBA process. Every Alloc-ID is connected to a T-CONT (Traffic Container) [21] identifier which defines the type of bandwidth. Type 1 is related to fixed bandwidth and mainly used for services sensitive to delay and with high priority like VoIP. Both T-CONT 2 and 3 are related to the guaranteed bandwidth and mainly used for video and data services of higher priorities. Type 3 also provides surplus bandwidth when it is available. Type 4 is of best-effort type and mainly used for data services such as Internet and services of low priority which do not require high bandwidth.

DBA algorithms that are adequate for use in the 10 Gbit PON standard [9] can also be utilized by the PANDA architecture.

4.2 PROPOSED FIBO DBA ALGORITHM

The proposed DBA algorithm relies on status reporting by ONUs that is appended at the back of each transmission grant. This info includes the current buffer (queue) size in bytes. The size of this queue depends on arrival rate of new data packets that will be transmitted later as upstream traffic. This info is valuable at the OLT site for improving bandwidth efficiency of the access network. Acquiring this knowledge, the OLT can apply an adaptive and efficient algorithm like the proposed one.

The Offline scheduling by the OLT entity is exploited by the proposed algorithm and it is a prerequisite. Avoiding starvation at the ONU side is achieved by giving the chance for transmission in every DBA cycle, even when there was zero buffer reported earlier. In this case, the lowest transmission size which is 8 bytes is granted, so if there are newly arrived data at an ONU, between the last report and the current time, there will be a time window to ease their transmission. Also, the largest grant scheduling policy is applied. This means that when the grant sizes are decided based on policy for the next cycle, a descending sorting will take place.

Since the traffic characteristics vary through time, there is need for DBA methods that can adapt to them. That way, performance improvement is feasible. When traffic data don't undergo extreme fluctuations as in constant bit rate (CBR), the DBA methods don't need to provide flexibility because they demonstrate acceptable performance. On the other hand, when the traffic is bursty, it is important for the method to adapt as fast as possible to fulfill the bandwidth requirements.

The proposed algorithm in Fig. 11.7 exploits this idea by using a sequence of Fibonacci numbers (every number in this sequence equals the sum of its two predecessors) to provide the required bandwidth. When the new bandwidth requirement is higher than the last served one for a specific Alloc-ID, a pointer increases its position and points to the next number of the sequence. When it is lower, the opposite happens. The number the pointer points to, is used as a multiplier for the default grant size. That way, when the traffic rate decreases, the pointer points to the first few Fibonacci numbers. When it increases, it points to the last numbers, providing higher grant sizes. The distances between the last numbers of the sequence is higher than the distance between the first numbers. So, when a burst suddenly arrives and disappears, there is fast adaptation to the high bandwidth demand and then fast adaptation to the low bandwidth demand that follows next.

The fast adaptation is feasible due to the differences the Fibonacci numbers have. For example, in the sequence 1, 1, 2, 3, 5, 8, and 13, the transition from the first numbers to the last is not linear and the

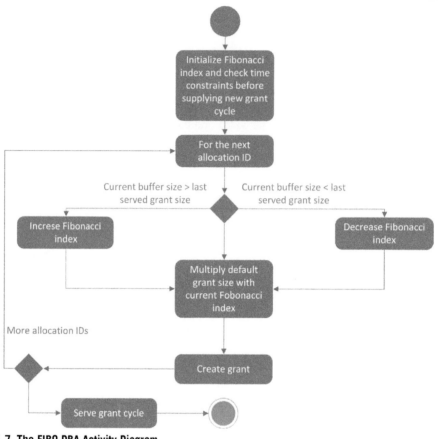

FIGURE 11.7 The FIBO DBA Activity Diagram

same applies to the opposite direction. When there is a sudden increase in traffic, the provided grant sizes increase rapidly. When there is a sudden decrease, the grant sizes fall accordingly.

The network environment that is used for the simulating procedure consists of an OLT that is connected to four ONUs through a splitter. The distances between OLT-splitter and splitter-ONUs are 40 and 8km respectively. The traffic traces were generated from [22]. An XG-PON simulator that is compatible with the PANDA architecture was developed in C++ 14 under the OMNeT++ simulating environment to demonstrate the performance of the FIBO DBA algorithm.

The Uncontrolled Excess DBA method [23] counts the underloaded and overloaded Alloc-IDs. The summary of the spare bandwidth of the underloaded IDs is then divided and provided to the overloaded ones, equally. The CBR method uses constant bitrate to provide bandwidth iteratively to the Alloc-IDs and exhibits low complexity with straightforward implementation. The next two figures depict the effect different default grant sizes have on the average delay time of the ONU data in their queues.

In Figs. 11.8 and 11.9, the average delay time is depicted according to the average grant sizes that are used. The adaptation of FIBO DBA is prominent when the grant sizes are low, which eventually

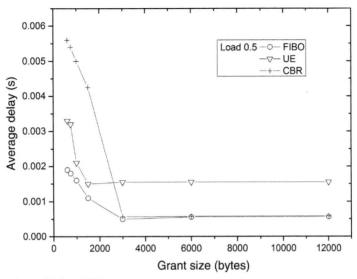

FIGURE 11.8 Comparison with Load 0.5

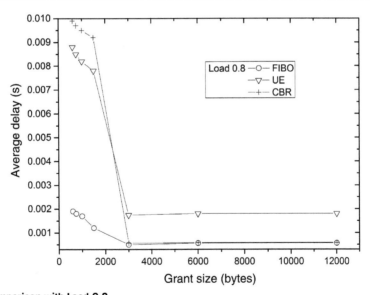

FIGURE 11.9 Comparison with Load 0.8

leads to higher performance. This happens because local queues get overloaded and the reports from the ONUs denote that high bandwidth is required. The multipliers increase with the positive side effect of providing high-bandwidth opportunities that decrease the average delay time. When grant sizes get higher (horizontal axis), the provided bandwidth windows to the ONUs are adequate to keep their buffers low. So in this range, FIBO DBA works like the CBR method since the Fibonacci pointer points mainly to the lower numbers of the array.

5 CONCLUSIONS

Concluding this chapter, a new and innovative optical network architecture that provides access to the end users was presented. Its scalability, geographical coverage, line rates, and the hybrid form provided by DSL were analyzed for the purpose of confronting the European 2020 agenda. At the same time, traffic aggregation and prediction techniques can coexist with this architecture for improving the overall performance. So, the basic ideas in this field that PANDA can exploit were presented. Finally, the DBA framework was introduced and a new adaptive DBA method was created and simulated. This method leads to decrease in the average delay time inside the queues of ONUs by taking advantage of the main property of a sequence of Fibonacci numbers.

ACKNOWLEDGMENT

This work has been funded by the NSRF (2007-2013) Synergasia-II/EPAN-II Program "Asymmetric Passive Optical Network for xDSL and FTTH Access," General Secretariat for Research and Technology, Ministry of Education, Religious Affairs, Culture and Sports (contract no. 09SYN-71-839).

REFERENCES

[1] Caragliu A, Del Bo C, Nijkamp P. Smart cities in Europe. J Urban Technol 2011;18(2):65–82.
[2] Yiannopoulos K, Varvarigos E, Klonidis D, Tomkos I, Spyropoulou M, Lazarou I, Bakopoulos P, Avramopoulos H, Heliotis G, Dimos LP, Agapiou G, Papastergiou G, Koukouvinos I, Orfanoudakis A, Oikonomou T, Kritharidis D, Spyridakis S, Dalakidis M, Synnefakis G, Reisis D, Papadimitriou GI, Sarigiannidis P, Liaskos C. PANDA: asymmetric passive optical network for xDSL and FTTH access. In: Proceedings of the 17th Panhellenic Conference on Informatics, ACM; 2013. p. 335–342.
[3] Zhang Y, Chowdhury P, Tornatore M, Mukherjee B. Energy efficiency in telecom optical networks. IEEE Commun Surv Tut 2010;12(4):441–58.
[4] Ismail MM, Othman MA, Zakaria Z, Misran MH, Said MM, Sulaiman HA, Zainudin MS, Mutalib MA. EDFA-WDM optical network design system. In: Procedia Eng 2013;53:294–302.
[5] Zhani MF, Elbiaze H, Kamoun F. Analysis and prediction of real network traffic. J Netw 2009;4(9):855–65.
[6] Sang A, Li SQ. A predictability analysis of network traffic. Comput Netw 2002;39(4):329–45.
[7] Johnson EL, Sivalingam KM, Mishra M. Scheduling in optical WDM networks using hidden Markov chain based traffic prediction. Photonic Netw Commun 2001;3(3):269–83.
[8] Yu X, Li J, Cao X, Chen Y, Qiao C. Traffic statistics and performance evaluation in optical burst switched networks. J Lightwave Technol 2004;22(12):2722–38.
[9] ITU-T, G.987.3, 10-Gigabit-capable passive optical networks (XG-PON): transmission convergence (TC) layer specification; 2010.
[10] Mirahmadi M, Shami A. Traffic-prediction-assisted dynamic bandwidth assignment for hybrid optical wireless networks. Comput Netw 2012;56(1):244–59.
[11] Luo Y, Ansari N. Limited sharing with traffic prediction for dynamic bandwidth allocation and QoS provisioning over Ethernet passive optical networks. J Opt Netw 2005;4(9):561–72.
[12] Morato D, Aracil J, Diez LA, Izal M, Magana E. On linear prediction of Internet traffic for packet and burst switching networks. In: Proceedings of tenth international conference on computer communications and networks, IEEE; 2001. p. 138–143.

[13] Wakabayashi N, Hirota Y, Tode H, Murakami K. Traffic prediction based wavelength resource management considering holding time. In: IEEE 2012 16th international conference on optical network design and modeling (ONDM); 2012. p. 1–6.

[14] Asensio A, Klinkowski M, Ruiz M, Lopez V, Castro A, Velasco L, Comellas J. Impact of aggregation level on the performance of dynamic lightpath adaptation under time-varying traffic. In: IEEE 2013 17th international conference on optical network design and modeling (ONDM); 2013. p. 184–189.

[15] Zhang J, Wang M, Li S, Wong EW, Zukerman M. Efficiency of OBS networks. In: IEEE 2012 14th international conference on transparent optical networks (ICTON); 2012. p. 1–4.

[16] Sang A, Li SQ. A predictability analysis of network traffic. Comput Netw 2002;39(4):329–45.

[17] Zhu Y, Jin Y, Sun W, Guo W, Hu W, Zhong WD, Wu MY. Multicast flow aggregation in IP over optical networks. IEEE J Selected Areas Commun 2007;25(5):1011–21.

[18] Bolla R, Bruschi R, Davoli F, Cucchietti F. Energy efficiency in the future internet: a survey of existing approaches and trends in energy-aware fixed network infrastructures. IEEE Commun Surv Tut 2011;13(2): 223–44.

[19] Shi L, Mukherjee B, Lee SS. Energy-efficient PON with sleep-mode ONU: progress, challenges, and solutions. IEEE Netw 2012;26(2):36–41.

[20] McGarry MP, Reisslein M, Maier M. WDM Ethernet passive optical networks. IEEE Commun Mag 2006;44(2):15–22.

[21] Han MS. Iterative dynamic bandwidth allocation for XGPON. In: 2012 14th international conference on advanced communication technology (ICACT); 2012.

[22] Kramer G. Ethernet passive optical networks. New York: McGraw-Hill; 2005.

[23] McGarry MP, Reisslein M, Maier M. Ethernet passive optical network architectures and dynamic bandwidth allocation algorithms. IEEE Commun Surv Tut 2008;10(3):46–60.

CLOUD COMPUTING SYSTEMS FOR SMART CITIES AND HOMES

12

T. Guelzim*, M.S. Obaidat†

*Department of Computer Science and Software Engineering, Monmouth University, West Long Branch, NJ, United States; †Department of Computer and Information, Fordham University, Bronx, NY, United States

1 INTRODUCTION

In today's information systems development model, cloud computing has become a de facto platform to enable content delivery to consumers. Cloud-based services have become more pervasive than ever: YouTube, Netflix, DropBox, Facebook, Amazon, and SoundCloud are being created and launched at a rapid pace. Evolving these platforms have allowed to reach mass user base all across the globe; creating communities and sharing key information-consuming patterns, delivering seamless user experience, and hiding all of the complexities behind such systems. Today, these systems have democratized access to information and made it available instantaneously. Everyday gadgets such as set-up boxes, wristband watches, athletic gear, and soon glasses to name a few have all been connected and data can be exchanged with a simple touch or a wink of the eye. This marks an initial experimenting era.

The term cloud computing has been used throughout the industry for more than a decade. In the early days, this term has been correlated with applications such as computing grid service, files storage, and early advanced email application. Later, cloud computing has reached a transition point in which every organization is considering cloud as a new cost cutter for its services and business offer [1]. Although considered a novel way of delivering computational resources, cloud computing is not a new technology. It is a delivery model based on internet infrastructure, computing, and storage. This new technological and economic model has attracted massive global investments in various sub areas such as performance, security, usability, and global accessibility [1].

It is undeniable that cloud computing has reached critical mass and become a very important aspect of any modern IT strategy or product development. In a recent study [2], the US government has suggested that 25% of the IT budget should be spent on cloud computing initiatives. It has been suggested that almost 30% of the cost would come from current infrastructure cost reduction.

In this chapter, we discover the basic concepts behind cloud computing and its application in the fields of smart homes and cities.

2 CLOUD COMPUTING FUNDAMENTALS
2.1 CLOUD COMPUTING OFFERINGS

There are three major offerings on cloud computing often described in the following pyramidal relation in Fig. 12.1.

Smart Cities and Homes. http://dx.doi.org/10.1016/B978-0-12-803454-5.00012-2

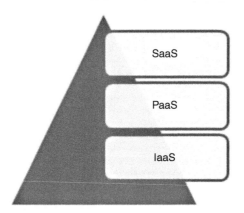

FIGURE 12.1 Cloud Computing Models

Fig. 12.1 shows the three types of cloud computing: SaaS, PaaS, and IaaS. These are the basic building blocks of any cloud computing service and are often used in combination. A description of each one is given subsequently.

2.1.1 SaaS: Software as a Service
Software as a service is the cloud offering that most people interact with today. Whether enterprise or consumer oriented, SaaS offers are accessed via the internet and often eliminates the need to download, install, and configure application on the client side. This deployment model makes it very attractive from a maintenance point of view. In addition, it allows seamless global reach of services to any users provided an internet access. In most cases, SaaS applications are managed by third-party vendors. Users access these services using an account based on a monthly subscription economic model.

2.1.2 PaaS: Platform as a Service
Platform as a service is a software environment used for the development and execution (run-time) of applications. PaaS is defined as a computing platform that allows the creation, testing, and implementation of SaaS applications or other applications without the complexities of setting up or buying expensive infrastructure. For PaaS vendors, many technical aspects need to be managed such as networking, storage, virtualization, middleware, etc. PaaS is bundled as installable units with names that differ from "cartridge" to "workable unit." PaaS is used to host applications that can later be consumed by a multitude of clients such as mobile, web, smart TV, and setup boxes. Application developers need not to worry about capacity and less about scalability. PaaS offerings allow them to request the allocation of new workable units on demand and also to scale down the system dynamically. Some well-known PaaS platforms are Google App Engine [3] and Microsoft Azule services [4].

2.1.3 IaaS: Infrastructure as a Service

IaaS is one of the most fundamental layers of cloud computing. It also interchangeably referred to as hardware as a service (HaaS). In this layer, the vendor offers virtualized infrastructure, networking as well as storage over the Internet. These resources are often available on demand and can scale linearly and dynamically if demand on the infrastructure increases. This makes it ideal for client and companies that require optimizing the cost of their IT infrastructure.

2.2 CHARACTERISTICS OF CLOUD COMPUTING ARCHITECTURE

Cloud computing, grid computing, and high-performance computing are all computing paradigms that belong to parallel computing [5,6]. To put simply, cluster resources are often located in a single internet domain whilst cloud computing relies on multiple data center that span multiple domains and geographical areas. This characteristic is important in order to optimize resource delivery from the closest and most optimal location. From a client point of view, cloud computing encompasses these characteristics [7]:

- dynamic and elastic computing
- rapid response and scalability
- self-managed, power efficient and self-repair
- weak consistency guarantees (CAP theorem)
- internet as a weak link in the chain
- consumption based billing
- low client investment in hardware for SaaS applications

From a technical perspective, Table 12.1 highlights the major system characteristics of cloud computing in contrast with grid computing:

Table 12.1 Main Differentiating Characteristics Between Cloud Computing and Grid Computing		
Characteristic	**Cloud Computing**	**Grid Computing**
Service-oriented paradigm	☑	☑
System loose coupling	☑	☒
System fault tolerance	☑	☒
Networking: TCP/IP stack	☑	☑
Infrastructure virtualization	☑	☒

Loose coupling is a fundamental system architecture quality attribute of cloud computing. Similar to software components, loose coupling in cloud computing ensures that components are well defined with specific responsibilities and clear interacting interfaces. Loose coupling in cloud computing infrastructure is accomplished partially through virtualization and recently through container-based deployment technology such as Docker. For it to work, infrastructure is separated into logical and physical parts. This separation ensures that the behavior of one part does not affect other parts of the system.

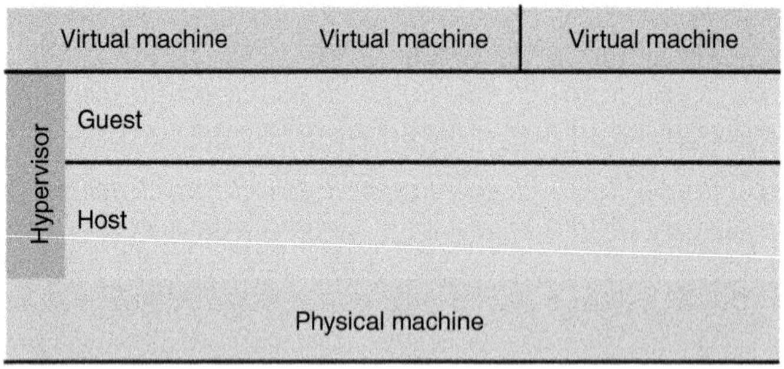

FIGURE 12.2 Loose Coupling in Cloud Computing

As shown in Fig. 12.2, loose coupling is achieved by separating the physical and the hypervisor layers on top of which sit the various virtual machines. In this schema, a client requests virtual machines that are backed up by infrastructure resources, that is, provisioning. The location and the capacity of resources are transparent to the client and it is possible to dynamically add new resources provided changes in the computational demand.

Another system characteristic of cloud computing is fault tolerance. In the basic definition, fault tolerance describes the ability of the underlying system to withstand errors (physical or software based). These errors can occur in two places as described in Table 12.2.

Table 12.2 Cloud Computing Fault Tolerance Model

Error Source	Description
Provider inner	In this scenario, the fault is fixed by substituting the failed part or by using redundancy mechanism.
Provider across	In this scenario, when different providers are aggregated to provide a service, the system attempts to redirect to healthy nodes or service providers in order to provide seamless runtime to clients. This can also be achieved using load balancing.

A last difference between cloud computing and grid computing is the reliance on virtualization technology. In grid computing, calculation intensive operations rely on physical hardware. Cloud computing, however, thrives on shared resources that can be easily virtualized using technologies such as VMware for enterprise or on other higher technologies such as OpenStack.

2.3 CLOUD COMPUTING MODELS

Cloud computing models can be categorized into three major types of clouds as shown in Fig. 12.3.

These models can be summarized as shown in Table 12.3.

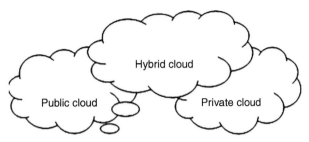

FIGURE 12.3 Cloud Computing Grids

Table 12.3 Description of Cloud Computing Types	
Cloud Type	**Characteristics**
Public	Offers pay as you go billing model Supports multiple tenants Services are either shared of dedicated Externally managed Externally designed
Private	Self-hosted Self-managed
Hybrid	Partner hosted Dedicated environment

2.4 CLOUD COMPUTING SECURITY

As consumers move their applications and data to cloud computing, it is primordial that the level of security at least matches that offered by the traditional IT departments [8]. Failure in doing so will result in higher costs and potential loss of data, business and consequently clients and thus eliminating the benefits of cloud computing. To do so, cloud providers establish policies and procedures to ensure that tasks are accomplished according to the same standard process. This leads to better governance and predictable overall performance. In a cloud computing migration endeavors, good and effective governance is key quality to preserve trust in IT infrastructure as well. Cloud providers and clients agree on SLA parameters so that each party can take appropriate security assessment, prevention, and control measures [8]. The split of responsibilities with the service provider requires the consumers to secure the operating system, the data put in the cloud and the network stack configuration. Fig. 12.4 illustrates the computing security pillars.

One additional issue in cloud governance is the jurisdictional protection of personally identifiable information (PII). There is a divergence across many countries on how to access PII data in case of investigation and enforcement. This is very problematic since cloud providers place datacenter in various geographical regions around the globe to optimize disaster recovery. This makes it difficult to know where the data actually resides and how to abide with international subpoenas on data since each piece of the data is governed by the laws of the country in which it resides. Since there is no clear solution on this matter, often governments and big corporations require that the data be hosted in servers inside their jurisdiction.

FIGURE 12.4 Cloud Computing Security Pillars

2.4.1 Effectively manage identities

Consumers need to ensure that the cloud provider has processes that control who can access their data and applications. This access shall be controlled and managed using regular standards. Some well-known standards are:

- Federated Identity Management (FIM)
- Identity Provisioning and Delegation
- Single Sign-On, Single Sign-Off (SSO)
- Identity and Access Audit
- Robust Authentication
- Role entitlement and Policy Management.

Cloud providers shall formalize processes for managing their own employees accessing the cloud infrastructure. It should also be possible to demonstrate to consumers that this is put in place upon demand.

2.5 KEY CONCERNS ABOUT CLOUD COMPUTING

Although cloud computing has evolved in many areas, there are many concerns that are considered drawbacks for many users:

- *Less control*: There are still many companies that are uncomfortable with the idea that owned information is stored elsewhere than the local infrastructure servers.
- *Data security*: Exchange of data across the network increases the risk of unauthorized exposure thus authentication and authorization mechanisms become important.
- *Reliability*: High availability is increasingly needed because businesses worry about loss of service and thus loss of customers.
- *Compliance*: In certain fields, regulatory compliance and audit is essential for certain IT infrastructures. Regulations such as HIPPA and SOX prohibit the use of cloud computing services to be used.
- *Security management*: Providers must provide easy control mechanisms in order to manage their PaaS of IaaS services.

2.6 MAJOR INDUSTRY PLAYERS

Table 12.4 summarizes the major cloud computing players.

Table 12.4 Cloud Computing Major Industry Players

Cloud Model	Industry Player	Description
IaaS	Amazon web services	AWS has been around for a while and is key played in elastic and dynamic resource allocation. It is widely used that it is also backbone for many SaaS providers such as spootify for example.
	Microsoft Azure	The Microsoft offer resembles to AWS in terms of elasticity and ease of use. In addition it boasts predictive analysis and disaster recovery. It supports major business such as Mazda, Lufthansa, and Mark & Spencer
	Google Drive	Google Drive is pioneer in cloud storage with "infinite" capacity offered to student and free 15GB offered to every subscribed user.
DaaS	Citrix	Citrix provides virtual desktop solutions along with host management file sync and shared services.
	VMware	VMware is long known for its hypervisor software that supports multiple OS flavors. VMware through the Horizon suite offers performance remote desktop which are hosted in cloud environments.
SaaS	Salesforce.com	Salesforce is the go to provider for CRM solutions. Its SaaS offer has also extended to PaaS on order to allow companies build other apps on top of its services.
PaaS	OpenShift	The RedHat Openshift offers major platform development software to be accessible and provisioned through simple interfaces. This is used by developers and enterprises to host applications in the cloud
	Heroku	Similar to Openshift, Heroku supports many programming languages and application servers.

3 CLOUD COMPUTING APPLICATIONS

Cloud computing is a technically dense topic that has many applications in various business domains. Here is a short list to name a few:

- custom relationship management
- online storage management
- collaboration tools
- financial applications
- human resources and employment services
- smart homes
- smart cities
- Big Data

The subsequent sections describe three major applications where cloud computing is an enabling technology.

3.1 BIG DATA AS AN ENABLING TECHNOLOGY FOR SMART HOMES AND CITIES

Big data is considered the backbone technology for many applications of cloud computing in the domain of smart cities and homes. The following graph in Fig. 12.5 was produced by searching web search content trends on Google Trends for the following key words: cloud computing, big data, smart homes, and smart cities from 2014.

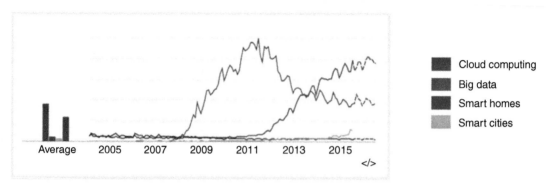

FIGURE 12.5 Google Trends Graph for Cloud Computing, Big Data, Smart Cities, and Smart Home Between 2014 and 2015

In Table 12.5, we explore the Google search index weights of cloud computing, smart homes, smart cities and big data by region.

To the reader, it is notable that Asia and North America are the two global players across all categories in addition to Germany and Spain in Europe. Although there are not too many experiences or players worldwide for smart homes or cities, this can be explained by their reliance on current advances on cloud computing and big data which we see thriving in multiple countries. We shall see major advances in smart homes and cities in the forthcoming years as the former technologies mature.

Table 12.5 Google Trends Search Index Weight for the Key Words: Cloud Computing, Smart Homes, Smart Cities and Big Data per Region of Interest

Cloud Computing		Smart Homes		Smart Cities		Big Data	
	Google search index		Google search index		Google search index		Google search index
Malaysia	19	UK	100	India	100	India	100
Australia	16	USA	41	Spain	41	Singapore	87
UK	14			UK	11	Hong Kong	69
USA	14			USA	7	Taiwan	48
Canada	11					South Korea	46
Indonesia	8					USA	44
Germany	8					Spain	38

3.1.1 Big data, data fusion, and data analytics

Cloud computing and big data is a compelling combination. In a nutshell, big data refers to huge data sets in volume, data that is diverse and that include structured, semi structured and unstructured data. In today's world, this flood of data is produced from a multitude of data sensors and gadgets such phones, RFID tags, homes, hospitals, roads, cars, public spaces, etc. This data in raw form is not easily exploitable and does not give value. In fact, what make it valuable are the insights and analytics it produces when it is analyzed. Current cloud computing systems have demonstrated large capabilities for moving data into the cloud, indexing and searching it as well as coordinating large scale cloud databases analysis. To do so, many supporting algorithms and technologies were developed and enhanced such as Map Reduce, Chord, and Dynamo. These algorithms have been adapted and optimized to serve in the cloud [9–11]. Fig. 12.6 illustrates the map-reduce algorithm in the cloud.

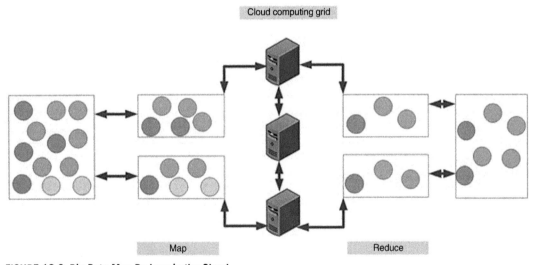

FIGURE 12.6 Big Data Map Reduce in the Cloud

In a growing number of enterprises that are requiring data analysis (Financial institutions, and scientific laboratories), Information Technology (IT) role is shifting from traditional computing grid based services to brokering cloud based big data analytics services [14] also known as Data as a Service or DaaS. Using cloud infrastructure to analyze big data makes sense for various reasons. To name a few [12]:

- Investing in big data analysis requires large IT budget. Cloud computing provides an appealing case when it comes to resource elasticity
- Data can come from internal as well as external sources. External data is often hosted in cloud stores so it makes perfect case to use the same infrastructure to analyze this data and keep the overall system coherent.
- Data services such as Analytics as a Service (AaaS) are needed to extract value from big data.

Cloud computing models are thus the next logical evolution in the field of scalable analytics solutions and we start to see many start-ups operating in this field. Organization using cloud to provide AaaS can weight many factors such as security, interoperability, workload when implementing such a solution. Often, a hybrid model is used where a private cloud is used to handle and manage in-house data while a public cloud is used as an extension to further provide scalability to the system [12].

3.1.2 Trends in big data as an enabling technology

From current experiences, it is undeniable that cloud computing is a cost-effective delivery model for big data and data analytics. Cloud will enable the enterprise as well as cities to deliver a new generation of agile and innovative solutions. The first generation big data applications were based on textual data. The second or next generation of big data analytics will aggregate data from multiple sources and encodings such as voice data, video stream, car flow and transportation data, hospital data, airline data, grid energy status, homes sensors, and user and objects tracking data, among others.

3.2 SMART CITIES

Many of the world cities have embarked on smart city projects, including Seoul, New York, Tokyo, and Shanghai. These cities might seem like cities of a future era but with the current advances in technology and especially cloud computing, they are only exploiting to a certain extent what current technology has to offer.

3.2.1 Smart city concept

A smart city is a concept of a knowledge, digital, cyber, and ecofriendly city. Based on the specificities of each city, the following two definitions emerge:

- "A city well performing in a forward-looking way in [economy, people, governance, mobility, environment, and living] built on the smart combination of endowments and activities of self-decisive, independent and aware citizens." [13]
- "A city that monitors and integrates conditions of all of its critical infrastructures including roads, bridges, tunnels, rails, subways, airports, sea-ports, communications, and water, power. Even major buildings can better optimize its resources, plan its preventive maintenance activities, and monitor security aspects while maximizing services to its citizens." [14]
- "A city that strategically utilizes many smart factors such as Information and Communication Technology to increase the city's sustainable growth and strengthen city functions, while guaranteeing citizens' happiness and wellness." [15]

Smart cities require a lot of planning in order to create coherence between city services. This can be achieved by many models and most notably a human-centric model that is based on ICT infrastructure. Fig. 12.7 describes this model.

The continuous evolution of the Internet and the ability to maximize user activities allowed accelerating the emergence of ideas that attempt to improve the quality of services for communities around the cities [16]. Smart cities are a new perception of what commonly provided services should be in the

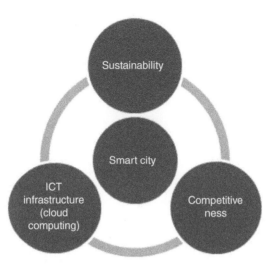

FIGURE 12.7 Cloud Computing Benefits in the Context of Smart City

age of the internet. Smart cities enclose services in diverse business and technological fields such as efficient use of natural resources as electricity, water, and air quality in addition to waste management. There are many examples of smart cities around the world. Each one of these cities is revolutionizing current processes in order to improve the quality of life of their citizens while optimizing the cost of these services [15]. One of the most cheerful and bold moves by the city of Berlin is to consider cloud computing as a natural resource [17].

For smart cities services to take shape, large amounts of data emerging from many sources must be collected, analyzed and synthesized in order to take informed actions and decision automatically and semi-automatically.

3.2.2 Smarter grid

According to the energy information administration, 62% of worldwide energy generation comes from gas and coal, 13% comes from nuclear, 16% from hydraulic systems and only 4% from renewable energies [18]. There is a constant rise in energy demand to levels above 80% until 2030 and by 2040, energy demand from large economic powers such as China will double that of the USA level. Having an outage such as that of 2003 in North America creates disturbances in businesses and the economies at large. These power grid failures could in most cases be prevented if the diagnostic information was available and ready in time.

In Fig. 12.8, we illustrate the smart grid ecosystem with energy efficiency at its center. Energy is stored and distributed to users. Excess energy can be exported to third party for example. There are many challenges and opportunities in smart grids that can be addressed by cloud computing [18,19]. Examples include dynamic energy pricing and shifting potential peak demand to a different time when the price of energy is low, real time massive data streaming and analysis from sensors plugged in the infrastructure [20]. To ensure proper coordination and efficiency in this area, ultraresponsive Supervisory Control and Data Acquisition (SCADA) systems can be used. This is illustrated in Fig. 12.9.

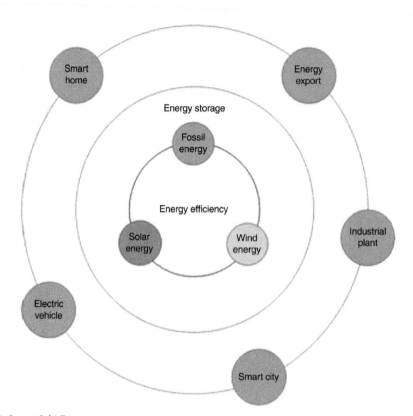

FIGURE 12.8 Smart Grid Ecosystem

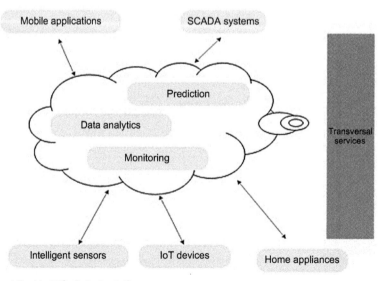

FIGURE 12.9 Cloud Enabled Big Data Analytics

New studies suggest that new paradigms need to be devised in order to support optimized production and consumption of power as well as to estimate a wide area state of the grid.

3.3 SMART HOME

3.3.1 Concept

The idea of moving home automation to cloud infrastructure is to provide a simple deployment model for client when deciding to install home automation devices [23]. From usability point of view, any new installation shall be easy to use, vendor agnostic as well as interoperable between devices providing the same or complementary data. IBM in [23] has defined three major characteristics of new smart home appliances:

- *Instrumented*: having the ability to sense and monitor changing conditions.
- *Interconnected*: having the ability to interact with people, systems and other objects.
- *Intelligent*: having the ability to make decision on data and produce a better outcome.

A smart home defines and offers many new capabilities. Here are few examples to name a few:

- entertainment and smart TVs
- energy management
- safety and security
- health and convenience
- user recognition and home profile management
- automatic contact of emergency services
- automatic virtual shopping
- practical data display

Cloud computing provides and intelligent platform to connect interoperating services [23].

As illustrated in Fig. 12.10, smart homes will be equipped with a variety of sensors such as power meters and monitoring devices. These devices will communicate together and contact services running in the cloud to provide the desired functionality. Table 12.6 illustrates how home devices will become smarter [23].

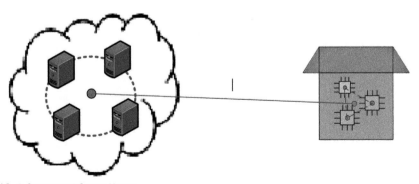

FIGURE 12.10 A Connected Smart Home

Table 12.6 Example of Smart Devices	
Home Device	**Description in a Smart Home Context**
Home energy distributer	On the basis of the activities in the home and the surrounding building, energy usage by other home appliances will be adaptive to optimize the cost of the KWH as well as the load of the grid. This adaptive information is extracted through continuous sensor data collection from the home as well as continuous behavior analysis and comparison with other relevant data from the cloud
TV set	Propose programs and content on the basis of user watch list history Propose targeted ads on the basis of content
Refrigerator	Adjust the thermostat on the basis of the volume of the food items it contains
Washer and dryer	Determine the water temperature in the wash/rinse/dry cycle on the basis of the load volume, dirt level, etc.
Water heater	Water heater that turns on to heat the water when the cost of energy is cheap and let the water cool off when this water is not needed
Air conditioner	It will consume and match usage and climate patterns, power cost and grid state in order to provide the most optimal temperature in the home at an optimal cost

3.3.2 Smart homes enabled by the cloud

Cloud computing platform offering or PaaS is ideal for providing the basic layer behind home automation. It allows dynamic allocation of resource applications. Provided the standardized web service interfaces, it is possible to enable dynamic composition of solutions in a plug and play mode. Relying on the cloud, smart home solution vendors can easily scale their solutions to millions of users worldwide. It is because we benefit from dynamic provisioning of resources. Using common PaaS and IaaS platforms allows devices to become connected and interoperable with other devices from different vendors. This is due to the expansion of industry wide standards as well as solid consortia of market leaders. The key benefits of using cloud computing to enable smarter home can be summarized as shown below:

- From a *consumer point of view*, the smart home devices are easier to use especially that the management has been moved to the cloud. This implies that no IT infrastructure needs to be put in place except a broadband network. Connecting smart home devices is only a matter of software adapters at the PaaS provider side. The extensibility of networked services along with the ability to link both existing and new devices provide reliable performance and increase service innovation.
- From a *device manufacturer point of view*, relying on industry standards through cloud allows creating innovative services as well as reaching large consumer base. Open standards prevent vendors and especially start-ups from being locked out of specific markets where device manufacturers have stroke specific deals. Once the cloud is the platforms, device and service manufacturers can concentrate on delivering added value business features and scale up or out in markets freely and easily.

- From a *service provider point of view*, providing services on top of standardized device interface allows a shorter time to market, and better pricing due to shared IT infrastructure. Along with device manufacturer, service providers can concentrate on added value services.

3.3.3 *Home cloud service delivery platform*

The service delivery platform, which sits on top of cloud computing infrastructure, enables the integration and monitoring of services and composites of services. The delivery platform concept has been developed by IBM and improved with deployments in the telecom IT industry in order to provide and reach content to consumers [21].

As illustrated in Fig. 12.11, the service delivery platform is a service-oriented Architecture (SOA)-based modular component services. It provides a controlled way to add new services and by aggregating and composing these services on the cloud and simplifying the consumer based side.

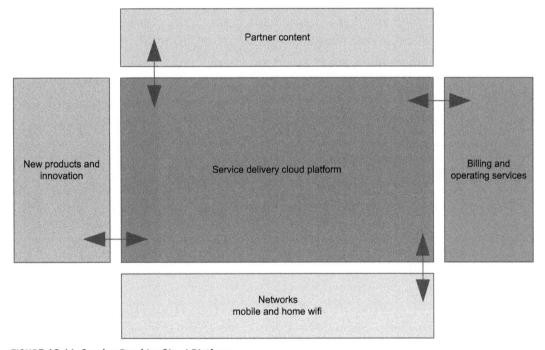

FIGURE 12.11 Service Provider Cloud Platform

3.3.4 *Emerging protocols for smart homes*

In order to successfully bring cloud based smart homes to market, it is necessary to ensure proper communication between the smart home devices and the cloud. This communication relies on the following protocols and standards [21]:

- *Home gate standard*: ISO/IEC 15045
- *Wired systems*: USB, Ethernet, IEEE 1395

- Low-level wireless protocols such as ZigBee, HomeRF, Wimax, Bluetooth
- Special level interoperability such as OSGi, TAHI, and Cablehome
- Home networking systems such as DVB, DLNA, and UPnP
- Power line such as DS2, X10, and HomePlug
- *Specifications*: HGI, ITU-T SG-5, ITU-T IPTV-GSI, and IEC TC100 gateway

4 CASE STUDY: SEOUL SMART CITY
4.1 PRESENTATION

Smart Seoul is one of the largest metropolitan cities in Asia and best known as one the most technological know-how cities in the world according to the UN smart cities survey [22]. Smart Seoul was announced in 2011 to promote Seoul's reputation as the world ICT leader by demonstrating the use of state of the art technologies. Strictly speaking, smart Seoul is not the first attempt of South Korea to use ICT for the development of a smart city. An initial attempt in 2004 was developed to use pervasive computing technologies in order to enhance the city's competitiveness and provide 'smart' services to citizens.

4.2 CLOUD BASED ICT INFRASTRUCTURE

Technological advances in large systems and specifically cloud computing allow the implementation of many concepts. The following are the three components behind Seoul smart city:

- *ICT infrastructure and Cloud computing*: With the massive amount of data that needs to be analyzed and processed as well as the increasing demand of computing resources in order to deliver services due to the dynamic nature of devices plugged in the smart city infrastructure, cloud computing is the most suitable technology stack.
- *Integrated management framework*: A management framework is essential to guarantee a cohesive and common layer for managing and monitoring all resources and services. Provided that smart city solutions are created by multiple vendors and in various domains, ICT infrastructure must adhere to common standards in order to ensure interoperability and provide simple interfaces to manage services in the city.
- *Smart users*: Having a comprehensive smart city plan relies also on tech-savvy users that are able to interact with smart services by increasing access to smart devices.

Provided the scale of such a city, cloud computing has been relied on to solve two major issues:

- Provision resources to public services in the cloud infrastructure
- Aggregate and synthesis the data to provide added value information

Fig. 12.12 depicts a high-level components architecture ICT stack of a smart city.

From a system architecture point of view, a smart city such as Seoul is a system of systems. This implies that there are many individual and independent systems that when combined create and form a meta-system which in turn becomes a sub system. Cloud computing can scale well to such a complexity. In such a system, mobile applications interact with the infrastructure using services through a public API which in turns hides all the complexity behind data source aggregation. This data is extracted using DaaS services. In order to ensure coherence and control across data access, city governance in the

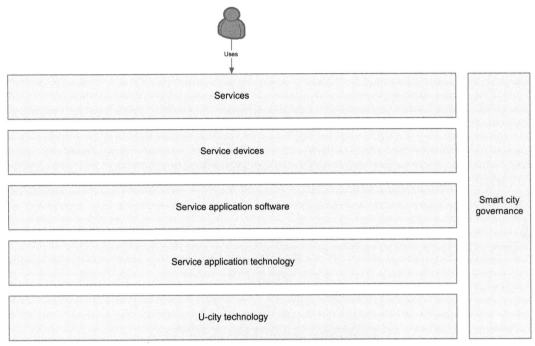

FIGURE 12.12 High-Level Components of Smart City ICT Ran on Cloud Computing

context of smart cities has been evolved. New monitoring tools, frameworks and interfaces are running on the cloud infrastructure to provide live feedback on city resources [23].

4.3 DATA AND SERVICE DELIVERY

With any content delivery system based on cloud infrastructure, strong communication networks are key components to deliver services on time and with high quality. Seoul smart city uses mobile and web technology in order to provide "smart" services to users. A wide range of information is available through mobile applications. Location based services pinpoint to public offices, hospitals, transportation live data, continuous air quality checks, emergency data, live crimes data and statistics, among others.

In order to provide Seoul citizens with useful information, one of the many challenges facing this system is how to aggregate this data from all sorts of databases, data sensors, live feeds, CCTV data, police reports, alerts, etc. It requires the use of unconventional IT systems and computing paradigms. Big data and data analytics are key technologies that are heavily used in this context.

4.4 OPEN APPLICATION PROGRAMMING INTERFACE (API) AND OPEN DATA

As described in earlier sections, Table 12.7 shows the evolution of the data sources in the Seoul smart city ICT system [23].

Table 12.7 Evolution of Data Sources in Seoul's Smart City Cloud Infrastructure

Classification	2011	2012	2013	2014
Accumulated number of databases	20	60	100	150
Proportion of total (target of 150 DBs at full system scale)	15%	40%	70%	100%

With such an increasing number of databases and unbounded number of connecting devices (either data producers or consumers), scalable data access architecture shall be put in place. A common paradigm is the use of public API interfaces, as shown in Fig. 12.13, which abstracts the complexity behind the data access.

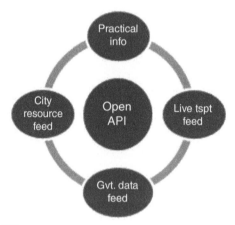

FIGURE 12.13 Smart City Open API

This data is accessible via the internet and can be used either in raw form for application developers or in the form of graphs, augmented maps and sheets to enhance and simplify understanding for citizens.

4.5 PaaS MOBILE APPLICATION DATA ACCESS

More than two thirds of Seoul's population is equipped with mobile phones and devices. Seoul government officials are leveraging this boom in technology as well as the private sector's activeness in order to accelerate the adoption of public apps and consumption of PaaS services [23]. Seoul metropolitan government has started a program to reward best apps developed by local companies or private developers. Fig. 12.14 depicts PaaS open API infrastructure in the context of smart city.

A mesh of content provider servers is coordinated to aggregate data and provides endpoint services. These servers are contacted via a common Open API service and mobile or web-accessing apps may represent the data in the adequate and convenient way to the citizen.

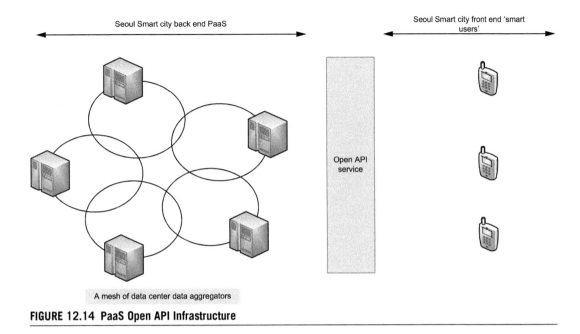

Seoul Smart city back end PaaS

Seoul Smart city front end 'smart users'

Open API service

A mesh of data center data aggregators

FIGURE 12.14 PaaS Open API Infrastructure

5 SUMMARY AND CONCLUDING REMARKS

Cloud computing is not a new topic. It is just an evolution of common paradigms in IT infrastructure and delivery based on virtualization and containers. Cloud computing promises agility, innovation, and lower cost by outsourcing traditional IT issues such as hardware maintenance, upgrades to specialized vendors with the benefit of acquiring the capacity that is needed. Although cloud computing has many benefits, it also has some shortcomings that still shall be addressed once the technology matures enough such as data security, compliance, and governance.

There is no doubt that cloud computing will dominate the IT landscape within the years to come. With the adoption of many protocols and standards, cloud providers erase the gap between enterprise expectations and technology maturity. This leads to the emergence of new applications in the consumer realm such as smart homes and smart cities. These applications rely on interconnected devices and software with an exchange of enormous data which need to be analyzed and interpreted in order to present it in a comprehensive form. In this vision enabled by cloud technology, an ecosystem is vital in order to develop new service delivery and business models. Consumer electronics manufacturers lead this movement so as to reach the widest consumer base possible. Once their devices are connected to the cloud, other vendors provide service delivery platforms to aggregate many services and provide valuable data and insights to consumers. In this setup, the cloud infrastructure constitutes the IaaS and the PaaS and other vendors or third party developers can manage in order to provide applications to offer a SaaS layer. All of these layers and technologies are interconnected in a seamless and simple manner.

REFERENCES

[1] Tripathi A, Mishra A. Cloud computing security considerations. In: Proceedings of the 2011 IEEE international conference on signal processing, communications and computing, ICSPCC'11; 2011. p.1,5.

[2] Kundra V. Federal cloud computing strategy. <http://www.dhs.gov/sites/default/files/publications/digital-strategy/federal-cloud-computing-strategy.pdf>; February 2011 [Online].

[3] Google, Google app Engine. <http://code.google.com/appengine>; March 2015 [Online].

[4] Microsoft, Windows Azure. <http://www.microsoft.com/windowsazure/windowsazure/>; March 2015 [Online].

[5] Gong C, Liu J, Zhang Q, Chen H, Gong Z. The characteristics of cloud computing. In: Proceedings of the 2010 international conference on parallel processing workshops, ICPPW'10; September 2010. p. 275,279.

[6] Buyya R, Yeo CS, Venugopal S, Broberg J, Brandic I. Cloud computing and emerging IT platforms: vision, hype, and reality for delivering computing as the 5th utility. In: Future generation computer systems. ACM 2009;25(6):599–616.

[7] Malcolm D. The five defining characteristics of cloud computing. <http://www.zdnet.com/article/the-five-defining-characteristics-of-cloud-computing/>; April 2009 [Online].

[8] Cloud Security Alliance. Security for cloud computing 10 steps to ensure success. <http://www.cloudstandardscustomercouncil.org/Security_for_Cloud_Computing-Final_080912.pdf>; December 2012 [Online].

[9] Bakken DE, Bose A, Hauser CH, Whitehead DE, Zweigle GC. Smart generation and transmission with coherent, real-time data. Proc IEEE 2011;99(6):928,951.

[10] Dean J, Ghemawat S. MapReduce: a flexible data processing tool. Commun ACM 2010;53(1):72,77.

[11] DeCandia G, Hastorun D, Jampani M, Kakulapati G, Lakshman A, Pilchin A, Sivasubramanian S, Vosshall P, Vogels W. Dynamo: amazon's highly available key-value store. In: Proceedings of the 21st ACM SIGOPS symposium on operating systems principles, SOSP '07, ACM; 2007. p.205–220.

[12] Intel. Big Data in the cloud: converging technologies. <http://www.intel.com/content/dam/www/public/us/en/documents/product-briefs/big-data-cloud-technologies-brief.pdf>; April 2015 [Online].

[13] Griffinger R et al. Smart cities—ranking of European medium-sized cities. <http://www.smartcities.eu/download/smart_cities_final_report.pdf>; October 2007 [Online].

[14] Hall R. The vision of a smart city. <http://www.osti.gov/bridge/purl.cover.jsp?purl=/773961-oyxp82/>; September 2000 [Online].

[15] Hernandez JF, Larios VM. Cloud computing architecture for digital services into smart cities. <http://smartcities.ieee.org/images/files/images/pdf/draft10-cloudieee.pdf>; April 2015 [Online].

[16] Birman KP, Chen J, Hopkinson KM, Thomas RJ, Thorp JS, Renesse R, Vogels W. Overcoming communications challenges in software for monitoring and controlling power systems. Proc IEEE 2005;9(5).

[17] Rohde F, Loew T. <http://www.hanse-parlament.org/images/images//pdf/Presenatations-Hanseatic-Conference-2012/t_parteka_smart_cities.pdf>; January 2011 [Online].

[18] U.S. Energy Information Administration. <http://www.eia.doe.gov>; April 2015 [Online].

[19] Misra FS, Xue G, Yang D. Managing smart grid information in the cloud: opportunities, model, and applications. IEEE Netw 2012;26(4):32–38.

[20] Banavar NG, Harrison C, Paraszczak J, Morris R. Smarter cities and their innovation challenges. IEEE Comput 2011;44(6):32–39.

[21] IBM. The IBM vision of a smarter home and technology. <http://www.ibm.com/smarterplanet/global/files/uk__uk_en__cloud__a_smarter_home_enabled_by_cloud_computing.pdf>; December 2010 [Online].

[22] UN-PAP. <http://www2.unpan.org/egovkb/global_reports/12report.htm>; December 2012 [Online].

[23] Seoul Open Data Square. <http://data.seoul.go.kr/index.jsp>; April 2015 [Online].

DESIGN AND MANAGEMENT OF VEHICLE-SHARING SYSTEMS: A SURVEY OF ALGORITHMIC APPROACHES

13

D. Gavalas*,‡, C. Konstantopoulos†,‡, G. Pantziou**,‡

*Department of Cultural Technology and Communication, University of the Aegean, Mytilene, Greece;
†Department of Informatics, University of Piraeus, Piraeus, Greece;
**Department of Informatics, Technological Educational Institution of Athens, Athens, Greece;
‡Computer Technology Institute & Press "Diophantus", Patras, Greece

1 INTRODUCTION

Sustainable principles in urban mobility urge the consideration of emerging transportation schemes including vehicle sharing as well as the use of electromobility and the combination of vehicle transfers with greener modes of transport, including walking, cycling, and public transportation.

Bike-sharing programs have received increasing attention in recent years aiming at improving the first/last mile connection to other modes of transit and lessen the environmental impact of transport [1]. Bike-sharing programs are networks of public use bicycles distributed around a city for use at low cost. The programs comprise short-term urban bicycle-rental schemes that enable bicycles to be picked up at any bicycle station and returned to any other bicycle station, which makes bicycle-sharing ideal for point-to-point trips. The principle of bicycle sharing is simple: individuals use bicycles on an "as-needed" basis without the costs and responsibilities of bicycle ownership [2]. The earliest well-known community bicycle program is launched in 1965 in Amsterdam, the Netherlands.

Current bike-sharing systems deploy bikes picked up and returned at specific locations (docking stations) and typically employ some sort of customer authentication/tracking (through the use of an electronic subscriber card) to avoid theft incidents [3]. Recent developments pave the way for next-generation bike sharing known as the "demand-responsive multimodal system" [2]. Such systems will emphasize on flexible docking stations (relocated according to usage patterns and user demands), incentivize user-based redistribution (by using demand-based pricing wherein users receive a price reduction or credit for delivering bicycles at empty dockings), enable integration of bike-sharing with public transportation and car-sharing locations (via smartcards, which support numerous transportation modes on a single card), and GPS-based tracking. Online services such as Social Bicycles (SoBi) [4] allows users to locate, reserve, and unlock a bike with a smartphone app, while also employing a rewarding scheme to motivate cyclists to return bikes to central stations/hubs.

Similar to bike sharing, car sharing is a model of short-term car rental, particularly attractive to customers who make only occasional use of a vehicle, enabling the benefits of private cars without the costs and responsibilities of ownership [5]. Car sharing first appeared in North America around 1994. Replacing private automobiles with shared ones directly reduces demand for parking spaces and decreases traffic congestion at peak times, thereby supporting the vision of sustainable transportation. Car-sharing operators typically allow cars to be picked up from designated stations (depots) with customers required to return vehicles to their original pick up locations (such schemes are referred to as two-way car-sharing systems). Most operators have been reluctant in introducing innovative features (eg, one-way rentals, ride sharing) due to added management complexities [6].

These complexities were responsible for the failure of Honda's Diracc system in Singapore, one of the best-known one-way car-sharing experiments in the world, after 6 years of operation (the system has been discontinued in 2008). Diracc failed mainly because it proved to be unable to maintain the quality of service (ie, car availability) required by customers due to one-way trips, leaving the system with significant imbalance in vehicle stocks. Indeed, during a typical day, the number of cars throughout a network shifts toward certain destinations; for instance, drivers commuting from the suburbs to downtown offices generate surplus of cars at certain stations, while depleting fleets at other stations. Nevertheless, some recent car-sharing initiatives—notably, Daimler's Car2Go[a] and BMW's DriveNow[b]—offer the option of one-way car sharing, as long as the customer drops off the car at any available public parking space within a designated operating area.

The design and management of a car-sharing system raise several optimization problems. First, optimal fleet sizes along with the location of the parking stations should be determined [7]. Further on that, operators allowing one-way rides need to develop strategies to reallocate the vehicles and restore an optimal fleet distribution among stations. Such a distribution could respond to the short-term needs at a particular station or be based on an historical prediction (ie, estimating future demand to proactively schedule relocations) [8]. Although bike-sharing operators typically employ dedicated vehicles for relocating bunches of bicycles to depots with depleted stock, vehicle relocation in car-sharing programs is more demanding. In particular, the activities of vehicle relocation can be carried out by the user itself or by the operator. In the first case, the user is incentivized to car pool or to choose another location or reservation time; in the second case, which is currently more common, the vehicles are physically transported using dedicated trucks or personnel.

A recent development in vehicle-sharing systems has been the employment of fully electric vehicles (EVs) as a means of lowering the environmental footprint of urban mobility. Further complicating things, the design of EV-sharing systems needs to consider two additional constraints: the availability of charging facilities on parking stations and the design of relocation strategies which take into account vehicles residual energy [9].

The aforementioned detailed challenges call for intelligent algorithmic solutions to support the long-term viability of vehicle-sharing systems. Such algorithmic approaches should aim at the highest possible quality of service for customers and reduced capital investment for operators with respect to both system deployment and operating expenditures. To achieve these objectives, the whole range of deployment and operational parameters inherent in vehicle-sharing systems should be carefully

---

[a]https://www.car2go.com/
[b]https://de.drive-now.com/en/

addressed: long-term strategic planning of systems, tactical decisions to enable user-based regulation to the benefit of the systems, and operational issues.

This chapter offers insights on research tackling the aforementioned main issues related to the design and operation of public bicycle/car-sharing systems. The focus is on mathematical models and algorithmic solutions developed so far, especially those that address cost and pricing models, depot location optimization, mobility and demand modeling, ways of balancing vehicle stocks across stations (ie, relocation strategies) in one-way vehicle-sharing systems. The objective is to identify the state-of-the-art along with possible paths for future developments in this field.

The remainder of the chapter is structured as follows. Section 2 elaborates further on the challenges and objectives relevant to the design of vehicle-sharing systems. Section 3 overviews models and algorithmic approaches for the design, operation, and management of vehicle-sharing systems. Section 4 presents algorithmic approaches for ride sharing. Finally, Section 5 provides insights on open issues and research challenges in the field whereas Section 6 concludes the chapter.

2 CHALLENGES AND OBJECTIVES IN THE DESIGN OF VEHICLE-SHARING SYSTEMS

Recent research analyzed the factors affecting the success of bike-sharing programs [10,11]. These factors range from the built environment (infrastructures, facilities at work, etc.) to factors related to the natural environment (topography, seasons and climate or weather), socioeconomic and psychological factors (attitudes and social norms, ecological beliefs, habits, etc.), and other factors related to utility theory (cost, travel time, effort and safety). Factors gaining growing interest involve bike station location, cycling network infrastructure (bike paths) and the operation of bicycle redistribution system [12]. The stations must be located in close proximity to one another and to major transit hubs and be placed in both residential (origin) and commercial or manufacturing (destination) neighborhoods, which makes bike-shares ideal as a commuter transportation system [1,13]. Existing examples show that the bike stations should not be located more than 300–500 m from important traffic origins and destinations. Given the complexity of bicycle facility planning and the importance of station distribution for operating bike-sharing programs, formal approaches are needed to model the problem variables and derive optimal solutions with respect to minimizing investment cost and maximizing utility for the users. Among others, optimal solutions should determine the number, location, and capacity (in bikes and docks) of the stations as well as the bicycle lanes needed to be setup.

On the other hand, equally important for bike-sharing systems success is to guarantee bicycle availability. Each rental station must carry enough bicycles to increase the possibility that each user can find a bicycle when needed. Therefore, measures of service quality in the system include both the availability rate (ie, the proportion of pick-up requests at a bike station that are met by the bicycle stock on hand) and the coverage level (the fraction of total demand at both origins and destinations that is within some specified time or distance from the nearest rental station). Due to the one-way rental policy, bikes are likely to get stuck in areas of lower individual mobility demand (cold spots) while needed in zones of higher demand (hot spots). To make the system more efficient and more profitable, this imbalance of supply and demand could be adjusted by applying different intervention (ie, relocation) strategies [14].

The need to ensure vehicle availability in high-demand areas is also acknowledged for car-sharing systems [15]. However, relocation of cars is more troublesome than that of bicycles (up to 60 bicycles can be transported altogether to hot spots on a bicycle carrier, contributing to cost and effort savings [16]). Some studies suggest the use of road vehicles (car carriers) with fully automated driving capabilities (typically moving along dedicated tracks), coordinated by centralized management systems, able to autonomously relocate to satisfy user demands [17]. Redistribution of vehicles may also be provided by a fleet of limited capacity tow-trucks located at various network depots; using such an approach the problem can be conveniently modeled as pickup and delivery problem [9]. However, dedicated transport trucks are of little use in most urban settings due to stations not easily reachable by heavy-duty trucks and the time consuming vehicle loading/unloading operations [18]. Thus, the scheme most commonly encountered in practice engages teams of employed drivers who undertake the relocation of vehicles thereby significantly increasing operational cost.

Recently, the decreased manufacturing cost of EVs along with their ecofriendly characteristics (fuel economy and lowered greenhouse gas emissions) has attracted the attention of car-sharing companies[c]. So far, the main body of EVs-relevant algorithmic research focuses on novel energy-efficient routing algorithms motivated by the unique characteristics of EVs (limited cruising range, long recharge times and the ability to recuperate energy during deceleration or when going downhill) [19,20].

EV-sharing systems are also unique with respect to their design and operational requirements. Specifically,

1. Sufficient battery availability at pick-up time should be ensured so as to travel reliably to user's destination [21].
2. Vehicle relocation policies should take into account the energy availability of vehicles at stations, in addition to physical availability [22].
3. Pick-up/drop-off locations are determined by the existence of charging stations (for instance, the 300 Car2Go vehicles and other EVs in Amsterdam have access to 320 charging stations in the city area).
4. The anticipated transformation of urban parking stations to charge-park stations in support to EV power demands is expected to create considerable load on the power grid, hence, intelligent approaches are in need to flatten the load peak, thereby deferring investments in grid enhancement [23].

3 MODELS AND ALGORITHMIC APPROACHES FOR OPTIMIZING VEHICLE SHARING SYSTEMS

Bicycle- and car-sharing systems are complex dynamical systems with stochastic demand whose modeling and performance analysis is very important for their implementation and performance as well as for ensuring an effective regulation of vehicle traffic flows. Different approaches and methodologies have been proposed in the literature for modeling and studying design, operational and management

[c]Among other operators, Car2Go has launched (as of November 2011) an EV car-sharing network currently covering San Diego and Amsterdam. Through a user-friendly web interface, users interested in driving a shared EV, Car2Go members are able to view the exact location of available EV along with their batteries state of charge and proceed to online reservations. If the battery performance sinks below 20%, the driver must end his/her trip at a charging station (found through an in-built navigation system.

issues of bicycle/car-sharing systems. Such approaches include mixed integer programming approaches [24,25], stochastic programming approaches [26], simulation methodologies [8,27,28]. Although Petri nets have been a tool used rather successfully in the literature for modeling and evaluating the performance of dynamic and complex systems in various domains (eg, traffic control of urban transportation systems [29–32] and planning [33–35]), very limited research work exists in Petri net models for modeling and performance analysis of bicycle- and car-sharing systems [36,37].

Besides OR approaches (using either mathematical programming or Petri nets and closed queuing networks) that support decision making in the design and management of bicycle and car-sharing systems, data-mining techniques have also received attention in the literature. Data mining is particularly suitable to analyze and predict the dynamics of such systems. The analysis of the temporal human mobility data in an urban area (using the amount of available bicycles/cars residing in the stations of vehicle-sharing systems) may offer insights on the system structure and operation; therefore, statistical and prediction models can be developed for the tactical and operational management of these systems. Some of the research works focusing on using data mining to analyze bicycle-sharing systems are the following:

- In Ref. [38] Froehlich et al. provide a spatiotemporal analysis of data collected for the number of available bikes and vacant bike stands from Barcelona's bicycle-sharing system. Stations are clustered according to the number of available bikes and an activity score assigned in the course of day. Then, visualization is used to identify shared behaviors across stations and show how these behaviors relate to location, neighborhood, and time of day. The authors show that fairly simple predictive models are able to predict station usage with an average error of only two bicycles and can classify station state (ie, full, empty, or in-between) with 80% accuracy up to 2 hours into the future.
- In Ref. [39] Kaltenbrunner et al. detect bike usage patterns in data from Barcelona's bicycle-sharing system. Their results are similar to those of Froehlich et al. [38]. They present a statistical model that predicts the number of free bikes and vacant bike stands at stations some minutes ahead in time.
- Borgnat et al. [40] use data mining to analyze the dynamics of bike movements in Lyon's bike-sharing system. Temporal patterns in the system-wide bike usage are examined. Weekdays show usage peaks in the morning, at noon, and late afternoon, whereas usage is concentrated in the afternoon on weekend days. A statistical model for the prediction of the number of rentals on a daily and hourly basis is proposed. Furthermore, spatial patterns are examined by clustering bike flows between stations. Spatial and temporal dependencies exist between stations of clusters interchanging many bicycles.

Vogel et al. [41] identify three main issues related to the design, management and operation of bicycle/car-sharing systems. The proposed design and management measures (aiming at alleviating imbalances in the availability of bicycles/cars) are distinguished into three separate planning horizons (see Fig. 13.1):

1. Strategic (long-term) network design comprising decisions about the location and the number of stations as well as the vehicle stock at each station.
2. Tactical (mid-term) incentives for customer-based distribution of bicycles/cars, that is, incentives given to users so as to leave their vehicle to a station different to that originally intended (this may be regulated through pricing schemes adaptable to the system state). For example, Fig. 13.2 illustrates two possible options offered to the user willing to move from a location *A* to a location

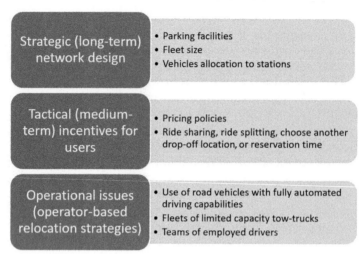

FIGURE 13.1 Main Issues Related to the Design, Management, and Operation of Vehicle-Sharing Systems

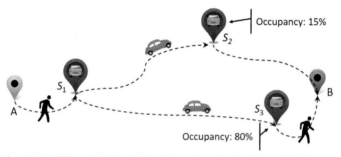

FIGURE 13.2 Illustration of the Different Options Offered to a User Moving From *A* to *B*

B: The $A \to S_1 \to S_3 \to B$ option (ie, the user leaves the vehicle at the station S_3 and walks to the destination location *B*) is the shortest time one, whereas the $A \to S_1 \to S_2 \to B$ option (ie, the user leaves the vehicle at the station S_2 and walks to the destination location *B*) is associated with the incentivized scheme.

3. Operational (short-term operator-based) repositioning of bicycles/cars on the basis of the current state of the stations as well as aggregate statistics of the stations' usage patterns. For example, Fig. 13.3 illustrates a relocation plan for a vehicle-sharing system, on the basis of the system data shown in Table 13.1.

In the sequel of this section, we overview algorithmic solutions proposed for the strategic design of vehicle-sharing systems (Subsection 3.1), present modeling approaches supporting pricing schemes and incentives for customer-based distribution of vehicles (Section 3.2) and summarize algorithmic approaches for the problem of operational repositioning of vehicles (Section 3.3).

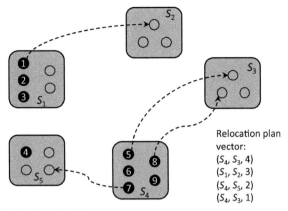

FIGURE 13.3 Illustration of a Relocation Plan

Circles filled with black color represent parked cars whereas empty circles represent empty parking spaces.

Table 13.1 Example Snapshot Showing the Capacity, Current/Targeted Occupancy and Surplus/Deficit in Vehicles

Stations	S_1	S_2	S_3	S_4	S_5
Capacity	5	3+	3	5	4
Current occupancy	3	0	0	5	1
Targeted occupancy	2	1	2	2	2
Surplus/deficit	+1	−1	−2	+3	−1

3.1 ALGORITHMIC APPROACHES ON THE STRATEGIC DESIGN OF VEHICLE-SHARING SYSTEMS

Integer-programming-based approaches. Lin and Yang [42] have been the first to investigate the problem of strategic design of bicycle-sharing systems. The problem investigated is the following: given a set of origins, destinations, candidate sites of bike stations and the stochastic travel demands from origin to destination, the problem's output comprises the location of bike stations, the bicycle lanes needed to be set up and the paths to be used by users from each origin to each destination, the objective being to minimize the overall system cost. The authors take an integrated view of the system cost, considering both the user's and the investor's point of view. In particular, the investor's cost comprises the facility costs of bike stations, the setup costs of bicycle lanes, bicycle stock and safety stock (for serving the demand at peak hours) costs. The level of service provided to the user is measured by the demand coverage level (defining penalty costs for uncovered demands) and travel costs (for both walking and cycling). The problem has been formulated as an integer nonlinear program.

Martinez et al. [43] formulated a mixed integer linear program (MILP) aiming to optimize the location of shared biking stations and the fleet dimension. This study also considered bike-relocation operations among docking stations (the relocation operations cost is considered as an additional system cost factor, yet not explicitly included as a decision variable in the MILP formulation). A general-model framework has been proposed, which computes several days of operation, maintaining the dimensioning data from previous iterations, recomputing the hour operation MILP model, and updating the system design, until the configuration reaches a net revenue equilibrium, producing a stable and optimal system configuration.

Correia and Antunes [25] addressed the optimization problem of selecting sites for locating depots in order to maximize the profits of a one-way car-sharing organization. Revenues are generated from renting the vehicles against some price rate while several types of expenses are considered (maintenance costs for vehicles and depots, vehicle depreciation costs and vehicle relocation costs). Relocation operations are only considered at the end of the day, unlike previous studies wherein the main emphasis was on optimizing such operations [24,26]. Three mixed integer programs (MIP) have been modeled which determine the optimal number, location, and capacity for the depots, each one corresponding to a different trip selection policy. According to the first policy, the operator is free to accept or reject trips in the period they are requested according to the profit-maximization objective; the second policy assumes that all trips requested by clients are approved; the third policy allows a trip request to be rejected in the case that there are no vehicles available at the pick-up depot. The optimization models have been tested in a case study involving the municipality of Lisbon, Portugal.

Boyaci et al. [44] proposed a generic model for supporting the strategic (number and location of required stations) and tactical (optimum fleet size) decisions of one-way car-sharing systems by taking into account operational decisions (ie, relocation of vehicles). The authors formulated a mathematical model (integer program) and conducted sensitivity analysis for different parameters. The objective function seeks to maximize the overall profit which considers the revenue generated from vehicle rentals in addition to user costs (proportional to the time required to reach the origin station from the start location and the end location from the destination station) and system costs (unserved customer cost, vehicle operating cost, station opening cost and relocation cost). The proposed model has been applied for planning and operating a station-based EV-sharing system in the city of Nice, France.

Heuristics. Recognizing the complexity of the bicycle-sharing system design optimization which precludes exact solutions for instances of realistic size, Lin et al. [45] approached the system's design as a hub location inventory problem[d] [46] that takes the coverage level into consideration and proposed a greedy algorithm for solving it efficiently. The greedy-drop heuristic iterates between locating bicycle stations given a collection of bicycle lanes, and locating bicycle lanes given a set of bicycle stations. In particular, all candidate stations and the bicycle lanes connecting them are initially marked as "open." The algorithm then iteratively removes the currently open station, which if closed, would result in the largest total cost reduction, until no further cost reduction is possible. Likewise, bicycle lanes are removed, as long as their removal results in cost reduction. The overall solutions cost is calculated utilizing the mathematical cost model introduced in an earlier study [42]. When testing the algorithm

---

[d]The hub location problem has been one of the important classic facility location problems. Hub facilities concentrate on flows to achieve economies of scale. Flows between origins/destinations and hubs and between pairs of hubs are consolidated into a smaller set of links rather than serving demand with direct links. The hub location problem involves determining the hub facilities and determining the links to connect origins, destinations, and hubs.

in test instances for which enumeration is possible, the heuristic solution has been found within a 2% gap from the optimal.

Kumar and Bierlaire [47] developed an optimization model to identify the most appropriate locations for establishing car-sharing stations such that the overall system performance is maximized (the main measure of stations performance is the average number of rides per day). The model considers car-sharing systems exclusively allowing round-trips. The model balances between the estimated attractiveness of key demand drivers in a locality and the locality's proximity to an existing station. The authors first build a linear regression model (applied on historical data of the Auto Bleue EV-sharing operator, in Nice, France) to identify the key demand factors that affect stations performance[e]. Then they formulate a mixed-integer quadratic (higher-order) program. The objective of the mathematical model is to maximize the combined performance of all selected stations (n stations are selected among k candidates; candidate stations are assumed to be located at the centroid of prespecified city localities). The main trade-off decision made by the model involves locating more stations in "highly attractive" localities versus locating new stations in "less attractive" but untapped localities (establishing too many stations at attractive locations does not increase the overall system performance as they tend to cannibalize each other's performance). Last, a heuristic is proposed to solve the problem. The heuristic first estimates the "best performing stations" on the basis of all parameters except distance and public transport ridership. In the first iteration, all k stations are assumed to be operational; the contribution of public transportation and distance is then computed. In the next step, the n best locations are picked to place the stations. Now the public transportation and distance contribution is recomputed assuming that only these n proposed stations are operational. On the basis of the changes in the objective function, the n best locations to place the stations are again selected. This process is repeated until the selected set of n stations remains unchanged.

The problem of determining the fleet dimension (size) and the distribution of vehicles among the stations of a car-sharing system was studied in relation to electrically powered one-person vehicles (personal intelligent city accessible vehicles, PICAVs), which enable accessibility for all in urban pedestrian zones [48]. This system allows one-way trips among stations (parking lots that offer vehicle recharging services) located at intermodal transfer points and near major attraction sites within the pedestrian area. The number, the location, and the capacity of the stations are not determined by the model. To cope with the imbalanced accumulation of the one-way system, this model enrolls a human supervisor. The task of the supervisor is to direct users that are flexible in returning the car to alternative stations, as to achieve a balanced operation and fulfil a maximum waiting time constraint. The cost minimization problem has been solved using a simulated annealing-based approach (the cost function takes into account both the transport system management and the customer cost, that is, the cost of vehicles and the total customer waiting time, respectively).

GIS-based approaches. Geographic information systems (GIS) represent a highly useful tool for determining bike station locations. Larsen et al. [49] presented a GIS-based grid-cell model to identify and prioritize cycling infrastructure investments using the example of Montreal, Canada. The main result is a grid-cell layer of the study region wherein high-priority grid-cells represent those areas most

---

[e]Stations performance have been found to increase with the share of high-income/education population (in the locality), the share of public transport ridership, the share of car usage to reach workplace, the presence of mobility attractors (mainly commercial centers, hotels and colleges), the population density, the presence of transit hubs; on the contrary, distance from customers residence decreases the performance of stations.

appropriate for bicycle infrastructure interventions (ie, the areas where new cycling facilities would provide the maximum benefit to both existing and potential cyclists). Rybarczyk and Wu [50] used GIS-based multicriteria decision analysis both to evaluate the quality of bicycle facilities utilizing supply- and demand-based objectives. Analyses were conducted at two levels: network (bicycle facility) level and neighborhood level. Network level analysis can address site-specific issues and provide detailed information for further improvements. By contrast, neighborhood-level analysis provides a strategic view of bicycle facilities in an urban area, and facilitates policy development and implementations.

Garcia-Palomares et al. [51] proposed a GIS-based method to calculate the spatial distribution of the potential demand for trips, locate stations using location-allocation models, determine station capacity and define the characteristics of the demand for stations. The authors follow a four-step approach: first, the distribution of the potential user demand is assessed (the number of trips generated and attracted for each transport zone is calculated on the basis of the population and employment associated with each building). The location-allocation models (p-median and maximum coverage[f]) are then applied defining obligatory bike-stations, candidate locations, the number of stations to be located and the type of solution chosen. Once bike-station locations and potential demand upon stations are obtained, the stations capacity (number of bicycles and docks) is calculated; also, the stations are characterized (as trip generators or attractors) making it possible to vary the number of bicycles according to the time of day, leading to more efficient bicycle redistribution systems. The final step is the analysis of stations in terms of accessibility (a measure of usefulness, which considers the volume of demand allocated to the station and its distance to the potential origin/destination stations of the users); this way, it is possible to prioritize stations within the bike-sharing program (eliminating those with poor accessibility).

Data-mining-based approach. Vogel et al. [41] use geographical information technology and data mining methods to gain insight into bicycle stations operations and try to incorporate this knowledge in the design of bicycle-sharing systems (strategic and operational planning). In a case study, collected data related to the activity of the bike stations are provided as input to a data mining phase where cluster analysis is used to group stations according to their pickup and return activity patterns. The analysis reveals spatiotemporal dependencies of pickup and return activities at stations which support the hypothesis of Vogel et al. that usage patterns at bike stations and the type of customers using certain stations depend on the stations' location. Note that if the hypothesis holds, then usage patterns for already existing stations can be mapped to potential stations based on their locations. Therefore, strategic decisions about the bicycle-sharing system can be supported.

The previous algorithmic approaches are summarized in Table 13.2 where a classification is also given according to whether they concern bicycle or car-sharing systems.

3.2 TACTICAL INCENTIVES FOR BICYCLES/CARS DISTRIBUTION

In Ref. [52], a bike-sharing system is modeled as a stochastic network and its steady state performance is analyzed using the mean field theory. Specifically, in this model, there are N bike stations, each of which can keep at most K bikes. Initially, there are s bikes in each station and therefore the total number

---

[f]In the maximum coverage location–allocation model, the stations are located such that as many demand points as possible are allocated to solution facilities within the impedance cutoff (200 m).

Table 13.2 Algorithmic Approaches on the Strategic Design of Vehicle-Sharing Systems

	Bike-Sharing Systems	Car-Sharing Systems
Integer programming – based approaches	Lin and Yang [42] Martinez et al. [43]	Correia and Antunes [25] Boyaci et al. [44]
Heuristic approaches	Lin et al. [45]	Kumar and Bierlaire [47] Cepolina and Farina [48]
GIS-based approaches	Larsen et al. [49] Rybarczyk and Wu [50] Garcia-Palomares et al. [51]	
Data mining–based approach	Vogel et al. [41]	

of bikes in the system is sN. It is also assumed that the arrival rate at each station is λ (symmetric case) and the travel time between any two stations follows the exponential distribution with parameter μ. The authors determine the proportion of problematic stations (empty or saturated) at steady state. Specifically, they prove that the proportion of problematic stations at steady state is minimal when $s = K/2 + \lambda/\mu$ and the minimum is equal to $2/(K + 1)$. This is not an encouraging result, since for achieving low proportion of problematic stations, large capacities in the stations are needed, which is not always feasible due to space constraints and construction costs. The authors also show that the situation does not improve even when the users are aware of the problematic stations and they always pick one of the remaining stations for getting and leaving a bike. The performance of the system is greatly improved if simple incentives for the users are adopted. Specifically, the authors test the case when the user selects a station at random for leaving its bike and then s/he finally selects the least loaded. The analysis and the simulation results clearly show that the proportion of problematic stations is now much lower. This also holds in the case that only a fraction of users accept to follow that policy. Then, the authors study the asymmetric case where there are two clusters of stations and the customer arrival rate at the stations of one cluster is higher than at the second cluster. In this case, the performance of the system is much worse than in the symmetric case when there is no regulation mechanism for the bike distribution across the stations. Even the aforementioned incentive of two choices is not that effective in this case. So, the authors propose bike repositioning using a number of tracks. Indeed, the simulation results demonstrate much higher performance in the steady state if the trucks regularly redistribute the bikes across the stations.

In Ref. [53], a bike-sharing system is presented where periodic redistribution of bikes across the stations is carried out by using a number of trucks and also incentives are given to the users to leave their bike to a different stations than the originally intended one. Incentives are regulated through a pricing scheme which is changing online according to the current state of the system. First, the authors use historic data for building user demand statistics of the bike-sharing system. Specifically, they determine the average arrival and departure rate of customers at each station for a number of time intervals on each day differentiating between working days and weekends. Then, periodically, each time for a fixed planning horizon, they determine the truck routes for optimal redistribution of bikes across the stations. For the problem formulation, the authors assume deterministic flows in the network and also define a utility function at each station which determines the benefit of removing or adding bikes at the station at the current time with respect to the increase of the percentage of users whose requests will be satisfied at this station in the near future. Then, they study the problem of finding the best route for the case when only one truck is used. They also assume that during each trip, the single truck can visit at most a small constant number of stations. Then, they use a greedy approach and they build a tree emanating always from a specific spot (named maintenance depot in the paper); a number of stations are added iteratively to the tree so that the increase in the utility function over the additional cost incurred for reaching the station is relatively high. Having constructed the tree, a separate optimization problem is solved for each different route starting from the root of the tree and ending at the tree leaves. This optimization problem which is in the form of a quadratic program, refines the truck loading actions across each route leading to a more effective solution. Then, the authors generalize their solution for the case of multiple trucks in the system. Essentially, they follow a sequential approach fixing the route of trucks one after another. Finally, the authors study the problem of determining the pricing scheme which will have the lowest monetary cost while keeping the bike distribution across the stations at an optimal level. The basic assumption in their approach is that the users are rational thinkers and when the system proposes to them an alternative nearby station to leave their bikes, the users weigh the monetary reward they are going to receive for this choice against the monetary cost of travelling additional distance. For determining the best pricing policy, the problem is formulated as a problem of model predictive control. More precisely, the best prices are determined for each different time step within a finite time horizon and then only the prices concerning the current time step are finally adopted. At the next time step, the problem is resolved since the system state may have changed in the meantime.

In Ref. [7], a vehicle-sharing system is modeled as closed queuing network. The authors make the simplifying assumptions that the users always find parking space at the destination station and also when they do not find a vehicle at the origin station, they simply leave the system. By regarding the vehicles as the customers of the closed queuing network system, each parking station is viewed as a single server node with FIFO service policy and the service time is equal to the interarrival time of users at that station. The user arrival at each station is modeled as a Poisson process. It is also assumed that the network of parking stations is complete and thus there is a direct link for each pair of parking stations. Each vehicle at the origin station leaves that station along a certain outgoing link with a specific probability. The traveling time between two stations i and j is exponentially distributed with parameter $1/\mu_{ij}$. Each link (i,j) of the station network is modeled as a node with infinite number of servers and with total service rate equal to n/μ_{ij} where n is the number of vehicles traveling along that link. The main objective in their analysis is to determine the optimal number of vehicles (fleet size) in the system such as the overall profit is maximized. In estimating this profit, the authors consider the revenue per unit time obtained from a vehicle rent by a user. They also take into account a maintenance cost per vehicle and an unavailability penalty when a customer cannot find vehicles available at a station. Then, they prove that the profit function is a concave function and its optimization derives two solutions at most. Further, they use mean-value analysis, for estimating these solutions.

In Ref. [54], the authors analyze a pricing scheme by modeling a vehicle sharing as a closed queuing network, basically following the approach in Ref. [7]. However, now, each station is assumed to have finite capacity and also the demand for each out-link of a station is elastic influenced by the price that should be paid for traveling along this link. It is also assumed that when a user picks a car at the origin station, the system ensures that there will be free parking space at the destination station by making reservation in advance. For determining the best pricing scheme, the time is partitioned into a number of time slots whose duration follows a certain distribution. In addition, the authors assume that the system has periodic behavior and the prices for each link should be determined only for the time slots within a single period of the system. Essentially, the whole problem is reduced to a Markov decision process wherein the set of actions applied at each moment should be determined. Apparently, this set of actions is the prices set for each link, which in turn affects the use of this link by the users of the system. Due to the huge state space of this process, the problem of obtaining the best pricing scheme cannot be solved in reasonable time. For this reason, the authors propose an approximation based on the fluid model where the stochastic demands are replaced by continuous flows with deterministic rate. Then, the problem is reduced to a continuous linear problem whose solution maximizes the sum of demands at each link of the station network.

In Ref. [55], the authors assume a vehicle-sharing system where stations have unlimited capacity and the travel time between any two stations is negligible. These two assumptions simplify the modeling of the vehicle-sharing system as a closed queuing network. Similarly to the previously discussed approaches, each station is a node of the closed queuing network where the jobs to be served are the cars at this station. The service rate of the server is equal to the rate of the customer arrival at that station which is modeled as a Poisson process. For each pair of stations there is a demand rate for the link connecting the corresponding nodes; this demand is leveraged by the price set for making a trip along this link. However, no method is proposed for adjusting these prices to maximize profit. Actually, the authors study the problem of finding the link demands which maximize the number of trips sold. Also, for each link, there is a separate upper bound for the demand passing through that link. This bound is implicitly determined by the lowest price that the system operator will set for the corresponding link. Then, the authors solve the maximum circulation problem on a flow network which results from the queuing network by viewing the upper bounds on the link demands as the edge capacities on this flow network. Note that in the maximum circulation problem, there is no source and sink node, and the objective is to maximize the circulated flow in the network without violating the capacity constraints. As the solution of this problem may yield zero flows for some links, the resulting flow network may be disconnected with a number of strongly connected components. Then, for each strongly connected component, the availability of each station at that component is determined, that is the probability that a new customer will find a vehicle at that station. Apparently, this probability is a function of the number of vehicles and the number of stations at the component. Given a specific distribution of vehicles across the different strongly connected components, the expected number of trips taking place in the system can now be calculated from the solution of the maximum circulation problem and the previously esti- mated station availabilities. Then, the authors give a greedy algorithm for determining the distribution of the vehicles across the strongly connected components mentioned previously, which maximizes the expected number of trips sold. They also prove that this greedy approach is actually optimal based on the fact that the expected number of trips is a concave function of the number of vehicles within each strongly connected component. Finally, they present some preliminary results about the approximation ratio of their approach. Specifically, they claim, without a complete proof, that the proposed policy is a tight $N/(N + M - 1)$ approximation on both static and dynamic optimal policies where N is the total number of vehicles and M is that number of stations of the vehicle-sharing system.

The deterministic version of the previous problem is also studied in Ref. [56]. In this setting, the trips planned to take place in a fixed horizon are known in advance. Similar to the previous approach, each link is associated with a fixed price to pay for following that link. In addition, for each trip, users set a maximum price they are willing to pay. A trip is cancelled, if the price of this trip's link is higher than the maximum price for that trip. The optimization problem in this scenario is to determine the prices at each link so that the total system revenue is maximized. The authors prove that this problem as well as a number of variants are all NP-hard problems.

In Ref. [37] a user-based solution for the vehicle relocation problem in car-sharing systems is proposed. In particular, an approach of using rental-pricing incentives is presented and assessed. Incentives are intended to influence the travel behavior of the users according to the system conditions, monitored in real time. The proposed solution is based on a model of an electric-car-sharing system developed in a timed Petri net (TPN) framework. Note that TPNs use graphical and mathematical descriptions to represent both the static and the dynamic aspects of the modeled system; the graphical representation enables a concise way to design and verify the model, while the mathematical description allows simulating the system in software environments, by considering different dynamic conditions [37]. The proposed vehicle relocation strategies have been applied to the real case of the electric-car-sharing system of Pordenone, Italy. The simulation results show that a system which ignores the operative conditions of the service suggesting always to its customers to return the vehicles as soon as possible, does not lead to the rebalancing of the number of vehicles parked in each station. On the other hand, giving incentives to the users which depend on the real-time monitoring of the system, can increase the number of served customers and, therefore, improve the overall system performance. The results also show that the effectiveness of the proposed solution decreases as the congestion level of the system grows highlighting the limits of such an approach.

In Ref. [57] the authors present a crowdsourcing mechanism that incentivizes the users in the bike repositioning process by providing them with alternate bike pick up or drop off locations in exchange for monetary incentives. The main component of the system is the incentives deployment schema (IDS) that handles the user's request through a Smartphone App. The IDS communicates with the bike-sharing system infrastructure to evaluate the current and predicted status of the stations, and decides whether to offer on not incentives to the user. In order to maximize the efficiency under given budget constraints, the authors design a dynamic-pricing mechanism using the approach of regret minimization in online learning that can learn over time about the optimal pricing policies. The users are considered as strategic agents who may untruthfully report information about their personal cost and location to maximize their profit. The pricing mechanism dynamic budgeted procurement using upper confidence bounds (DBP-UCB), is a dynamic variant of BP-UCB presented in Ref. [58]. The proposed system is evaluated through simulations using historical and user survey data. Finally, the system was deployed on a real-world bike-sharing system for a period of 30 days in a city of Europe, in collaboration with a large-scale bike-sharing company. According to the authors this is the first dynamic incentives system for bikes repositioning ever deployed in a real-world bike-sharing system.

The previous algorithmic approaches are summarized in Table 13.3 where a classification is also given according to whether they concern bicycle or car-sharing systems.

3.3 OPERATIONAL REPOSITIONING OF BICYCLES/CARS

In a bike-sharing system, there is a set of stations providing bicycles for rent, each with a specified capacity of allowed bicycles. A customer may rent a bicycle at a station, use it for a period of time and then leave it to another station. Since, the stations have a specified capacity and the number of bicycles available for rent is restricted, shortage events may occur. A shortage event occurs when a customer tries to rent a bicycle from an empty station or tries to return a bicycle in a full station [60]. To eliminate shortage events, hence customers' dissatisfaction, it is necessary to reposition bicycles using a fleet of

Table 13.3 Algorithmic Approaches on Tactical Incentives for Bicycles/Cars Distribution

	Bike-Sharing Systems	Car-Sharing Systems
Stochastic network modeling approach	Fricker and Gast [52]	
Model predictive control approach	Pfrommer et al. [53]	
Closed queuing network modeling approach		George and Xia [7] Waserhole and Jost [54], [55] Waserhole et al. [56] Briest and Raupach [59]
Timed petri net modeling approach		Clemente et al. [37]
Regret minimization approach	Singla et al. [57]	

dedicated vehicles. The repositioning can either be static [60], that is, it can take place during the night when no customer asks for bicycles or dynamic [61], that is, occur during the day in order to remove bicycles from full stations and transfer them to stations with lack of bicycles. Two main factors are considered in a repositioning process, the number of vehicles removed/transferred to a station to meet the customers' need and the operational cost of the fleet of vehicles performing the repositioning. In several applications the latter factor may be considered insignificant compared to the impact of a dissatisfied user and hence it may be disregarded.

Chemla et al. [62] study the static rebalancing problem in bike-sharing systems. The authors formulate the problem as a single-vehicle one commodity-capacitated pickup-and-delivery problem. In this formulation a single capacitated vehicle balances the stations transferring bikes from stations with excess of bikes to stations with shortage. The relocation is assumed to be static, that is, taking place during the night when there is no demand for bikes. The problem aims at producing a minimum cost vehicle route accompanied with loading/unloading number of bikes at each station. At route completion time all stations have to contain a predefined target number of bikes. The authors propose an intractable exact model for the problem. Then, the model is relaxed, obtaining an integer program with exponential number of constraints. This program is solved using a branch-and-cut algorithm, producing a lower bound for the solution. Apart from this approach, a tabu search heuristic is proposed to produce feasible solutions. The tabu search algorithm incorporates four different neighborhood structures. The tabu list contains a number of arcs previously removed during the execution of local search steps. Two different approaches are considered for the construction of the initial solution of the heuristic algorithm. In the first, a solution is constructed using a greedy heuristic procedure. In the second, an initial solution is obtained based on the solution of the integer program. The executed algorithms are the branch-and-cut algorithm for the integer program as well as the two versions (according to the construction of the initial solution) of the tabu search heuristic. The test instances used for evaluating the algorithms are based on the work in Ref. [63]. The experimental results indicated that tabu search incorporating the solution of the integer program produces higher quality solutions, that is, solutions with lower cost than the greedy initialization heuristic while the latter approach executes faster. Furthermore, the results indicate that the tabu search heuristic obtains, in general, solutions with cost close to optimal, achieving on average at most 5% gap.

Raviv et al. [60] study the static repositioning problem of bicycles performed during the night using a fleet of vehicles. The problem aims at producing vehicle routes for bicycle repositioning in order to minimize a cost function. The cost function considered is a weighted combination of a convex penalty function based on the expected number of shortage events per station of the next day and the travel cost of the vehicle routes. Two mixed integer linear-programming formulations of the problem are proposed, namely an arc-indexed formulation and a time-indexed formulation, each with different underlying assumptions. In the arc-indexed formulation a vehicle cannot visit a station twice, while no waiting is allowed at a station. These assumptions significantly reduce the number of decision variables and, hence, make the approach efficiently solvable. In the time-indexed formulation, the time period is discretized into small time periods and the decision variables taken into account extend the decision variables of the former formulation adding one more index, the time index. Furthermore, the restrictions of the arc-indexed formulation do not apply anymore. In this way the solution space of the latter formulation extends the solution space of the former. Since solving these programs would take a lot of computational time, a two-phase heuristic approach is considered. In the first phase the program relaxes the restriction of integer number of bicycles removed and transferred to stations, hence concentrating on the design of the vehicle routes. In the second phase, the program is solved with the integral restriction to the number of bicycles, with the decision variables concerning the vehicle routes treated as constants on the basis of the solution obtained from the first phase. The algorithms have been tested on data from Paris (Velib system) consisting of at most 60 stations and 2 vehicles and Washington DC (Capital Bikeshare) consisting of 104 stations and 2 vehicles. The results indicated that the arc-indexed formulation combined with the two-phase heuristic approach was the most efficient approach yielding higher quality results in the allowed 2 h execution time.

Weikl et al. [16] study the relocation problem of cars in free-floating car-sharing systems. The relocation strategies are categorized as user-based and operator-based. In the former, the relocation is performed by the customers. Incentives and bonuses are offered to the users to either change their destination, leaving the rented car in a station with shortage of cars or share a car with other customers with similar trips. In the latter, the relocation is performed by the employees of the system, transferring cars from stations with excess to stations with shortage. The first approach is very profitable for the system, since no cost for car transferring is added, however customers may refuse to changer their trip or share a car. The second approach adds cost to the system, requiring employees' actions and car movement. Nevertheless, it is more reliable. Then a user-based algorithm of Di Febbraro et al. [64] and an operator-based algorithm of Kek et al. [24] are described to illustrate the different approaches. Finally, a two-step algorithm for car relocation in car-sharing systems is introduced. In the first, of-fline step, a set of demand scenarios is produced based on real collected data. For each scenario, the optimum number of cars per station is computed and a set of relocation strategies is produced. In the second, online step, the number of vehicles currently placed in stations is compared to the optimum, computed in the current demand scenario. If these quantities differ, the appropriate relocation strategy produced in the previous step is applied.

The modeling of a car-sharing system as a closed queuing network is followed in Ref. [59], similarly to the works surveyed in Section 3.2. Again, the cars are considered as the pending jobs of the system and each parking station is viewed as a single server with the available cars at the station waiting in a queue for the next customer to come. Once more, the service rate of a server is essentially the inter-arrival rate of the customer arrival Poisson process at that server/station. The authors also assume that a customer picking a car at a station may drive to any other station with a certain probability. In addition, a redistribution policy is implemented wherein the staff of car-sharing company relocates cars so as to achieve maximum total profit. Specifically, a reward is credited when a customer uses a car for traveling between two stations, with the reward being proportional to the travelled distance. Similarly, a cost applies when a car is relocated by the company staff for achieving balanced car distribution across the stations. Again, this cost is proportional to the distance travelled for this relocation. It is also assumed that the reward value is higher than the relocation cost for the same travelled distance. Now, the overall objective is to determine the relocation policy which maximizes the total profit of the system. The problem is formulated as a linear programming problem and the optimal solution determines the average number of cars moving between each pair of stations due to customer requests and due to relocation which yields the highest net profit. On the basis of the optimal solution of this problem, a relocation policy is then determined. Namely, after a car arrives at a new station after completing a customer trip, the car is immediately relocated to a random target station according to a certain probability distribution. Specifically, the probability p_{uv} of relocating a car from a station u to a station v is equal to m_{uv}/y_u where m_{uv} is the average number of cars relocated from u and v and y_v is the average number of cars at node v after a customer request has been served at that station. The values of these two parameters derive from the optimal solution of the linear program discussed previously. Now, the authors prove that this relocation policy yields profit within a factor of 2 of the optimal policy's profit. They also prove, via a reduction from the set packing problem, that finding the optimal relocation policy in the car-sharing problem is an APX-hard problem, in general. Finally, they provide some preliminary results for the discrete-time version of the car-sharing problem where customers are not arriving according to a Poisson process but simultaneously at all nodes at regular intervals. Also, after each round of customer requests, a relocation policy may relocate all cars regardless of whether they were moved due to a customer request. Then, the authors study the problem assuming that the distance between any two stations is 1 and that a customer at a station will select the destination station uniformly. In this case, the optimal policy is proved to be not performing any car relocation. Then, the average fraction of nonempty queues in the system is determined and this is also the approximation ratio with regards to the optimal policy.

Gendreau et al. [65] tackle the problem of dynamically relocating emergency vehicles in order to cover the most possible population. For example, when a vehicle leaves its location for a service, the remaining vehicles are relocated to be able to cover as much population as possible. The problem is formulated as a maximal expected coverage relocation problem (MECRP). The input of the MECRP consists of n vehicles, a directed edge-weighted graph $G = (V, A)$, with V partitioned into two subsets, namely V_w, that represents the locations the vehicles may wait, and V_d, that represents the locations a service may occur. A vehicle in V_w covers a vertex in V_d if the vehicle can reach the vertex within a specified travel time. Each vertex $u \in V_d$ is associated with a demand that represents a measure of necessity for a vehicle covering the vertex. The objective is for each $k \le n$ to assign k vehicles in vertices of V_w, fulfilling a side constraint on the maximum allowed number of relocated vehicles, in order to obtain the maximum expected coverage. The calculation of all $k \le n$ assignments of vehicles to vertices in V_w is used in order to have all possible relocations known a priori, that is, when a vehicle leaves its position for a service the relocation of the remaining vehicles is already known. The expected coverage is the $\sum_{k=1}^{n} p_k, c_k$ where p_k denotes the probability exactly k vehicles to be available and c_k denotes the coverage obtained from the k vehicles. The problem is formulated as an integer linear programming problem. The approach was tested in real data from Montreal's medical services, with a small number of vehicles $3 \le n \le 6$ and the solution was calculated by an integer programming solver. The results indicate that the average response time would not exceed 10 min even in the case of only three vehicles. A drawback of the approach of Gendreau et al. is that although they calculate every possible assignment of $k \leq n$ vertices to V_w, they do not generate the actual routes to be assigned to vehicles when one of them leaves or becomes available. Furthermore, this approach cannot be used for large values of n, since the solution of the integer programming problem takes exponential time.

Contardo et al. [61] introduce the dynamic public bike-sharing balancing problem. The problem deals with the dynamic relocation of bicycles, that is, the relocation of bicycles during the day, when the demand for bicycles is not negligible. The aim is to derive vehicle routes for dynamically relocating bicycles in order to minimize the number of occurring shortages, that is, incidents wherein a customer requires a bicycle from an empty station or attempts to leave a bicycle in a full station. The time is discretized into periods and a space–time network is introduced. A node of the network denotes a station of the bike-sharing system at a specified time period, while an arc represents transition of a station at a specified time period to another station the expected time period. Regarding the number of bicycles loaded on a vehicle (during an arc traversal) as flow, a mixed integer linear programming (MILP) formulation is proposed. Since, the solution of the previous formulation would be computationally hard, a heuristic approach that produces lower bounds and feasible solutions is also proposed. In the latter, two decompositions of the problem are applied. Namely, Dantzig–Wolfe decomposition [66] is applied and the linear relaxation of the formulation is solved, creating a lower bound for an instance. Then a new formulation of the problem is proposed applying Benders decomposition [67]. Taking into account the process applied in the Dantzig–Wolfe decomposition, a feasible solution is obtained. Since no instances for the dynamic relocation of bicycles exist, Contardo et al. [61] created 120 test instances to test their approaches. The instances contained 25, 50, or 100 stations, the time span was set to 2 h, each time period was set to 5 or 2 min and the number of vehicles available for the relocation was considered to be 5. The MILP has been compared against the heuristic approach combining the two decomposition schemes. The former is solved using a commercial solver allowing 30 min execution time. The heuristic approach has been shown to clearly outperform the MILP approach in all test instances apart from the smallest ones. The lower bound produced by the heuristic is higher than the MILP solution, the solution obtained corresponds to lower cost, while the heuristic's execution time does not exceed on average the 6 min, even in the largest instances. On the downside, the problem formulation does not take into account the time spent for loading and unloading bicycles in stations. Since this time is not negligible in relation to the time length of routes and is usually proportional to the number of loaded/ unloaded bicycles [60], future work could focus in deriving new formulations of dynamic bicycle relocations incorporating loading/unloading times.

In Ref. [36] a bike-sharing system consists of a set $S = \{S_1, S_2, \ldots, S_N\}$ of N stations, where each station S_i has capacity C_i (ie, it is equipped with C_i bicycle stands). The system employs redistribution to transport bicycles from stations in excess of bicycles to stations that may run out of bicycles. The objective of the control system is for each station S_i, to maintain an (appropriate for this station) number of R_i bicycles ensuring that there are always bicycles available for pick up and also $C_i - R_i$ vacant stands available for bicycle drop off. The proposed Petri net model (initially defined only for three stations and generalized to any number of stations in the sequel) consists of three subnets (modules) representing three different functions: (1) the "station control" subnet, representing the control function of the stations to ensure availability of bicycles for pick up and vacant berths for bicycle drop off at every station; (2) the "bicycle flows" subnet representing the bicycle traffic flows between the different stations of the network; and (3) the "redistribution circuit" subnet representing the path to be followed by the redistribution vehicle in order to visit the different stations of the network. The proposed modular and dynamic Petri net model is validated through several simulations made for different interesting system configurations. Labadi et al. argue that Petri nets-based modeling is particularly useful to planners and decision makers in determining how to implement and operate successfully bicycle-sharing systems.

Krumke et al. [68] have studied the dynamic relocation problem in car-sharing systems. A customer may pick-up a car from a non-empty station and deliver it to another—not full—station. Similarly to bike-sharing systems, car relocations is necessary to counter the effect of stations with unbalanced vehicle stock. The relocation is assumed to take place using convoys, able to transfer a number of cars between the stations. For each relocation, the car system is charged with a cost depending on the number of convoys and the number of cars transferred as well as the distance covered. In the setting of this article a customer reserves in advance, that is, s/he requests a car rental from a specified station at a certain time to be returned to another specified station at a given time. Furthermore, each request is associated with a profit earned by the system. Based on the previous assumptions two variants of the relocation problem are tackled. In the first, all requests must be serviced and the goal is to minimize the cost of the relocation operations to meet all customers' demand. In the second, the goal is to decide which requests to service as well as to schedule the relocation tours of convoys to maximize the system's profit. An integer linear program is introduced for each of these variants. The solution approach incorporates a time-expanded network. The network consists of nodes representing stations at specified times. Network arcs represent feasible transitions of convoys with cars between station-time pairs. Arcs also model requests information, that is, a transition of a rent car from a station to another. Two kinds of flows are introduced: a flow representing the number of cars transferred between stations and a flow representing the number of convoys moved between stations. These two flows are related to each other, by introducing a constraint on the maximum number of cars transferred with a specified number of convoys, based on the capacity of the convoys. On the basis of these flow considerations, two integer linear programming formulations are proposed for the investigated problems. Notably, solving the proposed integer programs is highly inefficient, yet, no efficient algorithm is proposed by the authors to meet the requirements of the dynamic relocation scenario. Furthermore, the use of convoys for transferring cars may not be feasible in many urban settings and, hence, employed drivers may be restricted to transfer at most one car at a time.

Lee et al. [69] have studied the static relocation problem in EV-sharing systems. The authors assume concurrent relocation of EVs (ie, employed drivers are assumed to be at the stations where cars to be relocated reside at the time that relocation starts) without taking into account their residual energy. The relocation policy aims to restore a certain availability of vehicles at each station. They consider both even relocation schemes (stations end up having equal number of cars) and utilization-based relocation schemes (EVs are assigned to stations according to the demand ratio of each station, known a priori). Depending on the chosen relocation scheme, a relocation vector is determined, namely, the desired number of vehicles in each station after the completion of the relocation process. A heuristic algorithm is executed thereupon, deciding the destination of each vehicle through matching EVs from overflow stations to underflow stations so as to minimize the relocation cost, that is, the overall moving distance. For an EV, the preference to underflow stations depends on the distance to be travelled. Each EV to be reallocated has an index to its preference list of underflow stations, with the index initialized to the first (ie, nearest one). In addition, each underflow station is aware of its required number of EVs and maintains an allocation list (ie, list of EVs residing elsewhere and currently assigned to it). EV-station matching begins from the first

EV (among the ones scheduled to be relocated). The EV examines the option of relocating to the station marked by its local index within its preferences list. If the station currently holds within its allocation list less EVs than the required number e, the EV is added to the allocation list. Otherwise, if the allocation list is full, the EV which is farthest away from the station is removed. The removed EV then examines the next station, shifting ahead the index in its preference list. Having completed this iterative process, the allocation lists of underflow stations should be finalized. Such allocation, represented by (EV, station) pairs, should have the minimum relocation distance.

In a follow-up work, Lee and Park [70] designed a team-based relocation scheme for EV-sharing systems and proposed a genetic algorithm-based solution to obtain a reasonable quality relocation plan within a limited time bound. Each relocation plan, namely, the set of relocation pairs of EV from overflow to underflow stations, is represented by an integer-valued vector to run the genetic operators such as crossover, selection, reproduction, and mutation. Drivers performing vehicle relocations are assumed to move in teams of m members, wherein one of them drives a car following a route through a series of overflow stations. Upon arriving at a station, the rest of the team members drive m-1 EVs to the (same) planned underflow station, while the driver of the relocation vehicle follows them. This process is repeated until the relocation procedure completes. The experimental results have demonstrated that each addition of a service staff may significantly decrease the relocation distance.

The previous algorithmic approaches are summarized in Table 13.4 where a classification is also given according to whether they are static or dynamic, bicycle or car repositioning techniques.

Table 13.4 Algorithmic Approaches on the Operational Repositioning of Bicycles/Cars				
	Static Bike Repositioning	**Static Car Repositioning**	**Dynamic Bike Repositioning**	**Dynamic Car Repositioning**
One commodity capacitated pickup and delivery formulation + Heuristic approach	Chemla et al. [62]			
MILP formulation + Heuristic approaches	Raviv et al. [60]	Kek et al. [24]	Contardo et al. [61]	
Maximal Expected Coverage Relocation Problem/ILP approach				Gendreau et al. [65]
Petri net modeling approach			Labadi et al. [36]	
ILP approaches				Briest and Raupach [59] Krumke et al. [68]
Heuristic approaches		Lee et al. [69] Lee and Park [70]		

4 ALGORITHMS FOR RIDE SHARING

Ride sharing is promoted as a way to better exploit unused car capacity, thus lowering fuel usage and transport costs. In the context of a vehicle-sharing system, ride sharing can be used to maximize the profit of the system by further optimizing cars usage and minimizing the number of unsatisfied customer requests in the case that there are no available cars in certain pickup stations and/or parking slots in drop-off stations. In this section we summarize algorithmic approaches that deal with challenges arising in the domain of ride sharing, in particular the proper assignment of driver's offers and requests in ride-sharing applications. All techniques aim at fast running times to allow real-time applications.

In Ref. [71] Geisberger et al. provide practical algorithms to compute detours in the context of ride sharing. They consider the scenario where queries of users wishing to get from an origin s to a destination t should be matched to offers from riders going from s' to t'. Two types of possible matches are distinguished. In case of a perfect fit, the sources s, s' and destinations t, t' of driver and rider, respectively, are identical. In a reasonable fit, small detours and additional stops are allowed. The goal is to find the offer for which the detour is minimized. Formally, the goal is to minimize $d(s', s) + d(s, t) + d(t, t') - d(s', t')$. The authors present an algorithmic approach to find reasonable fits for a set of offers and a single incoming request by using Dijkstra's algorithm [72] to compute the detour for each offer, and return the offer with minimum detour.

Using a well-known speed-up technique called contraction hierarchies [73], Geisberger et al. are able to achieve query times that are faster than the straightforward approach described above. This alternative approach exploits the structure of search spaces in contraction hierarchies. The search space consists of two independent parts, namely the forward and the backward search space. More specifically, assuming that there are k offers, for every incoming s–t request, k queries from t to t_i' need to be run (one for each offer t_i'). However, all these queries have exactly the same forward search space, so the forward search space only needs to be computed once. In addition to that, the results of backward searches can be precomputed for each offer t_i'. Each vertex in the backward search space of t_i' gets a bucket assigned to store the corresponding distances. Experiments show that using these techniques allows to answer incoming queries several orders of magnitude faster than the straightforward Dijkstra-based approach.

In Ref. [74] Abraham et al. present a fast algorithm, HLDB, to compute shortest path distances using preprocessed data based on hub labels [75]. Hub labels are sets of "important" vertices of a graph $G = (V, E)$. Each vertex $v \in V$ has a forward label $L_f(v)$ and a backward label $L_b(v)$. For each hub vertex $h \in L_f(v)$, they precompute and store the distance $d(v, h)$ from v to h. An s–t distance query then checks for a hub $h \in L_f(s) \cap L_b(t)$ that minimizes the distance $d(s, h) + d(h, t)$. To preserve correctness, the labels must fulfill the *cover-property*, that is, for any pair $s, t \in V$, $L_f(s) \cap L_b(t)$ must contain a vertex on a shortest path from s to t. Precomputing labels that fulfill these properties can be accomplished using a technique based on Contraction Hierarchies [73]. Several heuristics are added to improve the performance of the algorithm.

One special property of the technique presented in Ref. [74] is that it works entirely with a database, using SQL queries. Although their basic case considers peer-to-peer shortest path queries, they consider several extended scenarios, such as POI-queries and ride sharing. The scenario mentioned above is examined, where queries and offers are to be matched. Again, the goal is to find an offer for which the detour is minimized, that is, that minimizes $d(s', s) + d(s, t) + d(t, t') - d(s', t')$. Using the HLDB approach, the authors show how to solve this problem efficiently with simple database operations. To answer queries, a table *offers* is created containing the four columns *id, source, target,* and *distance*. A second table *offers_labels* contains the four columns *id, hub_forward, hub_backward,* and *distance*. For every offer *(s', t')*, each combination *(h, h')* stores an offer *ID* in *id, h'* in *hub_forward, h* in *hub_backward* and the distance $d(s', h) + d(h', t') - d(s', t')$ in *distance*. Computing the minimum distance can now be done with loops over all possible combinations.

In order to allow for more flexible scenarios, Drews and Luxen [76] introduce multihop scenarios, where users can even transfer between cars of different drivers. Such transfers occur at designated stations $S \subseteq V$ (eg, parking lots). Their scenario extends the previous approaches by adding time dependency. Offers and requests are given as triples *(s, t, τ)*, where *s* and *t* are vertices, and *τ* is a departure time. Given all stations s_i at which transfers occur and possible waiting times $\omega_m(s_i)$ for a match *m*, the total duration of a journey equals

$$d(m) = \sum_{i=0}^{t} (\omega_m(S_i) + d(S_i, S_{i+1}))$$

One way to model the resulting time-dependent scenario is to use time-dependent graphs [77]. The authors extend this model by introducing the so-called "slotted time-expanded graphs." Here, the continuous time divided into equal-sized time slots, and departures are assumed only to happen at the end of such slots. This results in a directed acyclic graph, where finding the best fit for a certain request is done by running Dijkstra's algorithm. An A^*-variant is used to achieve speedup by about two orders of magnitude. The experimental study also evaluates the quality of their solutions and shows that request and offers are well matched by the proposed techniques.

5 RESEARCH CHALLENGES AND FUTURE PROSPECTS

The viability of bicycle/car-sharing systems largely depends on their effective strategic design, management and operation. Along this line, three separate planning aspects are identified: (1) strategic (long-term) network design comprising decisions about the location and the number of stations as well as the vehicle stock at each station; (2) tactical (mid-term) incentives for customer-based distribution of bicycles/cars, that is, incentives given to users so as to leave their vehicle to a station different to that originally intended (this may be regulated through pricing schemes adaptable to the system state); (3) operational (short-term) operator-based repositioning of bicycles/cars based on the current state of the stations as well as aggregate statistics of the stations' usage patterns.

However, each of the abovementioned planning aspects raises considerable research challenges:

- Strategic design should balance among the system's intended quality of service and investor cost. On one hand, the investor's cost includes the facility costs of bike docks or dedicated parking stations, acquisition and maintenance costs for vehicles, vehicle depreciation costs, and routine relocation costs. On the other hand, the level of service provided by the system mainly depends on vehicles' availability to satisfy user demand distribution in space and time.
- Incentives should be carefully designed so as to align the travel behavior of the users with the system's pursuit, which dynamically adapts to real-time demand patterns. In particular, the incentivization scheme should aim at increasing the number of served customers, offer meaningful and attractive alternatives to incentivized customers and minimize the revenue losses for the operator.

- The operational repositioning of vehicles differs considerable among bike and car-sharing systems. In the former, bikes are carried in bunches by dedicated vehicles whereas in the latter cars are relocated by groups of drivers. Both represent tough optimization problems, wherein the objective is to minimize relocation cost while satisfying user demand. Furthermore, the effectiveness of vehicle relocation schemes should ideally consider future demand so as to proactively ensure the availability of vehicles areas in which demand is expected to rise.

In the sequel of this section open research issues relevant to the abovementioned issues are identified.

Strategic design of bicycle-sharing systems. The design of bicycle-sharing systems is a particularly tough exercise as it involves several design decisions: the location and capacities of bike stations; the vehicles fleet size; the creation of bicycle lanes connecting bike stations. These design decisions should ensure sufficient coverage (ie, satisfy user demand with appropriate quality of service), while minimizing investment and maintenance costs. Despite the growing body of relevant literature, algorithmic approaches tackling this issue are very few. Given the problem's complexity and the metropolitan scale of realistic instances, heuristics represent a suitable method for efficiently deriving near-optimal solutions. Existing algorithmic approaches on hub location problems, maximal covering models and joint location inventory problems are expected to provide a suitable starting point for optimizing the design of bicycle-sharing systems. Along the same line, the mathematical modeling of bicycle-sharing systems should be refined so as to capture several system variables and constraints overlooked by existing models:

- Travel demands may largely vary over a day (eg, residential areas typically act as trip generators early in the morning and attractors late in the afternoon). It would therefore be helpful to develop a formal model incorporating demand variation and to evaluate the influence of demand variation on the system design and routing choices.
- Decisions on the establishment of new bicycle lanes between bike stations should take into consideration the existing street network structure (unlike existing models which simplistically consider direct links between stations [42]). Clearly, existing bicycle lanes infrastructure should be exploited in order to reduce facilities cost. Moreover, new lanes should be setup taking into account the attractiveness of alternative options with respect to distance, bicycle friendliness (eg, road segments with high motor traffic are less friendly to bikers than pedestrian zones), flatness, etc. Notably, OpenTripPlanner (OTP[g]), the leading open source platform for multimodal trip itinerary planning, already supports the provision of such information[h].
- The reallocation of bicycle stock is commonly practiced by shared bicycle system operators to enable balanced distributions among stations and allow coverage of anticipated demand. Given that bicycle reallocation largely contributes to maintenance cost, model formulations should consider this cost factor so as to influence the overall system design.

---

[g]http://opentripplanner.com/
[h]OpenTripPlanner relies on general transit feed specification (GTFS) (https://developers.google.com/transit/gtfs/) data to describe public transportation schedules and routes. It can use OpenStreetMap (http://www.openstreetmap.org/) or commercial data sources for data on sidewalks, bicycling infrastructure, and streets. It allows users to plan a trip that can combine multiple modes of transportation, such as cycling or walking to reach public transportation, while it can also incorporate several popular bike-sharing systems (http://wiki.openstreetmap.org/wiki/OpenTripPlanner).

Mobility on demand combined with dynamic incentives. It seems that incentives-based mechanisms represent the most promising way to deal with the major problem of car-sharing systems, that is, the fleet redistribution issue or asymmetric demand/offer. In effect, incentives motivate fleet redistribution and tackle the demand/offer asymmetry problem. This mechanism is based on real time bi-univocal information between the user and the system, allowing to modify, not only the drop off and pick up station, but also other trip parameters such as the routing options for moving from a location A to a location B the time for picking up or dropping off the car, suggest trip sharing with another user going along the same route, etc. Thus, it might occur that moving from a location A to a location B has different prices depending on the incentives or penalties offered and accepted by the user. Incentives must be managed in real time and the system should be adaptive and possess some kind of intelligence to infer/plan each user behaviors/tendencies, so that, it is able to offer a particular user the "right" incentive (ie, attractive enough for the user to modify his/her initial plan but adjusted enough to maximize the benefit of the fleet manager). Incentives may be offered in two forms: in kind or in price. "In kind" incentives refer to discount vouchers or special offers for services -directly or indirectly- relevant to mobility. For example, it could be a 15% discount on a restaurant, or free laundry service or allowance to top price range vehicles in the system. Of course, there should be previous agreement among cooperating establishments (offering these "in kind" incentives) and the fleet management authority. Price incentives refer to discounts on actual or future trip fares and exclusively involve fleet management services. Finally, taking the incentives scheme to the extreme, there could be a way to make it explicit to the users. When the asymmetric demand problem deteriorates, the fleet manager could "offer" to users - deliberately subscribed for this purpose - an attractive incentive to drive a vehicle from A to B. The user answering positively would earn future discounts or even monetary rewards for driving the car from A (place with low demand) to B (place with high demand). This option could be seen as a contractor-based redistribution system. However, the use of this incentive tactic should be implemented in severe asymmetry situations because of the high "redistribution" trip costs incurred by the fleet manager.

Vehicle relocation and effective reward schemes. Car relocation is deemed as a necessary instrument to restore the desirable allocation of vehicles among stations in car-sharing systems. Having to adapt to user demand dynamics, car relocation activities are typically needed several times on the course of a day; hence, relocation decisions are bound to time constraints. Given the complexity of the problem, heuristics represent a reasonable algorithmic tool to meet the strict time requirements. However, the algorithmic state of the art in dynamic vehicle relocation in car-sharing systems leaves a lot to be desired. For instance, the results obtained by the greedy approach of Lee et al. [69] could be significantly improved by approaching car relocation as a k-server problem (regarding the employed drivers of the car-sharing operator as servers that handle relocation requests). Moreover, the problem of optimally assigning employed drivers to cars to be relocated and transferring the drivers to the stations where those cars reside has not been studied, although being an essential part of the relocation process. The provision of incentives to customers has also been recognized as a cost-effective means of tackling the problem of unbalanced car distribution among stations in car-sharing systems. The benefit of incentive provision models has been evidenced by several simulation studies (see Section 3.2). In real-world systems, though, users indicating willingness to take advantage of a reward scheme would expect meaningful alternatives. For instance, a customer would consider delivering a car to a station further than that originally planned, under the condition that s/he could transfer to a transit service and reach his actual destination location with reasonably small delay. Furthermore, such meaningful recommendations should maximize the utility for the system (eg, incentivize the customer to undertake the most urgent, among pending, relocation) and should be derived in real time. Last, rewards (ie, rental discount) need to be adjusted so as to compensate the user enough for delaying his/her arrival time (or even having to pay for a transit service ticket), while minimizing revenue cost for the operator. To the best of our knowledge, no algorithmic methods have been proposed so far deriving concrete alternatives so as to effectively incentivize customers. Hence, this represents a particularly promising research topic.

Proactive vehicle relocation based on predicted demand. Contemporary vehicle-sharing systems take a reactive approach to handling user demand, wherein vehicles are relocated from station with surplus to those with shortage of vehicle stock, as soon as uneven vehicle distribution is detected. Given the highly dynamic nature of user demand, such relocations are likely to prove ineffective, for example, relocate vehicles to stations with relatively low stock and yet to remain unused. The use of historical data and demand prediction models may, however, give effect to more effective relocation strategies. For instance, a vehicle depot located nearby office premises with a few parked cars may be reasonable to supply before the end of the business hours. This relocation may be undertaken either by operator employees (relocators) or incentivized customers. In the special case of EV-sharing systems, demand prediction may be used to identify which vehicles (among those parked at a specific depot) should be relocated; for instance, vehicles with high battery level may be more appropriate to relocate to a station at a time that relatively long rides are expected to be requested. Furthermore, the limited range and the long charging of EVs give reasons to innovative incentivized schemes. For instance, in the event of a request issued at 20 pm for a 25 km ride towards a suburb where high user demand is not expected before 7 am, the customer could be incentivized to use a vehicle with battery status providing 35 km autonomy, which requires 8 h to be fully charged.

6 CONCLUSIONS

Vehicle sharing represents an emerging transportation scheme which may comprise an important link in the green urban mobility chain. One-way vehicle-sharing systems employ a flexible rental model (customers are allowed to pick-up a vehicle at any station and return it to any other station) which best suits typical urban journey requirements. However, the so-called demand-offer asymmetric problem (ie, the unbalanced offer and demand of vehicles) typically experienced in one-way-sharing systems severely affects their economic viability as it implies that considerable human (and financial) resources should be engaged in relocating vehicles to satisfy customer demand.

The design and management measures aiming at alleviating imbalances in the availability of bicycles/cars, are distinguished into three separate planning horizons [41]: (1) Strategic network design comprising decisions about the location and the number of stations as well as the fleet size at each station, (2) tactical incentives for customer-based distribution of bicycles/cars and (3) operational (operator-based) repositioning of bicycles/cars based on the current state of the stations as well as aggregate statistics of the stations' usage patterns. This chapter presents an extensive literature survey on models and algorithmic techniques for the design, operation, and management of vehicle-sharing systems. Different approaches applied either to bike or to car-sharing systems, are described and classified according to the involved solution method. Also, open research problems relevant to the abovementioned issues are identified highlighting important research issues that need to be addressed in the future.

ACKNOWLEDGMENT

This work has been supported by the EU FP7 Collaborative Project MOVESMART (Grant Agreement No 609026).

REFERENCES

[1] Midgley P. Bicycle-sharing schemes: enhancing sustainable mobility in urban areas. Background paper no. 8, CSD19/2011/BP8, Commission on Sustainable Development, Department of Economic and Social Affairs, United Nations; 2011.
[2] Shaheen SA, Guzman S, Zhang H. Bikesharing in Europe, the Americas, and Asia. Transport Res Rec 2010;2143:159–67.
[3] DeMaio P. Bike-sharing: history, impacts, models of provision, and future. J Public Transport 2009;12(4):41–56.
[4] Social bicycles (sobi). <http://socialbicycles.com/>; February 2014 [Online].
[5] Katzev R. Car sharing: a new approach to urban transportation problems. Anal Soc Issues Public Policy 2003;3(1):65–86.
[6] Shaheen SA, Cohen AP, Roberts JD. Carsharing in North America: market growth, current developments, and future potential. Transp Res Rec 2006;1986(1):116–24.
[7] George D, Xia C. Fleet-sizing and service availability for a vehicle rental system via closed queuing networks. Eur J Oper Res 2011;211(1):198–207.
[8] Barth M, Todd M. Simulation model performance analysis of a multiple station shared vehicle system. Transport Res Part C 1998;7(4): 237–259; Elsevier.
[9] Touati-Moungla N, Jost V. Combinatorial optimization for electric vehicles management. J Energy Power Eng 2012;6(5):738–43.
[10] Curran A. Translink public bike system feasibility study. Vancouver: Quay Communications Inc.; 2008.
[11] Heinen E, van Wee B, Maat K. Commuting by bicycle: an overview of the literature. Transport Rev 2010;30(1):59–96.
[12] Dell'Olio L, Ibeas A, Moura JL. Implementing bike-sharing systems. Proc ICE 2011;164:89–101.
[13] Martens K. Promoting bike-and-ride: the Dutch experience. Transport Res Part A 2007;41(4):326–38.
[14] Sayarshad H, Tavassoli S, Zhao F. A multi-periodic optimization formulation for bike planning and bike utilization. Appl Math Modell 2012;36(10):4944–51.
[15] Barth M, Han J, Todd M. Performance evaluation of a multi-station shared vehicle system. In: Proceedings of the 2001 IEEE intelligent transportation systems (ITS'2001); 2001. p. 1218–1223.
[16] Weikl S, Bogenberger K. Relocation strategies and algorithms for free-floating car sharing systems. In: Proceedings of the 15th international IEEE conference on intelligent transportation systems (ITSC'2012); 2012. p. 355–360.
[17] Parent M, Gallais G. Intelligent transportation in cities with cts. In: Proceedings of the 5th IEEE international conference on intelligent transportation systems (ITSC'2002); 2002. p 826–830.
[18] Bruglieri M, Colorni A, Lue A. The vehicle relocation problem for the one-way electric vehicle sharing. <http://arxiv.org/pdf/1307.7195v1.pdf>; 2013 [Online].
[19] Artmeier A, Haselmayr J, Leucker M, Sachenbacher M. The shortest path problem revisited: Optimal routing for electric vehicles. In: Proceedings of the 33rd annual German conference on AI (KI '10), lecture notes in computer science; 2010. p. 309–316.
[20] Sachenbacher M, Leucker M, Artmeier A, Haselmayr J. Efficient energy-optimal routing for electric vehicles. In: Proceedings of the 25th conference on artificial intelligence (AAAI-11); 2011.
[21] Kondo S, Ogura M, Takei A. Electric vehicle sharing system. US Patent No. 6,181,991; 2001.
[22] Chauvet F, Hafez N, Proth J-M. Electric vehicles: effect of the availability threshold on the transportation cost. Appl Stoch Model Bus 1999;15(3):169–81.

[23] Schuster A, Bessler S, Gronbaek J. Multimodal routing and energy scheduling for optimized charging of electric vehicles. e & i Elektrotechnik und Informationstechnik 2012;129(3):141–149.

[24] Kek AGH, Cheu RL, Meng Q, Fung CH. A decision support system for vehicle relocation operations in carsharing systems. Transport Res Part E 2009;45(1):149–58.

[25] Correia GHDA, Antunes AP. Optimization approach to depot location and trip selection in one-way carsharing systems. Transport Res Part E 2012;48(1):233–47.

[26] Fan WD, Machemehl RB, Lownes NE. Carsharing: dynamic decision-making problem for vehicle allocation. Transp Res Rec 2008;2063(1):97–104.

[27] El Fassi A, Awasthi A, Viviani M. Evaluation of carsharing network's growth strategies through discrete event simulation. Expert Syst Appl 2012;39:6692–705.

[28] Nourinejad M, Roorda MJ. A dynamic carsharing decision support system. Transport Res Part E 2014;66:36–50.

[29] Di Febbraro A, Giglio D, Sacco N. Urban traffic control structure based on hybrid petri nets. IEEE Trans Intell Transport Syst 2004;5(4):224–37.

[30] Julvez J, Boel.F R.K. A continuous petri net approach for model predictive control of traffic systems. IEEE Trans Syst Man Cybern 2010;40(4):686–97.

[31] Tolba C, Lefebvre D, Thomas P, El Moudni A. Continuous and timed petri nets for the macroscopic and microscopic traffic flow modelling. Simul Modell Pract Th 2005;13(5):407–36.

[32] List GF, Cetin M. Modeling traffic signal control using petri nets. IEEE Trans Intell Transport Syst 2004;5(3):177–87.

[33] Abbas-Turki A, Bouyekhf R, Grunder O, El Moudni A. On the line planning problems of the hub public-transportation networks. Int J Syst Sci 2004;35(12):693–706.

[34] Bouyekhf R, Abbas-Turki A, Grunder O, El Moudni A. Modelling, performance evaluation and planning of public transport systems using generalized stochastic petri nets. Transport Rev 2003;23(1):51–69.

[35] Takagi R, Goodman CJ, Roberts C. Modelling passenger flows at a transport interchange using petri nets. Proc Inst Mech Eng 2003;217(2):125–34.

[36] Labadi K, Benarbia T, Darcherif AM. Petri nets models for analysis and control of public bicycle-sharing systems. In: Petri nets-manufacturing and computer science. InTech; 2012.

[37] Clemente M, Fanti M-P, Mangini A, Ukovich W. The vehicle relocation problem in car sharing systems: modeling and simulation in a petri net framework. In: Application and theory of petri nets and concurrency, LNCS, vol. 7927. Berlin Heidelberg: Springer; 2013. p. 250–269.

[38] Froehlich J, Neumann J, Oliver N. Sensing and predicting the pulse of the city through shared bicycling. In: Proceedings of the twenty-first international joint conference on artificial intelligence (IJCAI-09) 2009. p. 1420–1426.

[39] Kaltenbrunner A, Meza R, Grivolla J, Codina J, Banchs R. Urban cycles and mobility patterns: exploring and predicting trends in a bicycle-based public transport system. Pervasive Mob Comput 2010;6(4):455–66.

[40] Borgnat P, Abry P, Flandrin P, Robardet C, Rouquier J-B, Fleury E. Shared bicycles in a city: a signal processing and data analysis perspective. Adv Complex Syst 2011;14(3):415–38.

[41] Vogel P, Greiser T, Mattfeld DC. Understanding bike-sharing systems using data mining: exploring activity patterns. Procedia 2011;20(0):514–23.

[42] Lin JR, Yang TH. Strategic design of public bicycle sharing systems with service level constraints. Transport Res Part E 2011;47(2):284–94.

[43] Martinez LM, Caetano L, Eiro T, Cruz F. An optimisation algorithm to establish the location of stations of a mixed fleet biking system: an application to the city of lisbon. Procedia 2012;54(0):513–24.

[44] Boyaci B, Geroliminis N, Zografos K. An optimization framework for the development of efficient one-way car-sharing systems. In: Proceedings of the 13th Swiss transport research conference (STRC'13); 2013.

[45] Lin JR, Yang TH, Chang YC. A hub location inventory model for bicycle sharing system design: formulation and solution. Comput Ind Eng 2013;65(1):77–86.

[46] Alumur S, Kara BY. Network hub location problems: the state of the art. Eur J Oper Res 2008;190(1):1–21.

[47] Kumar VP, Bierlaire M. Optimizing locations for a vehicle sharing system. In: Proceedings of the Swiss transport research conference; 2012.

[48] Cepolina EM, Farina A. A new shared vehicle system for urban areas. Transport Res Part C 2012;21(1):230–43.

[49] Larsen J, Patterson Z, El-Geneidy A. Build it. But where? The use of geographic information systems in identifying locations for new cycling infrastructure. Int J Sustain Transport 2013;7(4):299–317.

[50] Rybarczyk G, Wu C. Bicycle facility planning using {GIS} and multi-criteria decision analysis. Appl Geogr 2010;30(2):282–93.

[51] Garcia-Palomares JC, Gutierrez J, Latorre M. Optimizing the location of stations in bike-sharing programs: a {GIS} approach. Appl Geogr 2012;35(1–2):235–46.

[52] Fricker C, Gast N. Incentives and regulations in bike-sharing systems with stations of finite capacity. <http://arxiv.org/abs/1201.1178>; 2013 [Online].

[53] Pfrommer J, Warrington J, Schildbach G, Morari M. Dynamic vehicle redistribution and online price incentives in shared mobility systems. IEEE Trans Intell Transport Syst 2013;15(4):1567–78.

[54] Waserhole A, Jost V. Vehicle sharing system pricing regulation: a fluid approximation. OSP, HAL Id: hal-00727041. <https://hal.archives-ouvertes.fr/hal-00727041v4> 2012 [Online].

[55] Waserhole A, Jost V. Pricing in vehicle sharing systems: optimization in queuing networks with product forms. <http://hal.inria.fr/docs/00/85/87/10/PDF/VSS_CompactForm_sep2013.pdf>; 2013 [Online].

[56] Waserhole A, Jost V, Brauner N. Vehicle sharing system pricing regulation: deterministic approach, complexity results. <http://hal.archives-ouvertes.fr/docs/00/82/81/92/PDF/Waserhole_VSS_ScenarioBasedApproach_May2013.pdf>; 2012 [Online].

[57] Singla A, Santoni M, Bartok G, Mukerji P, Meenen M, Krause A. Incentivizing users for balancing bike sharing systems. Association for the Advancement of Artificial Intelligence (www.aaai.org); 2015.

[58] Singla A, Krause A. Truthful incentives in crowdsourcing tasks using regret minimization mechanisms. In: Proceedings of the 22nd international conference on World Wide Web; 2013. p. 1167–1178.

[59] Briest P, Raupach C. The car sharing problem. In: Proceedings of the 23rd ACM symposium on parallelism in algorithms and architectures. ACM; 2011. p. 167–176.

[60] Raviv T, Tzur M, Forma IA. Static repositioning in a bike-sharing system: models and solution approaches. EURO J Transport Logistics 2013;2(3):187–229.

[61] Contardo C, Morency C, Rousseau LM. Balancing a dynamic public bike-sharing system. <https://www.cirrelt.ca/DocumentsTravail/CIRRELT-2012-09.pdf>; March 2012 [Online].

[62] Chemla D, Meunier F, Calvo RW. Bike sharing systems: solving the static rebalancing problem. Discrete Optim 2013;10(2):120–46.

[63] Hernandez-Perez H, Salazar-Gonzalez J-J. A branch-and-cut algorithm for a traveling salesman problem with pickup and delivery. Discrete Appl Math 2004;145(1):126–39.

[64] Di Febbraro A, Sacco N, Saeednia M. One-way carsharing. Transp Res Rec 2012;2319:113–20.

[65] Gendreau M, Laporte G, Semet F. The maximal expected coverage relocation problem for emergency vehicles. J Oper Res Soc 2003;57:22–8.

[66] Dantzig GB, Wolfe P. Decomposition principle for linear programs. Oper Res 1960;8:101–11.

[67] Benders JF. Partitioning procedures for solving mixed-variables programming problems. Numerische Mathematik 1962;4(1):238–52.

[68] Krumke SO, Quilliot A, Wagler AK, Wegener J-T. Relocation in carsharing systems using flows in time-expanded networks. <http://hal.archivesouvertes.fr/docs/00/90/82/42/PDF/kqww_carshar_flows.pdf>; November 2013 [Online].

[69] Lee J, Park GL, Kang M-J, Kim J, Kim H-J, Kim I-K, Ko Y-I. Design of an efficient matching-based relocation scheme for electric vehicle sharing systems. In: Proceedings of the international conference in computer applications for modeling, simulation, and automobile, vol. 341. Berlin Heidelberg: Springer; 2012. p. 109–115.

[70] Lee J, Park G-L. Design of a team-based relocation scheme in electric vehicle sharing systems. Proc Int Conf Comput Sci Appl (ICCSA' 2013) 2013;7973:368–377.

[71] Geisberger R, Luxen D, Sanders P, Neubauer S, Volker L. Fast detour computation for ride sharing. In: Proceedings of the 10th workshop on algorithmic approaches for transportation modeling, optimization, and systems (ATMOS'10), vol. 14 of OpenAccess Series in Informatics (OASIcs); 2010. p. 88–99.

[72] Edsger W, Dijkstra. A note on two problems in connexion with graphs. Numerische Mathematik 1959;1:269–71.

[73] Geisberger R, Sanders P, Schultes D, Vetter C. Exact routing in large road networks using contraction hierarchies. Transport Sci August 2012;46(3):388–404.

[74] Abraham I, Delling D, Fiat A, Goldberg AV, Werneck RF. HLDB: location-based services in databases. In: Proceedings of the 20th ACM SIGSPATIAL international symposium on advances in geographic information systems (GIS'12). ACM Press; 2012. p. 339–348.

[75] Abraham I, Delling D, Goldberg AV, Werneck RF. A hub-based labeling algorithm for shortest paths on road networks. In: Proceedings of the 10th international symposium on experimental algorithms (SEA'11), LNCS, vol. 6630. Springer; 2011. p. 230–241.

[76] Drews F, Luxen D. Multi-hop ride sharing. In: Proceedings of the 5th international symposium on combinatorial search (SoCS'12). AAAI Press; 2013. p. 71–79.

[77] Pyrga E, Schulz F, Wagner D, Zaroliagis C. Experimental comparison of shortest path approaches for timetable information. In: Proceedings of the 6th workshop on algorithm engineering and experiments (ALENEX'04). SIAM; 2004. p. 88–99.

SMART TRANSPORTATION SYSTEMS (STSs) IN CRITICAL CONDITIONS

14

M. Cello*, C. Degano†, M. Marchese*, F. Podda†

*Electrical, Electronics and Telecommunication Engineering and Naval Architecture Department (DITEN)—University of Genoa, Genoa, Italy; †Research and Development Business Unit—Gruppo SIGLA S.r.l., Genoa, Italy

1 INTRODUCTION

In the last decades, we have seen a constant evolution of the ICT technologies that have brought us to an even more "connected" world and changed the way we use to interact with things and people. This evolution involves two main fields: computing and communication technology. Even if those two paradigms started independently, with the coming of the internet it is almost impossible to think about computation without networking and vice versa. Nowadays the two paradigms are strictly related and this bond will probably strengthen in the future.

Fig. 14.1 shows the evolution described previously and highlights the beginning of the next ICT revolution, the Internet of Things (IoT) [1], where devices speak to each other and provide services, storage, and computation in a heavily distributed environment.

As can be seen from Fig. 14.1, along with the ICT evolution, new areas of research have emerged and new kinds of innovative services have brought us to the so-called "digital era." With the technology development, we have changed our way of life and most aspects of it are constantly influenced by this continuous progress. We are now in the era of "smart things," where technology helps us to perform new tasks or to optimize actions and behaviors we already carry on. A new area of interest that represents this kind of progress and involves, at large scale, citizens and their way of life is the so-called "smart city."

The definition of a "smart city" is not univocal and various shades of this term can be found. In Ref. [2] there is a collection of different definitions of this term, but all of them agree with the fact that a smart city is composed of a collection of interconnected computing technologies that cooperate to handle, in a smarter way, different aspects of the urban spaces (eg, traffic and mobility, infrastructures, security and quality of life, etc.). Furthermore, almost all the emerging smart cities technologies and solutions massively use the elaboration of huge amount of data coming from different sensors/sources to offer services that make the city "smarter." For this reason, the term smart city is often related to the Big Data research field [3,4].

Smart city field has emerged in 2005 in the United States [5] and has immediately caught the attention of the scientific community. Later on, many industries and municipalities started to be interested in smart cities, and projects in this area started to increase. Probably, one of the first involved the city of Rio de Janeiro, where it was created a system for fasten the response of municipality to an emergency. In the following years other cities follow a similar path, and also the growth of infrastructures and the

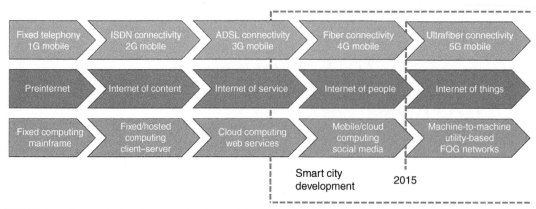

FIGURE 14.1 The Evolution of Communication and Computing Technologies

evolution of communication technologies have helped to find even more efficient and effective technology and solutions that have enriched smart city communities.

From the previous description, and looking at Fig. 14.1, it is clear why, currently, smart cities are considered as an emerging research topic: we have a very highly connected world, where computation is no more centralized, and where service-oriented architectures and IoT are opening the door to new interesting challenges in distributed and collaborative computing.

There are many research initiatives and projects that attest the interest of the scientific community in smart cities. In the last 3 years, the European Commission has allocated more than 350 Millions of Euro to the *Smart Cities and Communities* Work Programme and, concerning 2015, the call for projects on the same area in the Horizon 2020 framework has a budget of more than 100 millions of Euro. Furthermore, we have to consider that there are many other projects that are financed directly from the local municipalities or from the institutions (research ministries and so on). In Europe we can already find many cities that are making themselves smarter [6] with many pilot projects that involves all the aspect of the city: from the society to the infrastructure.

In Amsterdam [7], the "smart parking" project [8] aims to provide citizens a powerful instrument to book in advance a parking in a private or public space, preventing the so-called "search-for-a-parking traffic" and hence reduce the CO_2 emission and make roads safer. Another interesting project is the "Digital Road Authority" [9] that aims to shorten the response time of emergency services; The Digital Road Authority allows, for example, ambulances to arrive on time by linking them directly to the traffic control. In this way it is possible to clear traffic lanes for emergency services by controlling semaphores and roads access, providing emergency services the quickest route, also using the real-time traffic information.

In Barcelona, the "City OS" project [10] aims to build an entire platform for the smart city. This platform, that constitutes a sort of operating system of the city, will be the hub of all the information coming from the city (by sensors, municipal and non-municipal system databases, social media network data) and will offer a layer of functionalities serving as a basis or model for future city platforms.

In Italy, we can found a variety of initiatives, led by municipalities, devoted to collect, and coordinate research projects in this field. Examples are: *Genova smart city* [11], *Torino smart city* [12], and

Milano smart city [13]. Moreover, it is estimated that in Italy the already-done investments in the smart city research are of about 4 billions of Euro.

Another interesting aspect of the presence of municipalities in these activities is that research results and prototypes can be easily deployed and tested directly into the cities, with positive effects for the visibility of the research institutions and for the chance to put a product on the market with less effort also reducing the time-to-market. This convergence of interest has brought to the constitution of "Living Labs," open environments in the city for the design, test, and validation of new products and services devoted, for example, to Intelligent Mobility. In a Living Lab, users can interact and experiment prototypes, providing very important feedbacks in terms of refinement and potential improvements.

As said before, the smart city area covers different aspects of a city and involves different research topics, starting from citizens' life to environment preservation through transportation and mobility. Each of them represents a sort of branch of the main topic and they are strictly related to each other. Actually every aspect of a city must take into account the other ones to maximize the final impact on the city itself.

The topic covered in this chapter concerns mobility, because traffic management and public transportation is one of the most important problem of a modern city and involves many of the aforementioned aspects. When we talk about transportation systems, we need to consider how technologies have already improved them, and how new cutting-edge tools and paradigms can help to foster further progresses (Fig. 14.2).

ITS Category	Specific ITS Applications
1. Advanced traveler information systems (ATIS)	Real-time traffic information provision
	Route guidance/navigation systems
	Parking information
	Roadside weather information systems
2. Advanced transportation management systems (ATMS)	Traffic operations centers (TOCs)
	Adaptive traffic signal control
	Dynamic message signs (or "variable" message signs)
	Ramp metering
3. ITS-enabled transportation pricing systems	Electronic toll collection (ETC)
	Congestion pricing/electronic road pricing (ERP)
	Fee-based express (HOT) lanes
	Vehicle-miles traveled (VMT) usage fees
	Vatiable parking fees
4. Advanced public transportation systems (APTS)	Real-time status information for public transit system (eg, bus, subway, rail)
	Automatic vehicle location (AVL)
	Electronic fare payment (eg, smart cards)
5. Vehicle-to-infrastructure integration (VII) and vehicle-to-vechicle integration (V2V)	Cooperative intersection collision avoidance system (CICAS)
	Intelligent speed adaptation (ISA)

FIGURE 14.2 Intelligent Transportation Systems Category [14]

2 SMART TRANSPORTATION SYSTEMS

In the field of smart cities, a central role is covered by the solutions that aim to solve or improve different aspects of the urban mobility at varying levels, from pedestrian to freight. In the evolution of smart city technologies, we have seen a transition from the intelligent transportation systems (ITSs) to the STSs.

Intelligent transportation systems (ITS), also defined as advanced use of information and communication technology (ICT) in the transportation context [15], represent a major transition in transportation on many dimensions, improving the safeness, effectiveness and efficiency of surface transportation systems through advanced technologies in information systems, communications, and sensors.

Particularly, ITS can be defined as IT-based applications which enable elements within the transportation system (ie, traffic lights, roads, vehicles, etc.) to communicate with each other through wireless technologies, aiming to [14]:

- provide innovative services concerning different transport and traffic management modes;
- enable various users to be better informed;
- make safer, more coordinated, and "smarter" use of transport networks;
- improve transportation system performance reducing congestion and increasing safety and traveler convenience.

Nowadays ITS can be seen as an infrastructure platform able to provide product and services that can be grouped within five main categories [14]:

1. Advanced traveler information systems provide drivers with real-time information, such as transit routes and schedules, navigation directions, and information about delays due to congestion, accidents, weather conditions, or road repair work;
2. Advanced transportation management systems focus on traffic control devices such as traffic signals, ramp metering, variable message signs in order to provide drivers real-time messaging about traffic status;
3. ITS-enabled transportation pricing systems include systems such as electronic toll collection (ETC), congestion pricing, fee-based express (HOT) lanes, and vehicle miles traveled (VMT) usage-based fee systems;
4. Advanced public transportation systems deal with applications such as automatic vehicle location (AVL), allowing trains and buses to report their position so passengers can be informed of their real-time status, that is, arrival and departure information;
5. Fully integrated ITSs, such as vehicle-to-infrastructure integration (VII) and vehicle-to-vehicle (V2V) integration, enable communication among assets in the transportation system, for example, from vehicles to roadside sensors, traffic lights, and other vehicles.

The concept of transportation system is moving from intelligent transportation to smart transportation, also thanks to the advent of the IoT whose main characteristic is to link heterogeneous devices and technologies such as sensors, WiFi, RFID, smartphone, so giving the possibility to make transportations *cleverly smart.*

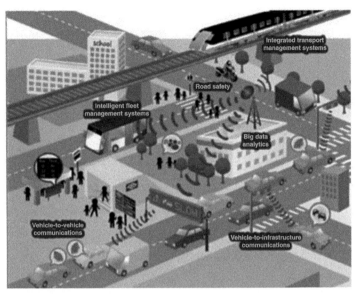

FIGURE 14.3 STS Scenario [16]

Fig. 14.3 shows a doable STS scenario in which:

- *Drivers and pedestrians* could get an increment of safety through the development of applications such as real time traffic alerts, collision avoidance, cooperative intersection collision avoidance, crash notification systems and, thinking to the smart concept, advanced safety vehicle (ASV) aimed at offering safer and smart driving via vehicle-to-vehicle communication;
- *Transportation networks* could improve the operational performance, particularly by reducing congestion through applications acting on real time traffic data in order to, for example, optimize traffic signal lights and ramp metering;
- *Personal mobility* could be enhanced through ICT-based applications able to provide the users (pedestrians and drivers) with real-time traveler information systems devoted to route selection and navigation capability;
- *Environmental benefits* may arise from the automotive point of view by equipping cars/motors with "eco-driving" applications that optimize driving behavior providing feedback to the driver on how to operate the vehicle [14].

A city that adopts an intelligent management of transportation and public mobility can benefit of different advantages, improve the quality of life, and preserve the environment.

According to the definition in Ref. [17], STSs use advanced and emerging technologies to deliver an end-to-end solution to transportation. A modern STS is composed of different actors shown in Table 14.1.

The next evolution of STSs takes into account the public transportation as an important part of mobility and is particularly devoted to the development of the new generation of the *advanced public*

Table 14.1 Principal Actors in a STS	
Sensors	Sensors and networks of sensors. They can be fixed (eg, traffic sensors, cameras, etc.) or mobile (eg, vehicles)
Decision center	One or *more control and decision* centres, where all the data are collected and where algorithms are executed and decisions taken
Actuators	A network of *actuators* (eg, semaphores, mobile barriers, etc.) to remotely control traffic
Information systems	Information systems to send messages to citizens/vehicles and public safety users (eg, Variable Messages Panels, SMS services, etc.);
Users	Citizens; police or other institutions that interact with the STS to provide information or to request data; STS services that control all the sensors and actuators, gathering information or sending specific commands; other actors that use STS services
Networks	Networks that interconnect all the elements mentioned previously and guarantee a certain level of quality of service

transportation systems (APTSs). APTSs are aimed at handling the main aspects of public and private mobility and at enhancing the concepts depicted in Fig. 14.3.

2.1 USERS AND MAIN APPLICATIONS IN AN STS

Different kinds of users can access the network and receive different advantages from an STS:

Private users: for ordinary users, to be aware of traffic conditions could allow them to plan the trip, and to avoid congested routes and/or temporarily closed roads. Thanks to the availability of real-time updates, private users will possibly change their route to avoid traffic jams created in the meantime, due, for example, to a road accident. In the following a possible list of user applications:

- intermodal route planner (APTS/private/mixed);
- availability/reservation/payment of parking areas;
- traffic information (accidents, congestion, etc.);
- information on road maintenance (eg, street cleaning);
- other information such as taxi areas, bus stops and timetables, car sharing, mo-bike;
- proximity services such as notification entry in limited traffic zones, availability of parking spots, information centers, pharmacies or points of public interest, events, museums;

Institutional operators: STS are able to offer a significant added value to city authorities in the field of traffic management. Advantages range from the management of traffic flows to urban access management, from public transport to urban logistics. In more detail:

- supervision and control of traffic;
- crisis management;
- innovative services for city logistics;
- information to local travelers, passengers, and logistics operators;
- collection of information on user behaviors and new mobility patterns;
- analysis user satisfaction level with respect to mobility metrics;
- information service for users such as visitors, workers, landed vehicles, of given areas (not only the city but also the port);

Public transport operators: STS may improve the following:

- planning (frequency, timetables, type of vehicle);
- rescheduling for abnormal events;
- real time information provision to drivers;

Logistics operators: STSs include a wide range of applications not only for passengers but also for the freight sector. In this area principal STS applications include: electronic tolling, dynamic traffic management (management of variable speed limits, reservations and parking guidance, support for real-time navigation), real-time information systems, driver assistance such as warning systems, stability and driving style control. STS can also make easier the connection of different modes of transport, for example by means of integrated multimodal travel planning tools or of monitoring services for the co-modal transport of goods.

Public safety operators: the knowledge, through the STS, of the urban traffic situation, enables significant advantages to security operators like Civil Protection, Fire Department, and Police Forces who can choose the fastest route in terms of traffic to successfully complete their operations. The following table summarizes the different advantages depending on the operator category:

- *Civil protection*: Knowledge of the territory and definition of the risk scenario; definition of the defense structure identifying, and mapping all the resources available to deal with the emergency; management of actions in case of emergency, through the operations room whose job is to coordinate the participation of municipal forces; provision of adequate information to citizens about the degree of risk exposure.
- *Firefighters*: They help provision during fires, uncontrolled releases of energy, sudden or threatening structural collapse, landslides, floods, and other public calamity, and, as in the previous case, definition of the defense structure identifying and mapping all the resources available to deal with the emergency;
- Law enforcement investigations of vehicles/people; support for intervention in the field.

STS services control all sensors and actuators, gathering information or sending specific commands.

It is clear from the previous list that the level of importance of the actors is not the same. At a first glance we can say that we can represent the users of the STS in a pyramid that defines the level of priority, as shown in Fig. 14.4.

We can reasonably argue that a citizen that requests an information to the STS service is less important than a police officer who is signaling some risk situations. Similarly, the operation of gathering information from sensors by the STS is typically more important than a routinely information request of an institution to the STS service. This can be a very first-level of granularity of the priority space that can be implemented in an STS. However, when we face with critical conditions, this schema is no longer feasible because it does not take into account the situation of the city when a request is made. Starting from this point, we can define a priority class for each actor, which represents its "importance" inside the STS. Moreover, since the "importance" may depend not only on the actor but also on the type of request and on the particular situation (critical, normal, etc.), we can define an extended concept of priority considering all these aspects. In more detail, the priority can be seen as a three-dimensional function as shown subsequently.

$$p = f(u,c,r)$$

Where p is the priority, u is the user that is making the request, c is the condition of the city, and r is the request type.

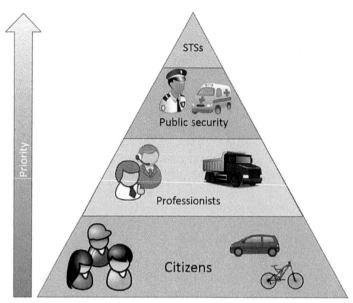

FIGURE 14.4 STS User Priority

Priority management helps provide different kind of service levels within different utilization scenarios. It is worth noting that such rules have a key role in emergency conditions, because if we do not have any problems in the infrastructure we can probably handle all the requests and services without any issue.

2.2 STS IN CRITICAL CONDITIONS

As said previously, the paradigm of IoT is improving the level of intelligence and information of transportation systems allowing intelligent recognition, location, tracking, and monitoring of mobility by exchanging information and communicating efficiently. This means that intelligent IoT–enabled transportation systems improve capacity, enhance travel experiences, and make moving safer and more efficient; for example, emergency and other police services can use sensor networks along with smart traffic management to gain citywide visibility, to help alleviate congestion, and to rapidly respond to incidents [17].

On the other hand, existing IoT devices and related infrastructures have to face vulnerabilities and weaknesses that can affect STS efficiency. Paramount to the success of any transportation system is the efficiency in terms of safety and security, integrity, confidentiality and availability of information and services. Smarter systems, as highlighted previously, can improve them. An STS can improve security by detecting and evaluating threats through the analysis of passenger information, electronic surveillance and biometric identification and ensuring in the same time that passenger and cargo data are only accessible to authorized personnel.

However, since all smart transportation solutions rely essentially on computing and networking, it is clear that the failure of one of these components represents a serious threat to the entire system. For this reason, resilience to failure is very important for a STS. It is a safety-critical system and wrong decisions can cause accidents and risks for the public security. In an STS, where almost all components are distributed, the failure of a single part is more than an exception. We can have different kind of failures depending on different causes; in this chapter, we consider failures depending on critical conditions.

We can define a critical condition as an exceptional situation where we have an emergency in a part of a city (or in the entire city) and where we probably have damages to infrastructures, things and people. The most popular critical conditions are caused by natural event such as floods, hurricanes and earthquakes, or by events such as terroristic attacks, blackouts, blazes, and environmental disasters. We have to take into account such events when implementing a STS in order to prevent situations where a damage of a part of our infrastructure can cause a general breakdown of the entire system. Furthermore, we have to consider that, in case of emergency or critical conditions, network loads may change unpredictably and quickly due to high amount of requests that users may perform, making the access to network services more difficult for users. So, we have to take into account these kinds of failures when we consider STSs:

Damages or failures of computational elements: Regarding computational elements, we have to consider that a service interruption due to a failure can lead to different problems. For example, if the service that regulates road traffic goes down for some reason (eg, hardware fault or blackout) we will probably have congestions and possible hazards for citizens. Countermeasures to these faults must be taken into consideration when we deploy the service. Concerning this aspect, we can reasonably say that today the trend is to host almost all important STS Services in a private datacenter or in a public cloud. Since modern datacenters massively use virtualization and data replication to ensure heavy fault-tolerance, one or more failures can be recovered without losing data or having service interruptions. Furthermore, datacenters are equipped with generators that guarantee electrical power even in case of blackouts and give the operators time to respond to these criticisms.

Damages or failures of sensors or actuators: Damages or failures of sensors and actuators generally do not represent a serious threat to the entire STS and can be resolved in different way) to replicate sensors in critical position to ensure that a failure of a sensor can be recovered) to temporary replace sensor or actuators with human beings (eg, the police) if the situation is critical.

Damages of network segments or failures of network devices: Differently from sensors and actuators, a failure of the telecommunication network could represent a serious problem to the STS. Since we have seen that all the STS elements rely on the network (Fig. 14.5), we can say that the communication infrastructure is a crucial point in the overall system, and its failure must be resolved as soon as possible avoiding connection interruptions and data loss. Damages to the network may occur in different places at different levels, due to the highly distributed nature of the network itself. Furthermore, damages at different levels cause different problems that a robust network must be able to tackle.

Examples of damage could be represented by a hurricane that gets down some antennas devoted to mobile communications, or a flood that damages the access network. In any case, we are facing with different problems and consequent different solutions. In some cases, reconfiguring the network to restore the service may be enough. Other situations may require to handle congestion due to the change of network topology and to the high network utilization. In other cases it may be important to guarantee to some "privileged" users the network access and ensure certain level of service.

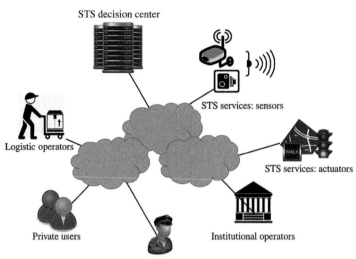

FIGURE 14.5 STS Actors Connected Through Telecommunication Networks

2.3 SECURITY ASPECTS IN STS

Two fundamental challenges arise in the context of STS: security and privacy. Security includes illegal access to information and attacks causing physical disruptions in service availability. As digital citizens are more and more instrumented with data available about their location and activities, privacy seems to disappear. Privacy protecting systems that gather data and trigger emergency response when needed are technological challenges that go hand-in-hand with the continuous security challenges [18–20]. To highlight the security/privacy problems in STS we describe the possible security challenges in three scenarios:

Private user that requires an intelligent route planner from the STS: The route planner request to the STS by a private user and the consecutive answer from STS (ie, the suggested route) constitutes sensitive information since a third party (ie, hacker) could trace the movement of the user. For these reasons there is a need of confidentiality requirement between the client device and STS system in order to prevent that sensitive information are accessible by third parties. The analysis over time of the requests from the same user could give information about usual route and itinerary of the user. This is serious risk of invasion of privacy. For this reason, STS will avoid to recognize user itinerary habits, except, as an option, the case of customized user experience through user authentication.

Emergency alerts to private users and logistics operators: Emergency alerts from institutional/public safety operators have not confidentiality or privacy requirements because of their broadcast nature. Given the importance of the service and the consequences, in terms of public safety, of any abusive emergency alerts, STS needs a strict access control for sending alerts. Emergency alerts have high priority because they have to alert the user to the presence of a critical conditions. In critical situation it is necessary that the messages are sent to users with a high degree of reliability, since the no reception of the warning message may pose a hazard to public safety.

Public safety operators that require/receive images/video: The exchange of videos between public safety operators and STS is an operation aimed at solving a critical situation. From the point of view of availability this kind of communications need to be highly available since is used by public safety operators. Whereas the information exchanged between users and the system is sensitive, it is necessary the use of appropriate tools (encryption algorithms) to guarantee its confidentiality.

Info mobility sensors: In a STS scenario there are also communications among remote sensors (ie, semaphores, cameras, traffic jam sensors) and STS operations centre. These kinds of communications are necessary to detect the state of the transport network, and to manage the transport network itself. The authentications of the sensor messages are very important here. Moreover, the communications between STS and sensors need to be reliable: it is necessary that the sensors provide a proper measure in order to avoid errors during data analysis and therefore errors in the output provided by the STS architecture. The availability plays an important role in these scenarios, since it is essential that the measures are available when required, especially in cases of critical condition.

3 NETWORK DESIGN FOR SMART TRANSPORTATION SYSTEMS IN CRITICAL CONDITIONS

3.1 THE LIMITS OF CURRENT NETWORK TECHNOLOGIES TO SUPPORT STS

Nowadays, networks and protocols are able to handle both failures and congestions, but the way they do does not completely respond to the requirements of traffic and services that run over an STS network. Moreover, the prioritization of the traffic and the quality of service (QoS) are not addressed, except for the service level agreements (SLAs) that can be signed with service providers that are generally not available for wide-scale consumers. An example of the requirements of applications and services in a STS is shown in Table 14.2. These requirements have been defined within the PLUG-IN research project, funded by the Italian Ministry of Research and devoted to study and develop an infomobility system to support citizens and professionals. In the PLUG-IN project we have defined different scenarios of interaction between users and STS platform and we have studied different requirements of network and services related to the criticism level and the priority of the particular scenario. Scenarios are primarily organized depending on the type of user, required services and level of critical condition. In particular, as far as the level of critical condition is concerned, we distinguish different cases:

- *low criticism*, which refers to a situation of normal use of the STS, where there are no emergency events and urban traffic is below or at normal levels;
- *middle criticism*, which corresponds to a situation where the traffic has exceeded normal levels or a particular event happens for which it is necessary to act so to create minimum inconvenience to citizens or to prevent any damage to people and/or things;
- *high criticism*, which refers to a particular situation in which public safety is endangered and where you need to act quickly to resolve the emergency situation so to avoid or minimize damage to people and/or things. It is important to emphasize that the level of criticality is determined by the operation center on the basis of information from the field about the status of roads, of areas subject to risks such as rivers and landslides, and of industrial areas.

A description of the main QoS parameters is shown in Table 14.2:

- Availability, which represents the percentage of time for which the service is available compared to the total time;

Table 14.2 Performance Requirements of Some STS Users[a][b]		
Scenario	Low/Middle Criticism	High Criticism
Private user that requires a path from A to B with an application on his/her smartphone	Availability: 99% Completion time: [5–15 s] Information loss: 0	Availability: ~90% Completion time: [25–40 s] Information Loss: 0
Logistic operator user that has to fulfil some deliveries and requires the optimum path	Availability: 99.5% Completion time: [2–5 s] Information loss: 0	
STS that sends alert messages to the smartphones of citizens/professionals to warn about a criticism in the city	Availability: ~99.9% Completion time: [10 s] Information loss: 0	Availability: ~99.999% Completion time: [5 s] Information loss: 0
Policeman that interacts with STS to resolve a criticism in the city	Availability: ~99.99% Completion time: [2–5 s] Information loss: 0	Availability: ~99.999% Completion time: [1–2 s] Information loss: 0
Institutional operators that want to send video/images to STS to document a particular situation Institutional operators that want to view some video/images to follow a particular situation	Availability: ~99.99% Video—Completion time: [2 s] Video—Information loss: ~10^{-2} Images—Completion time: [1–2 s] Images—Information loss: 0	Availability: ~99.999% Video—Completion time: [1s] Video—Information loss: ~10^{-3} Images—Completion time: [1–2 s] Images—Information loss: 0

[a]*ITU-T G.1010, End-user multimedia QoS categories, 11/2001.*
[b]*Optionally, it can also be included as an indication of the user/service priority. This priority is intended as a parameter to use in case of call admission control (CAC). For example, in case of network outage due to catastrophic events, STS could serve the requests from public safety users before those of private users due to the high priority characterization.*

- Time completion, also called "one-way delay," and defined as the time required to fulfill the user request;
- Information loss, expressed as the rate of packet loss at the application level;

The requirements have been fixed depending on the emergency level of each scenario. Achieving most demanding requirements with the already deployed network is not so easy, especially the ones at high criticism level. Generally, we can say that, in critical conditions, networks can suffer and could not be able to handle services and connections to guarantee the given requirements, unless there is a particular service level agreement (SLA) with the provider. Conversely, in normal conditions without any criticisms, we can argue that all aforementioned scenarios will respect requirements.

In the traditional approach to networking, most of the functionality of the network is implemented within network nodes (switch/router). Within them the majority of the functionality is implemented in a dedicated hardware, such as an application specific integrated circuit (ASIC), that is an integrated circuit designed to solve a specific computation task. Unfortunately:

- ASICs that provide network capabilities have evolved slowly.
- The evolution of the functionality of ASICs is controlled by the manufacturer of the switch / router.
- The operating systems on board of network nodes are proprietary.

- Each node is configured individually.
- Operations such as provisioning, change management and de-provisioning are time consuming and subject to errors.

In addition, the following features mean that the world of the implemented networks is somehow "crystallized":

- Currently used routing protocols (RIP, OSPF, BGP) were developed for the most part 20 years ago and did not evolve except for few differences.
- Mechanisms for traffic engineering are difficult to implement and the quality of service is difficult to achieve.

Current networks are difficult to manage and, although distributed control allows a certain scalability, they can be a source of problems when it is necessary to reconfigure them in order to achieve specific quality objectives (eg, traffic engineering for specific traffic flows). When a network device reveals a network topology change (ie, for a fault on a line or on a device), it can change its configuration to re-route incoming traffic through another path in the network by using distributed routing protocols such as BGP and OSPF. Although this mechanism is quite efficient (it takes few seconds to reconfigure a network), it can be improved. For example it does not take into account overall network topology and link utilization. It may happen that, in case of high traffic and network failure, the reconfiguration addresses the traffic to an already congested network portion. In this case, the congestion will worsen and many packets will be dropped.

Another problem with currently used networks is that generally all the traffic is routed following the *best-effort* strategy. In this situation the traffic is handled without any type of priority and if there is network congestion no priority level will privilege important traffic. Signing an SLA with service provider is the only way to handle this kind of problems, but, in an STS environment, where we have lot of different actors and different internet service providers (ISPs), this strategy is not applicable. Moreover, we have to take into account that SLAs are not thought to be dynamic and service levels can hardly be varied over time. A further problem is that with the current networks is practically impossible to quickly deploy new applications or protocols within the router since, as mentioned earlier, they implement proprietary operating systems.

3.2 TOWARD A NEW NETWORK ARCHITECTURE

Along with the evolution of ICT and services that massively use it, networks such as the internet have increasingly become critical infrastructures, because their failure compromise the entire systems that rely on them. Furthermore, with the IoT and its future evolution such as the Internet of Everything (IoE), the internet will become even more important and will represent a critical aspect of our life.

Keeping this in mind, we can consider STSs as critical systems that use a critical infrastructure, the network, which has to be designed to be fault-tolerant to critical conditions. Obviously, this network has to be as secure as possible, that is, the network must be able to contrast cyber attacks, such as distributed denial of service (DDoS), and traffic injection. We can summarize network fault tolerance and security requirements with the term survivability, which is the capability of a system to fulfil its mission, in a timely manner, in the presence of threats such as attacks or natural disasters [21].

The need to ensure QoS performance and adequately support the applications and services of a STS, pushes the use of network architectures that are different from those in place. This is true in particular regarding the dichotomy *data-plane/control plane*. In this sense, the main characteristics which a new generation architecture must meet are:

- Improving the management of the network: faster handling, targeted interventions to network elements and global network view in terms of topology, and real time traffic.
- Implementing end-to-end traffic engineering policies.
- Assuring the dynamic allocation of network resources for network function virtualization (NFV).
- Facilitating the evolution of faster network functionalities, based on the life cycle of software development.

In the subsequently sections, we will go deep in these problems trying to explore the innovative solutions that can respond to these needs, highlighting the problems that exist on the already-deployed networks and how to cope with them.

4 SOFTWARE-DEFINED NETWORKING

Current computer networks are complex and difficult to manage. They use with many kinds of devices, from routers and switches to middleboxes such as firewalls, network address translators, server load balancers, and intrusion detection systems. Routers and switches run distributed control software that is typically closed and proprietary. On the other hand, network administrators typically configure individual network devices by using configuration interfaces that vary across vendors and even across different products from the same vendor. This operation mode has slowed innovation, increased complexity, and inflated operational costs to run a network.

SDN is a new paradigm in networking that is revolutionizing the networking industry by enabling programmability, easier management and faster innovation. These benefits are made possible by its centralized control plane architecture, which allows the network to be programmed by the application and controlled from one central entity.

The defining feature of SDN is its large scale adoption in industry [22,23], especially as compared with its predecessors. This success can be attributed to a perfect storm of conditions among equipment vendors, chipset designers, network operators, and networking researchers [24]. Here the principal reasons of the transformation from designed networks to programmable networks [25]:

- *Networks are hard to manage* whereas computation and storage resources have been virtualized by creating a more flexible and manageable infrastructure, networks are still notoriously hard to manage. In facts network administrators need large share of sys-admin staff.
- Networks are hard to evolve especially compared with the ongoing and rapid innovation in system software (eg, new languages, operating systems, ...). On the other hand, networks are stuck in the past. For example routing algorithms change very slowly, as highlighted before, and network management is extremely primitive.
- Network design is not based on formal principles. Whereas operating systems courses teach fundamental principles like mutual exclusion and other synchronization primitives (eg, files, file systems, threads, and other building blocks), networking courses teach a big bag of protocols with no formal principles, and just general design guidelines.

4.1 BRIEF HISTORY

The problems described before exist from the dawn of networking. Making computer networks more programmable makes innovation in network management possible and lowers the barrier to deploy new services. This section is dedicated to the review of early efforts on programmable networks. This section follows in large part the work published in [24].

SDN has gained significant traction in the industry. Although the excitement about SDN has become more palpable fairly recently, many of the ideas underlying the technology have evolved over the past 20 years. SDN resembles past research on active networking, which articulated a vision for programmable networks, albeit with an emphasis on programmable data planes. SDN also relates to previous work on separating the control and data planes in computer networks.

4.1.1 Active networking

Active networking represented a new radical approach to network control by envisioning a programming interface through network API that exposed resources (eg, processing, storage, and packet queues) on individual network nodes and supported the construction of functionalities to apply to the packets inside the router. Active networking community pursued two programming models:

- the capsule model, where the code to execute at the nodes was carried in-band in data packets [26];
- the programmable router/switch model, where the code to execute at the nodes was established by out-of-band mechanisms. [27,28].

Active networks offered intellectual contributions that relate to SDN. In particular the research in active networks pioneered the notion of programmable networks as a way of lowering the barrier to network innovation. Moreover the need to support experimentation with multiple programming models led to work on network virtualization. Finally, early design documents cited the need to unify the wide range of middlebox functions with a common, safe programming framework.

4.1.2 Separating control and data planes

As the Internet flourished in the 1990s, the link speeds in backbone networks grew rapidly, leading equipment vendors to implement packet-forwarding logic directly in hardware, separate from the control-plane software. Moreover, the rapid advances in commodity computing platforms meant that servers often had substantially more memory and processing resources than the control-plane processor of a router deployed just one or two years earlier. These trends catalyzed two innovations:

- an open interface between control and data planes, such as the ForCES (Forwarding and Control Element Separation) [29] interface standardized by the IETF and the Netlink interface to the kernel-level packet-forwarding functionality in Linux [30];
- a logically centralized control of the network, as seen in the RCP (Routing Control Platform) [31,32] and SoftRouter [33] architectures, as well as the PCE (Path Computation Element) [34] IETF protocol.

Moving control functionalities off the network equipment and into separate servers (selecting better network paths, minimizing transient disruptions) represented a paradigm-shift from the Internet's design making sense because network management is, by definition, a network-wide activity. Logically centralized routing controllers [31,33,35] were made possible by the emergence of open-source routing software [36–38] that lowered the barrier to create prototype implementations. To broaden the vision of control- and data-plane separation, researchers started exploring clean-slate architectures for logically centralized control: 4D project [39] and Ethane project [40].

4.2 WHY DID IT NOT WORK OUT?

Although active networks articulated a vision of programmable networks, the technologies did not see widespread deployment due to the paradigm shift of active networks research compared to the Internet community. Moreover the lack of an immediately compelling problem or a clear path to deployment was a block for the large adoption of the mechanisms proposed by active networks research.

Concerning the separation of control plane from data planes, the dominant equipment vendors had little incentives to adopt standard data-plane APIs such as ForCES, since open APIs could attract new entrants into the marketplace. At the end, although industry prototypes and standardization efforts made some progress, widespread adoption remained elusive.

The ideas underlying SDN faced a tension between the vision of fully programmable networks and pragmatism that would enable real-world deployment. OpenFlow, described later, is a balance between these two goals by enabling more functions than earlier route controllers and building on existing switch hardware. Although relying on existing switch hardware did somewhat limit flexibility, Open-Flow was almost immediately deployable, allowing the SDN movement to be both pragmatic and bold. The creation of the OpenFlow API [41] was followed quickly by the design of controller platforms such as NOX [42] that enabled the creation of many new control applications.

4.3 THEORETICAL ASPECTS

In networking we can envision two "planes" [43]:

- *Data plane*: devoted to process and deliver packets by using the information of local forwarding state (eg, routing entries);

$$\text{Forwarding state} + \text{packet header} \rightarrow \text{Forwarding decision}$$

- *Control plane*: devoted to compute the status in routers. For example, it determines how and where packets are forwarded, takes decisions about traffic engineering, firewalls, …

In current network devices (eg, routers and switches), see Fig. 14.6, the control plane is implemented in distributed protocols or directly through a manual configuration of network devices.

FIGURE 14.6 Stack of Current Network Devices

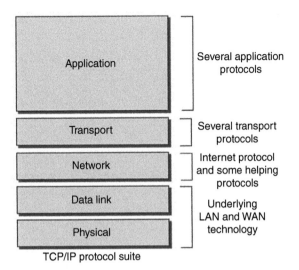

TCP/IP protocol suite

FIGURE 14.7 TCP/IP Protocols Stack/Suite

The two different planes (data and control) would require different abstractions. As far as data plane is concerned, the abstraction is well-known and is the protocol stack (ISO/OSI or TCP/IP). Applications are built on reliable (or unreliable) transport (eg, TCP, UDP) that, in turn, are built on a best-effort global packet delivery protocol (eg, IP). IP is then built on best-effort local frame delivery that uses a local physical transfer of bits. TCP/IP protocol stack is shown in Fig. 14.7.

Many mechanisms belong to the control plane functionalities. Some of them are access control lists (ACLs), Virtual LAN, firewall, traffic engineering mechanisms (adjusting weights, MPLS). Unfortunately, there is no abstraction for the control plane.

Control plane must compute forwarding state. To accomplish its task, the control plane must:

- figure out what network looks like → *topology discover*;
- figure out how to accomplish the goal on a given topology → *accomplish the goal*;
- tell the switches what to do → *configure forwarding state*.

Currently, when we want design and implement a new control plane functionality we view the three tasks as a natural set of requirements and we require each new protocol to solve all three. Obviously *two of these tasks can be "reused"* for any new control plane functionality we want to implement. In particular:

- determining the topology information
- configuring the forwarding state on routers/switches

In other words, if we implemented the mechanisms able to determine the topology and configure the forwarding state in network devices we could reuse the same mechanisms for any new control functionality. This is the theoretical core idea of SDN:

"SDN is the use of those two control planes abstractions."

Abstraction 1—*Global Network View*:

- the global network view provides information about the current network and is then devoted to the topology discover;

- its implementation is "network operating system" that runs on a centralized server in network (replicated for reliability);

    Abstraction 2—*Forwarding Model*:

- the forwarding model provides a standard way to define the forwarding state inside network devices;
- a common implementation is OpenFlow (described later).

In current networks, traditional control mechanisms (eg, routing, traffic engineering, multicast) run in a distributed fashion as shown in Fig. 14.8.

In SDN, as shown in Fig. 14.9, Network OS that runs on a server, communicates with each SDN device through a standard mechanism (Forwarding model), builds the Global Network View (a graph) and makes visible the resources of the network (in terms of a graph) to all control programs that run on Network OS.

With SDN, the control plane functionalities are not distributed control mechanisms but simple pieces of software that run on graphs (built by network OS).

A clean separation of concepts is reached through SDN.

- Control programs (eg, router mechanisms, TE schemes): to express their goals on a Global Network View (not a distributed protocol but a graph algorithm);

Distributed algorithm running between neighbors
Complicated task-specific distributed algorithm

FIGURE 14.8 Traditional Control Mechanisms

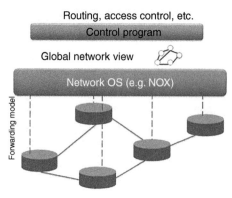

FIGURE 14.9 SDN "Layers" for Control Plane

- Network OS: to build the Global Network View through the communication with switches/routers. It also conveys configurations from the control program to switches;
- Router/switches: which merely follows the orders from NOS;

There is a clean separation of control and data planes that will not be packaged together in proprietary boxes. SDN enables the use of commodity hardware (network devices) and 3rd party software (network OS and control programs). Obviously, abstractions don't eliminate complexity, but now every system component is tractable. Network OS are still complicated pieces of code but Network OS is reusable for every control program.

Other aspects and details can be found in [25,44–46].

4.4 SDN IN PRACTICE

As said before, SDN is a new networking paradigm that is revolutionizing the networking industry by enabling programmability, easier management and faster innovation. These benefits are made possible by its centralized control plane architecture (network OS), which allows the network to be programmed by the application and controlled from one central entity.

SDN architecture is composed of both switches/routers and of a central controller (SDN controller or network OS), as in Fig. 14.10. The SDN device processes and delivers packets according to rules stored in its flow table (forwarding state), whereas the SDN controller configures the forwarding state of each switch by using a standard way. The controller is also responsible to build the virtual topology representing the physical one. Virtual topology is used by application modules that run on top of the SDN controller to implement different control logics and network functions (eg, routing, traffic engineering, firewall state).

4.4.1 Implementation aspects: forwarding model

An implementation of the forwarding model and a standard de facto in SDN is Openflow (OF) [47]. In OF the forwarding model is implemented as a <match-actions> as shown in Fig. 14.11.

In more detail, the OpenFlow architecture is illustrated in Fig. 14.12 [48]. The forwarding device, or OpenFlow switch, contains one or more flow tables and an abstraction layer that securely communicates with a controller via the OpenFlow protocol. Flow tables consist of flow entries, each of which

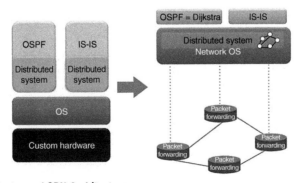

FIGURE 14.10 Legacy Router and SDN Architecture

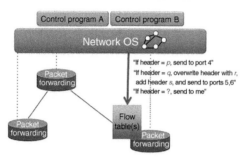

FIGURE 14.11 Basic of Forwarding Model in OF

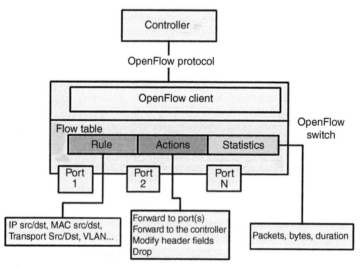

FIGURE 14.12 OF Device

determines how packets belonging to a flow will be processed and forwarded. Flow entries typically consist of:

1. match fields, or matching rules, used to match incoming packets; match fields may contain information found in the packet header, ingress port, and metadata;
2. counters, used to collect statistics for the particular flow, such as the number of received packets, the number of bytes and the duration of the flow;
3. a set of instructions, or actions, to be applied upon a match; they dictate how to handle matching packets.

Upon a packet arrival at an OpenFlow switch, packet header fields are extracted and matched against the matching fields portion of the flow table entries. If a matching entry is found, the switch applies the appropriate set of instructions, or actions, associated with the matched flow entry. If the flow table look-up procedure does not result on a match, the default action is to send the packet to the SDN controller that will take the decision and will install the rules on OF devices. The communication between controller and switch happens via OpenFlow protocol, which defines a set of messages that can be exchanged between these entities over a secure channel. Using the OpenFlow protocol a remote controller can, for example, add, update, or delete flow entries from the switch's flow tables. That can happen reactively (in response to a packet arrival) or proactively. Figs. 14.13 and 14.14 contain, respectively, main current available commodity switches by makers and current software switch implementations, both compliant with the Openflow standard.

4.4.2 Implementation aspects: controller

The SDN controller has been compared to an operating system in [42], in which the controller provides a programmatic interface to the network that can be used to implement management tasks and offer new functionalities. A layered view of this model is illustrated in Fig. 14.15. This abstraction assumes the control is centralized and applications are written as if the network is a single system. It enables the SDN model to be applied over a wide range of applications and heterogeneous network technologies and physical media such as wireless (eg, 802.11 and 802.16), wired (eg, Ethernet) and optical networks.

As a practical example of the layering abstraction accessible through open application programming interfaces (APIs), Fig. 14.16 illustrates the architecture of an SDN controller based on the OpenFlow protocol. This specific controller is OpenDaylight controller [Opendaylight]. In this figure it is possible to observe the separation between the controller and the application layers. Applications can be written in Java and can interact with the built-in controller modules via a REST API (topology manager,

Maker	Switch model	Version
Hewlett-Packard	8200zl, 6600, 6200zl, 5400zl, and 3500/3500yl	v1.0
Brocade	NetIron CES 2000 Series	v1.0
IBM	RackSwitch G8264	v1.0
NEC	PF5240 PF5820	v1.0
Pronto	3290 and 3780	v1.0
Juniper	Junos MX-Series	v1.0
Pica8	P-3290, P-3295, P-3780, and P-3920	v1.2

FIGURE 14.13 Main Current Available Commodity Switches by Makers, Compliant With the Openflow Standard

Software Switch	Implementation	Overview	Version
Open vSwitch	C/Python	Open source software switch that aims to implement a switch platform in virtualized server environments. Supports standard management interfaces and enables programmatic extension and control of the forwarding functions. Can be ported into ASIC switches.	v1.0
Pantou/OpenWRT	C	Turns a commercial wireless router or Access Point into an OpenFlow-enabled switch.	v1.0
ofsoftswitch13	C/C++	OpenFlow 1.3 compatible user-space software switch implementation.	v1.3
Indigo	C	Open source OpenFlow implementation that runs on physical switches and uses the hardware features of Ethernet switch ASICs to run OpenFlow.	v1.0

FIGURE 14.14 Current Software Switch Implementations Compliant With the Openflow Standard

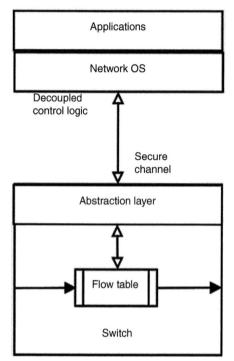

FIGURE 14.15 The Separated Control Logic Can Be Viewed as a Network Operating

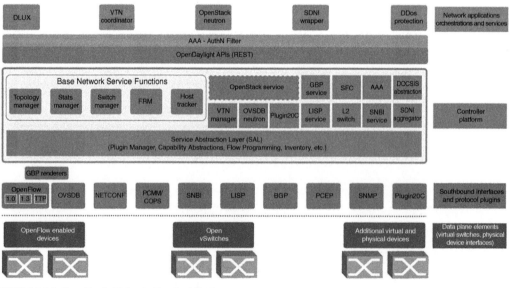

FIGURE 14.16 OpenDaylight Controller Architecture

stats manager, …). The controller, on the other hand, can communicate with the forwarding devices via the OpenFlow protocol through the abstraction layer present at the forwarding hardware. While the aforementioned layering abstractions accessible via open APIs allow the simplification of policy enforcement and management tasks, the bindings must be closely maintained between the control and the network forwarding elements. The choices made while implementing such layering architectures can dramatically influence the performance and scalability of the network.

5 QoS APPLICATIONS FOR STS

As said before, many applications nowadays rely on QoS guarantees. Some of them are telemedicine, tele-control, tele-learning, telephony, video-conferences, online gaming, multimedia streaming and applications for emergencies and security. Each application, having very different characteristics, needs a specific degree of service, defined at the application layer. As described previously, applications and services in STS mainly rely on the satisfaction of QoS requirements.

In order to guarantee specific QoS requirements, QoS management is strongly necessary. QoS management functions are aimed at offering the necessary tools to pursue this objective. A possible classification of the QoS Management functions is shown in Fig. 14.17, from [49]. Others classifications can be found in [50–52].

Traditionally QoS management has been left to the end-hosts by virtue of "end-to-end principle" [53] by using end-to-end congestion control (TCP) [54]. In-network solutions in legacy networks are proposed in MPLS-TE [Awduche01, 7], in OFPS-TE [55] and Segment Routing [56] but these approaches rely on distributed control architectures that may limit the control power of the entire network.

As described before, the ability of the SDN controller to receive (soft) real-time information from SDN devices and to make decisions based on a global view of the network, coupled with the ability of "custom"-grained flow aggregation inside SDN devices, makes Traffic Engineering (TE) one of the most interesting use cases for SDN networks.

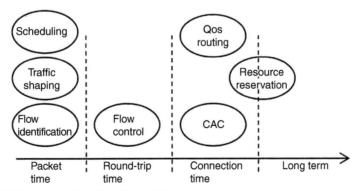

FIGURE 14.17 QoS Management Functions Versus Time

5.1 STS SCENARIO

In a STS scenario (Fig. 14.18) different actors (described in Section III) use the network (here simplified and composed by 8 SDN devices). In case of *middle criticism* conditions, we need to give higher priority to certain actors. As clear from the scenario in Fig. 14.18, public safety operators, institutional operators, actuators, sensors and STS decision center generate *high priority flows* (in red), logistic operators generate *mid priority flows* (in yellow) and private users generate *low priority flows* (in green).

Each SDN device is configured with multi-queues: in each OF switch one or more queues can be configured for each outgoing interface and used to map flow entries on them. Flow entries mapped to a specific queue will be treated according to the queue's configuration in terms of service rate.

In particular, let us suppose that for each interface of each OF device, there are two assigned queues, each of them dedicated to a specific type of traffic with a predefined service rate. The first queue (q_0) of each OF device is assigned to *high priority flows* while the second one (q_1) is dedicated to *low priority flows. Mid priority flows* are assigned to the first queue at first instance (Fig. 14.19). If a mid priority flow traversing q_0 suddenly increases its rate (ie, they are no longer conformant with QoS constraints on which the OF device has been configured), because, for example, the user wants an added value service, the queue will start to grow and it could end up losing packets (Fig. 14.20). The performance of the other flows (high priority flows) traversing the queue will be affected by this event, both in terms of delay and packet loss.

A possible solution that copes with a limitation of OF is traffic re-routing (TR). By TR we can change the path/route of a flow (or a subset of them) in order to reduce imminent congestion, to avoid link disruption and to improve the QoS. A key factor of TR is time: time that elapses between the

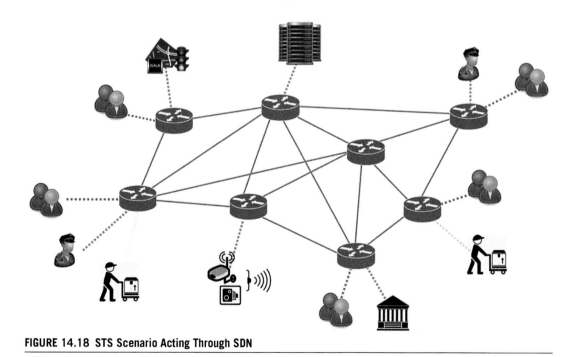

FIGURE 14.18 STS Scenario Acting Through SDN

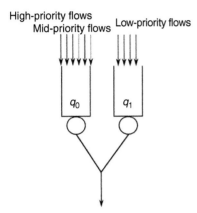

FIGURE 14.19 Queue Model of Device

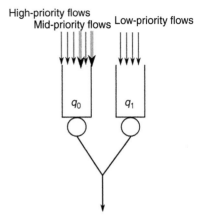

FIGURE 14.20 Some of Mid-Priority Flows That Increased Their Rate Are No Longer Conformant

congestion event and the reroute of the flow should be as small as possible. This problem has been tackled in [57,58].

We propose a strategy that limits this effect, ensuring that flows that prove to be compliant with the QoS constraints can exploit the needed bandwidth without suffering from any performance degradation. This solution has the task to identify the traffic which is not conformant to the rate constraints and re-route it (or drop it, if needed), in order to avoid the degradation of the quality experienced by other flows traversing the network. Since we want to devise a solution which is compatible with any underlying hardware, we design and implement the strategy inside the SDN controller. We chose Beacon [59] as SDN controller. Beacon is a multi-threaded Java-based controller that relies on OSGi and Spring frameworks and it is highly integrated into the Eclipse IDE. In spite of a specific choice of the controller, our modifications can be implemented in any controller. The principal modifications of Beacon are:

Statistics polling: Beacon periodically sends statistic requests to the switches. The statistics requested are flow, port and queue statistics. In addition to the basic statistics that the OpenFlow protocol 1.0 makes available, we added specific functions to the controller, which allow Beacon to exploit the collected data in order to compute the parameters useful to apply the chosen strategy. The statistics computed by the controller are shown in Table 14.3. The main extracted feature is the estimated rate (ER) for ports, queues, and flows.

Routing: This module has been modified to the purpose of implementing the proposed algorithms. When a switch receives a new flow, it contacts the controller in order to know where to forward the traffic. When the controller has to assign each flow to a specific queue, it checks a variable that identifies the algorithm to run. Beacon performs a routine to select the correct queue based on the chosen strategy and then notifies the node through the installation of a flow modification.

The scenario in which we present our solution involves a class-based system in which flows are identified by traffic descriptors. The issue we want to investigate deals with a flow characterized by a

Table 14.3 BeaQoS Statistics

Statistics Available in OpenFlow 1.0		Statistics Computed by BeaQoS
Tx bytes per flow	$\rightarrow$	Estimated rate per flow
Tx bytes per port	$\rightarrow$	Estimated rate per port
Tx bytes per queue	$\rightarrow$	Estimated rate per queue
Flow match		
Flow actions	$\rightarrow$	Flows per queue
Queue ID		

specific rate limit, which, for some reason, violates this constraint. At this point our system recognizes the problem and re-routes the flow in a more suitable queue, in order to avoid traffic congestion and quality degradation. We suppose two main types of traffic:

- *High priority (HP) and mid priority (MP)*: Characterized by a rate not exceeding 100 kbit/s
- *Mid priority (MP) not conformant*: Characterized by a rate that sometimes can exceed 100 kbit/s
- *Low priority (LP)*: Displaying a rate greater than 100 kbit/s, most of the time.

We introduce and implement a solution that will be called *Conformant*, to the purpose of re-establishing the correct routing of flows, based on their rates. This scheme assigns incoming flows to the queue associated to a specific traffic descriptor. q_0 is dedicated to process HP and MP flows, whereas q_1 is devoted to serve LP flows. Furthermore, the controller periodically checks the statistics related to the flows belonging to q_0 in order to figure out if a flow is not compliant to its constraint. When Beacon finds a MP flow which is violating its traffic descriptor, it re-routes it to the HR queue q_1 in order to be able to serve the traffic without causing congestion. If the newly re-routed flow overcomes a pre-defined threshold while traversing q_1, this traffic will be dropped by the Beacon controller. In the present simulation we set this threshold to 700kbit/s. We implemented this part of the strategy inside a specific Beacon module aimed at collecting statistic data.

We ran the performance analysis on a PC with Mininet (version 2.1.0) [Mininet]. The scenario is composed of two hosts connected to a SDN switch. The chosen implementation of the switch is Open vSwitch 2.0.2 [Ovs], managed by an instance of Beacon running on the same machine. Each port of the switch is configured with two queues, q_0 and q_1. We tested our strategy with a set of simulations involving 50 flows generated using *iperf*. Queue configuration and traffic characteristics are shown in Table 14.4.

This test is aimed at comparing the performances of our *Conformant* algorithm (as described before) with the *Dedicated* strategy. This last scheme consists in assigning each traffic class to the corresponding queue based on the traffic descriptor upon flow arrival and then take no further actions independently of the flow behaviour.

The results in Fig. 14.21 show that, while the *Dedicated* strategy produces a 6.8% packet loss, the *Conformant* one allows to completely avoid the packet loss of High priority flows. This benefit is obtained together with the fact that the quality experienced by Low priority flows is not affected. It is

Table 14.4 Queues Configuration and Traffic Characteristics

Queue ID	Service Rate	Buffer Size
0	4 Mbit/s	1000 packets
1	16 Mbit/s	1000 packets
Traffic Descriptor	**Rate**	**Percentage**
HP/MP	40-60 kbit/s	40%
MP no conformant	200-800 kbit/s	20%
LP	200-800 kbit/s	40%

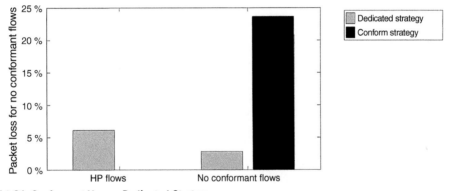

FIGURE 14.21 Conformant Versus Dedicated Strategy

worth noting that the loss of no conformant flows increases, but this is acceptable since these flows are not compliant with the constraints.

Being our proposal a programmable solution, it is however possible to tune the threshold that defines the behavior of the strategy in order to cope with different needs and situations. This parameter can be set through an external properties file, making the customization of the scheme even more flexible. In conclusion, we showed the results obtained in performance tests in which we compared the alternative QoS approaches. Our cases of study show that the proposed QoS solution allows getting good results when applied to the current OpenFlow environment. Future developments could consist in testing the network environment with a larger amount of traffic in order to test the scalability of our solutions. We also plan to devise alternative approaches such as exploiting the low rate queue in order to improve the quality perceived by high rate flows. Furthermore we plan to run our algorithms in other scenarios set with different queue configurations. Finally we hope to be able to conduct testbed measurements with commodity hardware routers in order to avoid the problems related to the software emulation of this type of devices.

6 INCREMENTAL DEPLOYMENT OF STS–SDN SOLUTION

In previous sections we have seen how the SDN paradigm can improve networks in terms of reliability and programmability. Software Defined Network can also address many of the requirements of a next-generation STS because it offers a way to deploy a network that is flexible, fault-tolerant and QoS-aware. Since STSs are typically deployed in large cities with dense population and involve different actors at different levels as said before, we cannot think to build an ad-hoc network to handle all STS generated traffic, but we need to use already-existent networks and services offered by a set of operators/service providers. Nowadays, service providers and network operators do not have SDN-compliant networks, but they are planning to evolve their infrastructures toward this new paradigm.

One of the major problems in SDN is that deploying a SDN network means to change the entire network structure. This can be easily done in relatively small environments such as data-centers and campuses, where we can already find a variety of deployments, such as Google Andromeda [60] used for the Google's data centers or OpenStack Neutron [61] used by Rackspace's cloud services. However, from the network operator point of view, the process to adopt SDN in its infrastructure is very costly, because a very large amount of devices at different levels are involved in this change. Furthermore, an operator has to train its programmers and personnel to interact with such new paradigm, and this requires investments in terms of money and time. For all these reasons, operators could be reluctant to do this upgrade. A possible solution that has been considered by large-scale operators, such as AT&T [62], is an incremental deployment of SDN into the already-existent infrastructure, since SDN nodes can be configured to be transparent to the other "legacy" nodes. In this way, deployment can be done without affecting network functionalities and preserving network access. Another advantage of the incremental deployment is that such new functionalities can be tested and evaluated before making major changes to the infrastructure. Of course, with this method, the time-to-market of the overall network is considerably higher but this problem is covered by the other advantages. Doing this incremental deployment, we introduce a sort of hybrid SDN network where traditional forwarding methods (called Comercial Off-The-Shelf Networks or CN) and newest one coexist.

As shown in [63], we can have different types of hybrid SDN networks:

- Topology-based: in this model we have a topological separation between SDN and CN with adapters to make the different zones to interact. This kind of deployment introduces SDN at regional-level, and could be adopted if we want to extend SDN starting from particular kind of zones (eg, large cities). Operators can take advantage from this model by means of tests and by isolating major failures on their new deployments.
- Service-based: in this model we use SDN only for a subset of network services and forwarding types, leaving CN to handle all the others. For example, we could use SDN at the edge nodes to improve load balancing and traffic engineering and leave all core-network functionalities to CN. With this model we can strategically place SDN nodes in the network and incrementally deploy new services as they will be ready to be handled by SDN.
- Traffic-based: in this model SDN handles only a certain class of traffic, while CN handles the other ones. In this way we can initially forward by SDN only the lowest priority traffic, and incrementally switch the other traffic classes from CN to SDN where the model is better consolidated. With traffic-based model we need to have many SDN-compliant nodes (eg, nodes that are OpenFlow capable), since we put both CN and SDN paradigms in all the nodes of the network.

- Integrated model: in this model there are no SDN nodes, but SDN-like working is obtained by controlling the CN nodes and transferring the control plane at the controller node. In this way the behaviour of all distributed routing protocols such as BGP, OSPF, etc. is managed directly by the controller that sends to CN nodes all the messages to create, for example, a particular forwarding path or other SDN-like behaviours. This method has a clear advantage: no SDN nodes are required and so this kind of hybrid model can be easily deployed into operators' networks. Obviously this solution offers a SDN-like network where we cannot have all innovative functionalities that are proper of such paradigm, and the complexity of the controller could be not negligible. The choice of the solution depends on different factors. First of all operators need to decide where to make the first changes and define a sort of "road-map" of the next steps. Doing this is not a trivial process, because operators have to decide on which level of the network to do the deployment considering that each level has different peculiarities, with obvious impact over network performance and deployment cost.

Summarizing, operators must take into account, for each kind of intervention:

- the number of involved devices;
- the cost of the devices substitution/update;
- the benefits to have an SDN node in such place;
- the interaction between the "legacy" nodes.

With these parameters operators can define an effective incremental deployment of the SDN-compliant network, choosing the proper model that fits its needs and requirements.

7 CONCLUSIONS

In this chapter we have analyzed the emerging technologies in the fields of the Smart Cities, focusing on STSs. After that, we have discussed about the possible infrastructures on which STS relies, founding that telecommunications networks are one of the most critical aspects because emergency conditions in the cities (hurricane, earthquakes, etc.) could damage the network (or part of it) resulting in interruption of service that, for a STS, may result in putting citizens and city infrastructures in danger. Facing with network survivability we have seen that current network paradigms and protocols are not able to guarantee sufficient level of service in case of disaster, and we saw how a new emergent network paradigm such as SDN could cope with this problem. Furthermore, we have proposed a simple example on how an application in a STS environment could take advantage from the SDN paradigm in terms of Quality of Service. Finally, we have considered how to deploy the SDN solution over operator networks and discussed related issues and open questions.

REFERENCES

[1] The IoT: the next step in internet evolution. <http://www2.alcatel-lucent.com/techzine/iot-internet-of-things-next-step-evolution/> [Online].
[2] Chourabi H, Nam T, Walker S, Gil-Garcia JR, Mellouli S, Nahon K, Pardo TA, Scholl HJ. Understanding smart cities: an integrative framework. In Proceedings HICSS; 2014.
[3] Kitchin K. "The real-time city? Big data and smart urbanism". Springer GeoJ 2014;79(1):1–14.

[4] Batty M. Big data, smart cities and city planning. Sage Dialogues Human Geogr 2013;3(3):274–9.

[5] IBM. Cisco and the business of smart cities—how two of the IT industry's largest companies plan to rewire urban living. <http://www.information-age.com/industry/hardware/2087993/ibm-cisco-and-the-business-of-smart-cities> [Online].

[6] European Smart Cities. <http://www.smart-cities.eu/index.php?cid=7&ver=4> [Online].

[7] Amsterdam smart city. <http://amsterdamsmartcity.com/> [Online].

[8] Amsterdam smart city–Smart Parking. <http://amsterdamsmartcity.com/projects/detail/id/64/slug/smart-parking> [Online].

[9] The Digital Road Authority–Incident Management. <http://amsterdamsmartcity.com/projects/detail/id/76/slug/the-digital-road-authority-incident-management> [Online].

[10] Barcelona smart city. <http://smartcity.bcn.cat/en/city-os.html> [Online].

[11] Genova smart city website. <http://www.genovasmartcity.it> [Online].

[12] Torino smart city website. <http://www.torinosmartcity.it> [Online].

[13] Milano smart city. <http://www.milanosmartcity.org> [Online].

[14] Ezell S. Intelligent transportation systems. ITIF The Information Technology & Innovation Foundation 2010.

[15] Sochor JL. User perspectives on intelligent transportation systems. PhD Thesis in Transport Science Stockholm, Sweden. <http://www.diva-portal.org/smash/record.jsf?pid=diva2%3A621461&dswid=5631>; 2013 [Online].

[16] Smart Mobility 2030—Singapore ITS Strategic Plan. <http://www.lta.gov.sg/content/ltaweb/en/roads-and-motoring/managing-traffic-and-congestion/intelligent-transport-systems/SmartMobility2030.html> [Online].

[17] Lopez Research. Smart cities are built on the internet of things. <http://www.cisco.com/web/solutions/trends/iot/docs/smart_cities_are_built_on_iot_lopez_research.pdf> [Online].

[18] Zhao M, Walker J, Wang C-C. Security challenges for the intelligent transportation system. In: Proceedings of the First International Conference on Security of Internet of Things; 2012.

[19] ETSI TR 102 893. Intelligent transport system (ITS), security, threat, vulnerability and risk analysis (TVRA), V1.1.1.(2010-03).

[20] Papadimistratos P, Gligor V, Hubaux J-P. Securing vehicular communications–assumptions, requirements, and principles. In: Proceedings of 4th Workshop on Embedded Security in Cars; 2006.

[21] Sterbenz JPG, Hutchison D, Çetinkaya EK, Jabbar A, Rohrer JP, Schöller M, Smith P. "Resilience and survivability in communication networks: strategies, principles, and survey of disciplines". Elsevier Comput Netw 2010;54(8):1245–65.

[22] Jain S, Kumar A, Mandal S, Ong J, Poutievski L, Singh A, Venkata S, Wanderer J, Zhou J, Zhu M, Zolla J, Hölzle U, Stuart S, Vahdat A. B4: experience with a globally deployed software-defined WAN. In: Proceedings of ACM SIGCOMM; 2013.

[23] Nicira. It's time to virtualize the network. <http://www.netfos.com.tw/PDF/Nicira/It%20is%20Time%20To%20Virtualize%20the%20Network%20White%20Paper.pdf> [Online].

[24] Feamster N, Rexford J, Zegura E. The road to SDN: an intellectual history of programmable networks. In: ACM SIGCOMM Comput Commun Rev 2014;44(2):87–98.

[25] Shenker S. A gentle introduction to software defined networks. Technion Computer Engineering Center. <http://tce.technion.ac.il/files/2012/06/Scott-shenker.pdf>; 2012 [Online].

[26] Wetherall D, Guttag J, Tennenhouse D. ANTS: a toolkit for building and dynamically deploying network protocols. In: Proceedings of IEEE OpenArch; 1998.

[27] Bhattacharjee S, Calvert KL, Zegura EW. An architecture for active networks. In: Proceedings of the IFIP TC6 HPN; 1997.

[28] Smith J et al. SwitchWare: accelerating network evolution. Technical report MS-CIS-96-38, University of Pennsylvania; 1996.

[29] Yang L, Dantu R, Anderson T, Gopal R. Forwarding and control element separation (ForCES) framework. Internet Engineering Task Force, RFC 3746. <https://www.rfc-editor.org/rfc/rfc3746.txt>; 2004 [Online].

[30] Salim J, Khosravi H, Kleen A, Kuznetsov A. Linux Netlink as an IP services protocol. Internet Engineering Task Force, RFC 3549. <https://tools.ietf.org/html/rfc3549>; 2003 [Online].

[31] Caesar M, Feamster N, Rexford J, Shaikh A, van der Merwe J. Design and implementation of a routing control platform. In: Proceedings of Usenix NSDI; 2005.

[32] Feamster N, Balakrishnan H, Rexford J, Shaikh A, van der Merwe K. The case for separating routing from routers. In: Proceedings of the ACM SIGCOMM workshop on future directions in network architecture; 2004.

[33] Lakshman TV, Nandagopal T, Ramjee R, Sabnani K, Woo T. The SoftRouter architecture. In: Proceedings of ACM Workshop HotNets; 2004.

[34] Farrel A, Vasseur J-P, Ash J. A path computation element (PCE)-based architecture. Internet Engineering Task Force, RFC 4655. <https://tools.ietf.org/html/rfc4655>; 2006 [Online].

[35] van der Merwe J, Cepleanu A, D'Souza K, Freeman B, Greenberg A. Dynamic connectivity management with an intelligent route service control point. In: Proceedings of ACM SIGCOMM Workshop on Internet Network Management; 2006.

[36] BIRD Internet routing daemon. <http://bird.network.cz/> [Online].

[37] Handley M, Kohler E, Ghosh A, Hodson O, Radoslavov P. Designing extensible IP router software. In: Proceedings of Usenix NSDI; 2005.

[38] Quagga software routing suite. <http://www.quagga.net/> [Online].

[39] Greenberg A, Hjalmtysson G, Maltz DA, Myers A, Rexford J, Xie G, Yan H, Zhan J, Zhang H. A clean-slate 4D approach to network control and management. ACM SIGCOMM Comput Commun Rev 2005;35(5):41–54.

[40] Casado M, Freedman MJ, Pettit J, Luo J, McKeown N, Shenker S. Ethane: taking control of the enterprise. In: Proceedings of ACM SIGCOMM; 2007.

[41] McKeown N, Anderson T, Balakrishnan H, Parulkar G, Peterson L, Rexford J, Shenker S, Turner J. "OpenFlow: Enabling innovation in campus networks". in ACM SIGCOMM Computer Communication Review 2008;38(2):69–74.

[42] Gude N, Koponen T, Pettit J, Pfaff B, Casado M, McKeown N, Shenker S. NOX: towards an operating system for networks. ACM SIGCOMM Comput Commun Rev 2005;38(3):105–10.

[43] Wu C, Feng TY, Lin M-C. Star: a local network system for real-time management of imagery data. IEEE Trans Comput 1982;C-31(10):923–33.

[44] Shenker S. The future of networking, and the past of protocols. Open Network Summit. <http://www.opennetsummit.org/archives/oct11/shenker-tue.pdf>; 2011 [Online].

[45] Shenker S. Software-defined networking at the crossroads. Standford, Colloquium on Computer Systems Seminar Series (EE380); 2013.

[46] McKeown N. ITC keynote. <http://yuba.stanford.edu/~nickm/talks/ITC%20Keynote%20Sept%202011.ppt> [Online].

[47] Open Networking Foundation. OpenFlow Switch Specification, Version 1.5.1. <https://www.opennetworking.org/images/stories/downloads/sdn-resources/onf-specifications/openflow/openflow-switch-v1.5.1.pdf>; 2015 [Online].

[48] Nunes BAA, Mendonca M, Xuan-Nam Nguyen, Obraczka K, Turletti T. A survey of software-defined networking: past, present, and future of programmable networks. IEEE Commun Surveys Tutor 2014;16(3):1617–34.

[49] Marchese M. QoS Over Heterogeneous Networks. 1st ed. Wiley; 2007.

[50] Aurrecoechea C, Campbell AT, Hauw L. A survey of QoS architectures. Elsevier Multimedia Systems 1998;6(3):138–51.

[51] Campbell A, Coulson G, Hutchison D. A quality of service architecture. ACM SIGCOMM Comput Commun Rev 1994;24:6–27.

[52] Hong DW-K, Hong CS. A QoS management framework for distributed multimedia systems. Int J Netw Manage 2003;13:115–27.

[53] Saltzer JH, Reed DP, Clark D. End-to-end arguments in system design. In: Proceedings of ICDCS; 1981.

[54] Afanasyev A, Tilley N, Reiher P, Kleinrock L. Host-to-host congestion control for TCP. In: IEEE Commun Surveys Tutor 2010;12(3):304–42.

[55] Katz D, Kompella K, Yeung D. Traffic engineering (TE) extensions to OSPF Version 2. Internet Engineering Task Force, RFC 3630. <https://tools.ietf.org/html/rfc3630>; 2003 [Online].

[56] Filsfils C, Previdi S, Decraene B, Litkowski S, Shakir R. Segment Routing Architecture. Draft Internet Engineering Task Force, draft-ietf-spring-segment-routing-03. <https://www.ietf.org/id/draft-ietf-spring-segment-routing-03.txt>; 2015 [Online].

[57] Boero L, Cello M, Garibotto C, Marchese M, Mongelli M. Management of non-conformant traffic in OpenFlow environments. In: Proceedings of SPECTS; 2015.

[58] Boero L, Cello M, Garibotto C, Marchese M, Mongelli M. BeaQoS: quality of service support in OpenFlow, draft.

[59] Erickson D. The beacon Openflow controller. In: Proceedings of ACM SIGCOMM HotSDN; 2013.

[60] Enter the Andromeda zone—Google Cloud Platform's latest networking stack. <http://googlecloudplatform.blogspot.it/2014/04/enter-andromeda-zone-google-cloud-platforms-latest-networking-stack.html>; 2014 [Online].

[61] OpenStack Neutron description. <https://wiki.openstack.org/wiki/Neutron> [Online].

[62] SDN and NFV will come to life in the operator network, eventually. <http://searchsdn.techtarget.com/news/2240215704/SDN-and-NFV-will-come-to-life-in-the-operator-network-eventually> [Online].

[63] Vissicchio S, Vanbever L, Bonaventure O. Opportunities and research challenges of hybrid software defined networks. ACM SIGCOMM Comput Commun Rev 2014;44(2):70–5.

OPTIMIZATION CLASSIFICATION AND TECHNIQUES OF WSNs IN SMART GRID

15

M. Naeem*,†, M. Iqbal*, A. Anpalagan†, A. Ahmad*, M.S. Obaidat**

*Department of Electrical Engineering, COMSATS Institute of IT, Wah Campus, Wah, Pakistan;
†Department of Electrical and Computer Engineering, Ryerson University, Toronto, ON, Canada;
**Department of Computer and Information, Fordham University, Bronx, NY, United States

1 INTRODUCTION

The recent state-of-the art developments in various fields, namely efficient wireless transmission, efficient onboard computational capabilities, and miniaturized electromechanical systems have enabled the widespread use of sensors in many application areas [1]. Peculiar characteristics including but not limited to increased fault immunity, improved accuracy, extended coverage area, and extraction of localized features make wireless sensor network (WSN) a preferable choice as compared to traditional sensing [2]. WSNs can be deployed at various locations in smart grid network that includes substations, overhead lines, generating station, and smart homes. In the preceding, the existing literature has been analyzed to show the trend of the research community in this area with respect to the research emanating from different geographical areas of the world. Fig. 15.1 shows a snap shot of the year wise distribution of articles included in this chapter with respect to the affiliation of the authors from different geographical areas of the world.

It can be inferred from the figure that the research relating to WSNs in smart grid is emanating mainly from North America and Asia. The regions of North America intend to meet their most of the energy requirements through renewable resources by the end of 2050 [3] which can be facilitated by the deployment of smart grid. On the other hand China is focusing to reduce her dependency on fossil fuel based energy generation and to minimize carbon dioxide emissions resulting from ever increasing industrial growth [4]. The other significant contributions are from the researchers belonging to Australia and Europe. Different researchers throughout the globe have used different types of sensors and technologies to target various applications in the smart grid paradigm. Some common sensors with their application in smart power systems are as follows:

1. *Cognitive radio sensors (CRS)*: CRS have been proposed in Ref. [5] to be used in CRS network for smart grid applications. The authors argued that the proposed network can be used to achieve quality of service for different type of traffics in sensor network.
2. *Phasor measurement units (PMUs)*: PMUs can provide GPS-synchronized precise, temporally correlated and much denser measurements of voltage and current phasors unlike those of

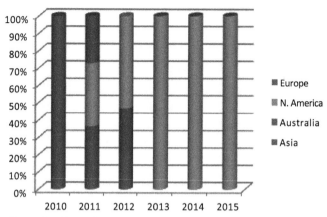

FIGURE 15.1 Research Emanating From Different Areas of the World

traditional sensors [6]. A reactionary smart grid model has been proposed in Ref. [7] which is equipped with the PMUs to distinguish between the normal and abnormal operating conditions. An outage identification and state estimation model has been proposed in Ref. [8] by using PMUs in the smart grid system. In Ref. [9], a multiobjective optimization formulation has been proposed for the optimal placement of PMUs in the smart grid paradigm.

3. *Thermal/hygro sensors*: Thermal/hygro sensors have been used in Ref. [10] to develop a case study building in order to analyze thermal efficiency of the building and find the cost-optimal level for the French single family building typology. Thermal sensors have been used in Ref. [11] to develop a framework for maximum power point tracking in order to maximize the power generated from a photovoltaic system. Thermal sensors have been used in Ref. [12] to facilitate the framework for the control of heating, ventilation, air conditioning, and the charging of plug-in hybrid electric vehicles.

4. *Voltage and current sensors*: Voltage and current sensors have been used in Ref. [13] for implementing the self-healing features in the distribution systems. Self-healing is one of the integral characteristics of smart grid. In Ref. [14], a load monitoring scheme to identify the individual appliances has been proposed by using current and voltage sensors. A nonintrusive load monitoring capability is one of the inherent features of a smart house connected to a smart grid through advanced metering infrastructure.

5. *Luminosity sensors*: Luminosity sensors have been proposed in Ref. [15] to optimize the renovation of the smart building in order to minimize the energy consumption, cost and the life-cycle environmental impact of the building materials.

6. *Rechargeable/energy harvesting sensors*: Limited battery power available with the conventional WSNs can shorten the life of the network considering that the network may be deployed in hard to reach hostile environment. Radio frequency energy harvesting sensors have been proposed in Ref. [16] for monitoring the smart grid. The proposed scheme enabled the high priority sensors to get maximum radio frequency energy. Energy replenishment strategy for the rechargeable sensors has been proposed in Refs. [17,18] to prolong the lifetime of the network.

7. *Relay sensors*: Relay sensors have been proposed in Ref. [19] for the wide area monitoring and control application in smart grid communication. The nodes are deployed for the detection and isolation of faulty relays.

8. *Magneto-resistive sensors*: A magnetoresistive-sensors based approach has been proposed in Ref. [20] for the monitoring of transmission line in smart grid. The proposed method involves measuring emanated magnetic field from a line conductor at the ground level and then calculating the source position and current inversely.

9. *Infrared sensors*: Infrared sensors have been proposed in Ref. [21] in order to facilitate the heating, cooling and air conditioning of smart buildings in the smart grid arena.

10. *Wireless multimedia sensors*: Smart grid can be visualized as an intelligent control system consisting of sensors and communication platforms. Wireless multimedia sensors can provide better surveillance for detecting the grid failure and recovery, monitoring of subsystems and security of the assets. In Ref. [22], authors have used cognitive radio networks to handle the traffic from multimedia sensors deployed in the smart grid.

11. *Generic sensors*: In addition to the aforementioned specific sensors, the literature is brimming with the use of generic wireless sensor nodes to provide information and communication infrastructure for the smart grid. For example in Ref. [23], a cyber physical system consisting of WSN has been proposed for improving the resiliency of the smart grid. The authors suggested a Markov based model for the cluster-tree WSN topologies that leverages the stability of the network. An information and communication infrastructure consisting of WSN has been proposed in Ref. [24] for managing distributed generation more effectively and to alleviate solar energy congestion in the smart distribution grid. The use of WSN based on Zigbee for facilitating the implementation of web services on a central computer has been proposed in Ref. [25]. The proposed approach can provide a web based interface to remotely interact with smart home elements in a smart grid paradigm. In Ref. [26], the authors have proposed a three layer communication architecture that consists of the backbone network using optical fiber, access network using cognitive radios and the terminal network based on WSN. Issues of data-prioritization and delay-sensitive data transmission for WSN in smart grid environment have been addressed in Ref. [27]. The authors argued that the proposed approach could provide robust communication protocols for smart grid monitoring applications. Zigbee based WSN has been considered in Ref. [28] to be used in electricity distribution layer in the smart grid. WSN architecture has been proposed in Ref. [29] for cost efficient residential energy management in the smart grid.

These devices are equipped with limited computational power, low battery power and very low onboard memory space [30–33]. These inherent limitations of the sensors make it difficult to realize a sensor network with reliable and prolonged operational life [34–36]. Therefore, it becomes imperative to efficiently utilize the scarce resources available with the sensor nodes. The optimal utilization of these scarce resources heavily depends on the opted optimization framework and algorithms paradigm [37–44]. To design a good optimization framework and proposing efficient algorithm, there is a need of in depth knowledge of optimization principles and its classification. In the following sections first give a brief primer on the optimization and then elaborates some common objectives and corresponding optimization types related to sensor networks in smart grid paradigm.

2 OPTIMIZATION PRIMER

Sensor optimization is used in smart grid for various purposes. It has wide canvas of applications that includes monitoring, identification, detection, prevention and many more. Before getting into sensor optimization, we will briefly give a primer on the optimization. Optimization is used to get best possible solution from a given set of inputs and constraints. A general notation of any optimization problem is

$$\min_{X} \quad \underbrace{f(X)}_{\text{objective function}}$$

subject to

$$C1: \quad \underbrace{g(X) \leq 0}_{\text{inequality constraints}}$$

$$C2: \quad \underbrace{h(X) = 0}_{\text{equality constraints}}$$

$$C3: \underbrace{X^{\min} \leq X \leq X^{\max}}_{\text{box constraints}}$$

Figs. 15.2 and 15.3 show self-explanatory pictorial view of a simple constraint optimization problem and optimization problem classification respectively [45,46]. Fig. 15.2A clearly shows feasible and infeasible regions, objectives, side and behavioral constraints, local minimum and global minimum. The objective function in the figure is a multimodal nonlinear nonconvex optimization problem with multiple local minimums. Any solution outside the (shaded) boundary is infeasible. Fig. 15.2B, shows a 3D view of constraint multimodel optimization problem with linear constraint. Based on the nature/ structure of the problem, the optimization can be classified into various categories. An optimization problem can be deterministic or stochastic. In deterministic optimization problem, the input data is known whereas stochastic optimization deals with uncertain data. Robust optimization is used if the uncertain data has certain bounds.

Another classification of optimization problem depends on the nature of objective function. Most of the optimization problems for engineering in general and as smart grid in particular have a single objective function. A rare type of optimization problem exists that does not have any objective function. The problems without objective function are known as feasibility problems. In feasibility problems, main goal is to find the values of decision variables that satisfy certain constraint with minimization or maximization of objective function. A number of times in optimization, one has to deal with multiple objective functions [47,48]. These objective functions, in general, may conflict with each other—that is, maximization of one objective may lead to minimization of the second objective.

Unlike single objective optimization, in multiobjective optimization there exist multiple optimal solutions and the decision maker selects one of the feasible solution depending upon the order of preference given to different conflicting objectives. A multiobjective optimization problem can be tackled using different strategies depending on when the decision maker enforce preference on various conflicting objectives. The most commonly used approach is to combine multiple objectives to one figure of merit by assigning different weights to different objectives and then perform single objective optimization algorithm. Weights can be assigned to multiple conflicting objectives through direct assignment, eigenvector method, entropy method, and minimal information method, etc. Few other

(A)

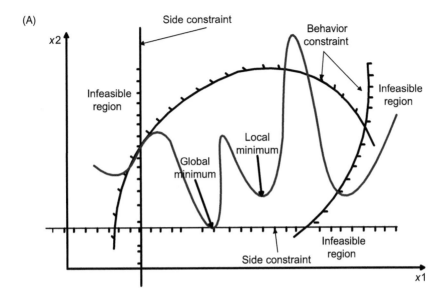

(B)

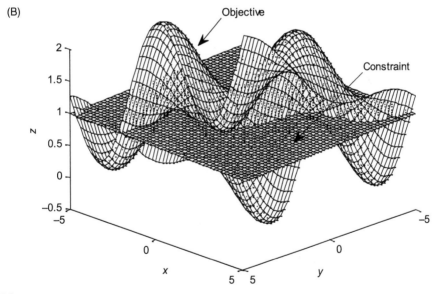

FIGURE 15.2

(A) Geometrical view of constraint optimization problem. (B) 3D Geometrical view of multimodel constraint optimization problem

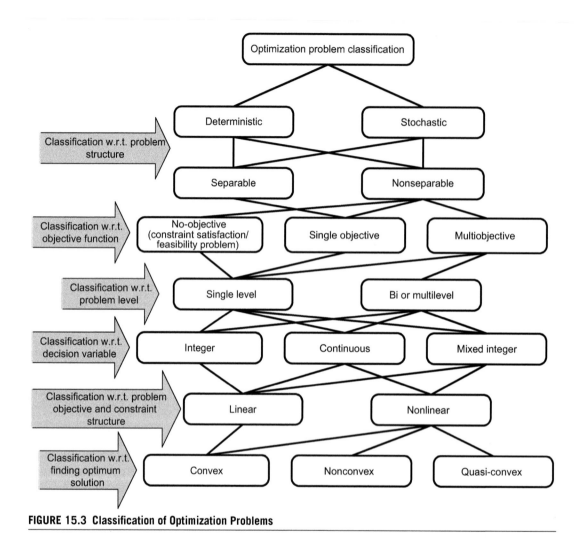

FIGURE 15.3 Classification of Optimization Problems

commonly used multiobjective handling techniques are Min–Max, Pareto, Ranking, Goals, Preference, Gene, Subpopulation, Lexicographic, Phenotype sharing function and Fuzzy. The algorithms used to solve multiobjective optimization problems try to determine best possible tradeoff between different conflicting objectives. Classification of any optimization problem with respect to decision variable and objective/constraint structure reveals to the researchers what kind of solution they can get and what will be the complexity of their solutions. If the problem is linear or convex, then we can get the optimum solutions.

If problem is not convex then we may or may not get the optimal solutions. A generic mathematical way to write an optimization problem is presented in Eq. (15.1). It shows a unified structure of single/bi-level, single/multiobjective, mixed integer programming problem with box, inequality and equality constraints.

$$\text{Outer level} \begin{cases} \min_{\substack{X \in R, \ Y \in Z^+ \\ \text{Continuous \ Integer}}} \overbrace{\underbrace{F_1(X,Y), F_2(X,Y), \cdots, F_M(X,Y)}_{\text{single objective}}}^{\text{Multiobjective}} \\ \textit{Subject To} \\ \quad \underbrace{G(X,Y) \le 0,}_{\text{inequality constraint}} \\ \quad \underbrace{H(X,Y) = 0,}_{\text{equality constraint}} \\ \quad \underbrace{X^{\min} \le X \le X^{\max}, Y^{\min} \le Y \le Y^{\max}}_{\text{box constraints on X,Y}} \\ \text{Inner Level} \begin{cases} Y = \arg\min_{\substack{X, W \in R, Y \in Z^+ \\ \text{Continuous \ Integer}}} \overbrace{\underbrace{f_1(W,X,Y), \cdots, f_N(W,X,Y)}_{\text{single objective}}}^{\text{Multiobjective}} \\ \textit{Subject To} \\ \quad \underbrace{g(W,X,Y) \le 0,}_{\text{inequality constraint}} \\ \quad \underbrace{h(W,X,Y) = 0,}_{\text{equality constraint}} \\ \quad \underbrace{W^{\min} \le W \le W^{\max}, X^{\min} \le X \le X^{\max}, Y^{\min} \le Y \le Y^{\max}}_{\text{box constraints on W,X,Y}} \end{cases} \end{cases} \quad (15.1)$$

After this brief primer, in the next sections, we will present different types of objectives, constraints and optimization problems used in the smart grid WSN.

3 TYPES OF OBJECTIVES AND OPTIMIZATION TECHNIQUES IN SMART GRID PARADIGM

The detail of objectives and corresponding optimization techniques being considered while optimizing the resource constrained sensor networks in smart grid applications has been depicted in Fig. 15.4. For example, an optimization formulation based on linear programming (LP) has been used in Refs. [5] and [49] to maximize the overall traffic through CRS network in smart grid applications. A framework for integrated sensor and equipment monitoring has been proposed in Ref. [50] by using nonlinear (NL) programming techniques to maximize the gross error detection in sensor measurements. Optimal placement of sensors in the power system has been addressed in Ref. [8] by using mixed integer convex programming to minimize the probability of error in detecting the real-time state of the system. An optimization formulation has been used in Ref. [51] to minimize the error between the actual power system frequency and that of the estimated by using nonlinear programming approach. Energy management in hybrid renewable energy generation system has been proposed in Ref. [52] by using integer programming to maximize the profit. Energy management in a smart home with wireless network based on Zigbee has been addressed in Ref. [25] by using dynamic programming to minimize the energy expenses. The authors proposed a secured web based approach to remotely control the home appliances for optimal energy management. Maximization of quality of service for smart grid applications has been proposed in Ref. [7] by using linear programming. The authors suggested WSN by optimally allocating the bandwidth for different smart grid applications. In Ref. [53], the quality of service maximization has been achieved by using linear programming for optimal spectrum management of

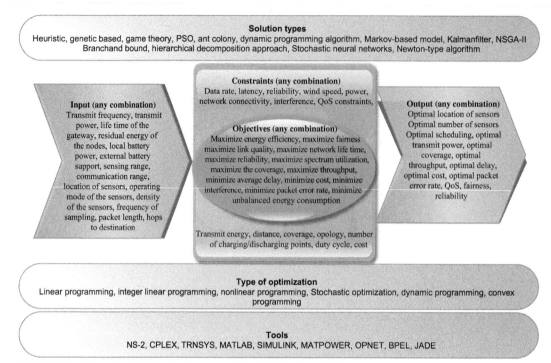

FIGURE 15.4 A Generic Optimization Framework for WSN in Smart Grid

cognitive radio based smart grids. Delay is critical for different smart grid applications and has been considered in various optimization formulations related to WSN in smart grid system. For example, in Refs. [23] and [54], the authors have proposed an optimization formulation based on nonlinear programming to minimize the delay in WSN for different smart grid applications sensitive to delay. A micro grid energy management framework has been proposed in Ref. [55] by using linear formulation to minimize the power purchase expenses and to maximize the profit of each stakeholder in the system. A nonlinear formulation has been used in Ref. [56] to optimize demand response for plug in electrical vehicles in the smart grid. Dynamic maintenance strategies have been proposed in Ref. [57] for multiple transformers to minimize the cost of operation and maintenance. Fault recognition in smart grids has been formulated as nonlinear optimization problem in Ref. [58] to minimize the total cost of inspection and repair. An integer linear programming formulation for the monitoring of the overhead transmission line in smart grid has been proposed in Ref. [19] to minimize the cost of the system.

A linear programming technique has been proposed in Ref. [29] for cost-efficient residential energy management in smart grid to minimize the total cost of electricity used at home. Energy conservation in smart grid is one of the critical issues and has been considered in various articles. For example, in Ref. [59], a smart grid energy management system has been proposed for distributed response management, control and optimization. A linear programming formulation has been used in Ref. [60] to adaptively optimize power conservation in the smart grid. Optimal utilization of the onboard battery of the sensor node is critical to prolong the life time of the network.

In Ref. [29], a WSN based cost efficient in hoe energy management (iHEM) scheme is presented. The main aim is to minimize the home energy expenses by applying optimization based residential management (OREM). The proposed OREM scheme employs a wireless sensor home are network. A similar approach is proposed in Refs. [39,40]. The OREM is capable of employing home energy management in the presence of self-generation of electricity, electricity from grid and time of use tariff (TOU). In time of use tariff, any service provider can charge differently at different times in any time span. The optimization problem can be written as follows:

$$\text{Minimize } U_{\text{OREM}}(X)$$
$$\text{subject to}$$
$$C1: f_{\text{CWI}}(X) \leq \Delta_t$$
$$C2: f_{\text{LOC}}(X) = D$$
$$C3: f_{\text{CON}}(X) = 0$$
$$C4: f_{\text{SCH}}(X) = 1$$
$$C5: X \in \{0,1\}^N$$

Where $U_{\text{OREM}}(X)$ is the objective utility which minimizes the total billing cost. Constraint C1 ensures that scheduling task should be complete in without interruption (CWI), C2 and C3 constraints ensure that each task should complete in required interval with consecutive time slots and C4 is the assignment constraint. Authors use ILOG CPLEX tool to solve the aforementioned optimization problem.

In Ref. [61], an optimization formulation has been suggested to maximize the nodes battery utilization, distribute power effectively from the energy harvester and optimize the distance between mobile charger and sensor nodes. Stochastic optimization has been used in Ref. [12] for the home energy management to minimize the discomfort level of the customer. Energy management in smart grid has been tackled in numerous ways. One of them is vehicle to grid (V2G), where electric vehicles are used to feed the stored energy to the grid when and where needed. In Ref. [62], the authors have proposed an optimization formulation to minimize the route of electric vehicles destined to charge different stations. The problem of energy management in direct current grid powered light emitting diode based lightening system has been considered in Ref. [63] to minimize the energy consumption. A photo voltaic system has been considered in Ref. [11] under partial shade and an optimization formulation has been used to maximize the power generation. Due to lack of transmission lines, energy transfer capability, and transmission capacity there can be congestion of energy. In Ref. [56], an advanced metering infrastructure based (AMI) based distributed demand response of plug-in-vehicle in smart grid is presented. AMI is a kind of sensing device that not only senses the power flow but also communicates with central control center for real-time update in tariff rates. In the system model, we have multiple residences and each residence has plug-in-vehicle. A user is the residence can sell or buy electricity from utility company. Four different kind of loads are in the system namely, base load, schedulable load, PEV load and the distributed generation. Main aim of the optimization is to reduce overall cost of the system subject to scheduling and charging/discharging constraints. Authors apply alternating direction method of multipliers to solve the cost minimization problem.

The energy congestion problem has been tackled in Ref. [24] by maximizing the number of connected load units. The estimation of energy usage in smart buildings has been formulated by using nonlinear programming in Ref. [64] to minimize the expected cost of displaying wrong signal about energy consumption. In any smart house, to have better demand side management, load demand identification is critical. In load identification procedure, without harming the existing signals, demand side manager nonintrusively identify the load and extract its transient features. A time-frequency analysis based nonintrusive load monitoring and load identification is investigated in Ref. [14]. Linear integer programming formulation has been proposed in for nonintrusive load monitoring to identify the load demand and to maximize the profit. Authors proposed a multidimensional knapsack framework for load identification in conjunction with multiresolution S-transform for transient feature extraction.

$$\text{Maximize } U_{\text{LDI}}(X)$$
$$\text{subject to}$$
$$C1: \quad f_{\text{RC}}(X) \le A + \varepsilon$$
$$C2: \quad X \in \{0,1\}^{N}$$

In the aforementioned formulation, $U_{\text{LDI}}(X)$ is the load identification objective and is equal to $U_{\text{LDI}}(X) = \sum_{j=1}^{N} p_j x_j$. The constraint C1 is resource constraint, in this constraint A is the available quantity and ε is the tolerance. Authors uses different household appliance for identification that includes vacuum cleaners, electric boiler, microwave oven, and hair dryer. A swarm intelligent ant colony algorithm is applied for load identification.

A building energy consumption model based on stochastic Markov models has been proposed in Ref. [65] to predict the potential energy saving benefits from building retrofit. Integer linear programming based optimization formulation has been used in Ref. [16] for differentiated radio frequency power transmission to charge WSN nodes in smart grid environment. This approach gives the energy harvesting as an alternative to extend the lifetime of wireless sensor nodes for long term smart grid monitoring and surveillance tasks. The main aim of that differentiated approach is to maximize the power received by the highest priority sensor node. A CPLEX-based optimization solver is used to optimal solution is reasonable time. Dual convex optimization has been proposed in Ref. [66] to optimize the power flow in smart grid networks. The distributed convex optimization method is applied to constraint multi-agent system with large number of sensor nodes. The author models the optimization problem as separable optimization problem so that function can be decompose into small elementary objective functions. A Langrangian duality based decentralized iterative approach is applied to solve the constraint multi-agent problem. Linear programming has been used in Ref. [67] for optimizing the dynamic state estimation of the power system. Authors proposed and applied dynamic modeler to conserve energy in smart grid assisted electric transmission system. The modeler, in the beginning identify the controlling parameters and then solve the linear optimization to set the states of these controlling parameters under the constraint that service level will be maintain all the time. The authors argued that the proposed model can capture the accurate runtime models of large power systems and is able to adapt itself with the change in the size or structure of the system.

A congestion aware smart metering communication is proposed in Ref. [24]. Congestion in power system may be due to unexpected surplus renewable power production. In their work, authors formulated the congestion control problem in smart grid as knapsack problem by disconnecting some solar units in order to maintain the reliability of the smart distribution grid. Their communication model is based on a disconnection process with the help of smart meter infrastructure. A greedy solution is provided to control the congestion in the smart grid system. In the proposed framework, the authors assume that instead of stand-alone system, the solar unit in each home is connected to the main distribution grid. We will only consider a unit as a connected unit if the energy generated by the solar unit is more than the demand of its connected load. Main aim of congestion control is to maximize the number of connected solar units under the constraint of network power transmission capacity. A generic congestion control optimization problem can be written as

$$\text{Maximize } U(X)$$
$$\text{subject to}$$
$$C1: \quad f_{\text{ATC}}(X) \le P_{\text{ATC}}$$
$$C2: \quad X \in \{0,1\}^{N}$$

where $U(X) = \sum_{n=1}^{N} x_n$, $x_n \in \{0,1\}$ is total number of connected solar units and P_{ATC} is available power transmission capacity. Constraint C1 is the network power transmission power capacity constraint. The variable x_n is a binary selection variable

$$x_n = \begin{cases} 1 & \text{if the nth solar unit is connected to grid} \\ 0 & \text{otherwise} \end{cases}$$

The maximization of $U(X)$ ensure a large number of connected solar units to the grid. That gives monetary benefits to consumers by selling surplus energy to other consumers. A greedy strategy is proposed to solve the connected solar load maximization problem.

To take care of quality of service (QoS) for mesh WSN deployed on smart grid network, authors in Ref. [32] proposed and analyzed QoS aware guaranteed time slot allocation scheme named as QGA. The proposed scheme decreases the end to end latency for delay sensitive data. In this scheme, smart grid monitoring infrastructure decides whether data is delay critical or delay tolerant. Delayed tolerant data is buffered and is transmitted at suitable time. The system model includes three kinds of devices in the WSN infrastructure. These devices are reduced function devices (RFD), full function devices (FFD) and sink as shown in Fig. 15.5. RFD can only communicate the information but cannot perform routing. Any FFD can communicate with its neighboring RFDs and can also perform routing.

The main goal of this time slot allocation is to reduce the time delay of data from FFD to the sinks. The constraint optimization problem can be written as

$$\text{Minimize } U_{\text{Delay-FFD}}(X)$$
$$\text{subject to}$$
$$C1: f_{\text{packet-T}}(X) \le 0$$
$$C2: f_{\text{packet-R}}(X) \le 0$$
$$C3: f_{\text{packet-Rate}}(X) \le 0$$

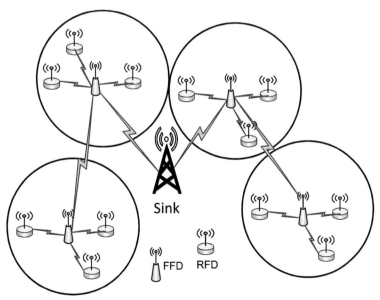

FIGURE 15.5 Differentiated Traffic Optimization in Smart Grid Mesh WSN

The aforementioned problem is linear optimization problem. Constraints C2, C2, and C3 are box constraint on packets that can be transmitted, number of packets received and arrival rate of packets respectively. Authors apply interior point method to solve aforementioned optimization problem.

Maximization of life time of the network has been addressed in Ref. [68] by using convex programming. The proposed scheme efficiently combined the two media namely, radio frequency and power line communication in order to improve the life time and the reliability of the network. A convex optimization technique has been suggested in Ref. [69] for the selection of sensor nodes in order to optimize the energy consumption and to maximize the network life time. The main aim of the sensor selection is to increase the lifetime and reduce the energy consumption under orthogonal and coherent multiple access schemes between wireless sensor node and fusion center. The proposed optimization framework is binary integer program having NP-hard characteristics. An opportunistic sensor selection scheme similar to water filling is applied by relaxing the binary integer constraints. Maximization of the sensor network utility has been addressed in Ref. [70] by using convex programming.

In Refs. [5] and [49], authors investigate a QoS aware cross layer design for CRSN in smart grid environment. The proposed method is useful for smart grid as existing path determining techniques under QoS constraint cannot simultaneously handle data traffic of different characteristic—for example, information about real-time pricing, time of use tariff, smart meter readings and observations, fault diagnostic information and power system control signaling. The problem is formulated as weighted network utility maximization problem under the assumption that wireless sensors are randomly deployed and each sensor can only occupy one channel at one time. The network utility $U(\beta,\tau,\alpha)$ is function of data rate, latency and reliability. A generic network utility maximization for smart grid can be written as

$$\text{Maximize } U(\beta,\tau,\alpha)$$
$$\text{subject to}$$
$$C1: f_{\text{Rate}}(\beta,\tau,\alpha) \leq 0$$
$$C2: f_{\text{Reliability}}(\beta,\tau,\alpha) \leq 0$$
$$C3: f_{\text{Delay}}(\beta,\tau,\alpha) \leq 0$$
$$C4: f_{\text{Network Flow}}(\beta,\tau,\alpha) = 0$$

Where β the data is rate, τ is delay, α is reliability. The constraints C1, C2, C3, and C4 are rate, reliability, delay, and network flow constraints. The data rate is in bits per seconds, latency is in packet delay till destination and bit error rate is used to ensure reliability. Authors proposed a cross layer heuristic to solve the problem.

A sustainable wireless rechargeable sensor network has been proposed in Refs. [17] and [71] which employed mobile chargers to charge multiple sensors from several landmark locations. The authors used integer programming to find minimum number of landmarks according to the locations and energy replenishment requirements of the sensors. Replenishment of the energy of sensor nodes will prolong the operational life of the network and will strengthen its reliability. The reliability and the robustness of the power system can be leveraged by ensuring real-time observing ability of the system. In order to maximize real-time monitoring of the power system, PMUs have been proposed in Ref. [9]. The authors have used integer programming to minimize the number of PMUs and to maximize the system state monitoring. A convex optimization formulation has been proposed in Ref. [72] for optimal sensor placement to minimize the vulnerability of the system from cyber attacks.

Accurate and cost effective selection of sensors plays important role in smart grid real-time operations. In Ref. [69], authors presented a sensor selection scheme for parameter estimation in energy constraint WSN for smart grid to monitor and control their environment. Main aim of wireless sensor selection in smart grid is to increase the network life time and minimize network energy consumption. Authors simply minimize the mean square error to get the best estimate of received signals. Mathematically:

$$\text{Minimize } U_{\text{MSE}}(X)$$
$$\text{subject to}$$
$$C1: f_{\text{Sel}}(X) \le K$$
$$C2: X \in \{0,1\}^N$$

Where K is the maximum possible sensors that can be selected at one time. The aforementioned problem is binary integer programming problem and valid for both orthogonal and coherent multiple access between wireless sensors and fusion center. The $U_{\text{MSE}}(X)$ is a utility function having the expression $\left(\sum_{n=1}^{N} x_n \dfrac{g_n}{\sigma_n^2 g_n + \xi_n^2} \right)^{-1}$, where x_n is a binary indicator used for selection on the nth sensor, g_n is the channel gain of the nth sensor σ_n^2 is the noise variance. The authors proposed a relaxed version of this problem by relaxing the binary constraint and apply convex optimization techniques and water filling to get the optimal results.

An optimization formulation based on linear programming has been used in Ref. [21] for optimum building energy retrofits under technical and economic uncertainty to minimize the loss. Optimal deployment of smart meters has been addressed in Ref. [71] by using dynamic programming for efficient electrical appliance state monitoring. A gradient based optimization formulation has been proposed in Ref. [73] for minimization of the peak load in smart grid without compromising the discomfort of the customers. In Ref. [44], a hybrid sensor network (HSN) is presented to increase the network life time for home energy management system. The HSN consists of wireless sensor nodes and power-line sensor nodes. Wireless sensor nodes will be used for data collection—for example, temperature measurement and appliance energy measurement while power-line sensors will be installed on appliances and used as complements of wireless sensor nodes. The mail goal of optimization is to maximize the network lifetime subject to routing. MAC and physical constraints on sensor nodes using cross layer design paradigm. The proposed network life time maximization problem is mixed integer convex optimization problem. The optimization problem can be written as

$$\text{Maximize minimum } U(T)$$
$$\text{subject to}$$
$$C1: f_{\text{Life time}}(T) \le Z$$
$$C2: f_{\text{Data rate}}(T) \ge R$$
$$C2: f_{\text{In}}(T) = f_{\text{Out}}(T)$$

In the aforementioned optimization $U(T)$ is the life time of worst sensor node. Constraints C1, C2, and C3 are life time, rate and flow conservation constraints. Authors apply Karush–Kuhn–Tucker (KKT) conditions to derive analytical expression to get life time maximization.

4 CONSTRAINTS CONSIDERED WHILE FORMULATING OPTIMIZATION PROBLEMS IN SMART GRID PARADIGM

Real-life problems of sensor network design, planning, deployment, operation and management in smart grid paradigm are influenced by many practical constraints. Table 15.1, shows some constraints which have been considered while formulating an optimization problem relating to sensor networks in smart grid environment. For example, data rate, latency and reliability constraints have been considered in Refs. [5] and [49] while formulating an optimization formulation to maximize the sensor network traffics.

Wind speed has been considered as constraint in Ref. [52] to optimize the energy management for hybrid renewable energy generating systems. An optimization formulation has been proposed in Ref. [7] to ensure quality of service for different competing users of the network while considering the latency constraint for different smart grid applications. Power congestion problem has been addressed in Ref. [24] by maximizing the number of connected units while satisfying the maximum power transfer capacity constraint of the system. Optimal power flow problem has been considered in Ref. [55] to find the optimal active and reactive powers in order to reduce the overall system cost while considering the reliability constraints. In Ref. [16], the authors have tackled the problem of differentiated radio frequency power transmission for WSN deployment in the smart grid to maximize the power received by the high priority nodes while satisfying the constraint of number and availability of charging stations. Minimization of users bills has been addressed in Ref. [56] where authors have considered the constraint of charging and discharging energy/rate. The constraint of the electric current that flows in each half line has been considered in Ref. [58] to formulate an optimization problem for the minimization of the total cost of inspection and repair. Minimization of overall cost of the communication infrastructure for the monitoring of smart grid has been addressed in Ref. [19] while satisfying latency and coverage constraints of the network.

In Ref. [53], the minimization of the number of leased channels has been tackled while considering the constraints of dropping and blocking probabilities. Maximization of network life time has been addressed in Refs. [68] and [69] by optimizing the energy consumption while considering the energy constraints of the network. An optimization formulation has been proposed in Ref. [70] to maximize the network utility while satisfying the power constraints. Problem of finding optimal number/location of charging points in the sustainable wireless rechargeable sensor networks for the smart grid has been tackled in Refs. [17] and [18] while considering the constraints of receiver power. In Ref. [12], an optimization formulation has been suggested for home energy management to minimize the discomfort level of customers while satisfying the cost and peak power constraints. Constraints of data rate and latency have been considered in Ref. [22] to ensure the provision of quality of service for cognitive radio communication infrastructure based smart grid applications. Cost minimization has been proposed by using WSN based home energy management in Ref. [29] while satisfying the constraint of reliability.

5 OPTIMIZATION SOLUTION TYPES AND RELATED SIMULATION TOOLS IN SMART GRID PARADIGM

Diversified natures of real world problems relating to the application of sensor networks in smart grid have culminated a plethora of optimization algorithms and simulation tools. Table 15.2 highlights optimization solution types and related simulation tools in smart grid paradigm. For example in Ref. [5], a heuristic algorithm has been proposed to meet the quality of service requirements in CRS networks for

Table 15.1 Different Constraints Considered While Formulating Optimization Problems in Smart Grid Paradigm

References	Objectives	Constraints									
		Data Rate	Latency	Reliability	Wind Speed	Blocking Problem	Power	Charge/Discharge Energy	Current	Charging/Discharging	Duty Cycle
[5,49]	Traffic maximization	X	X	X							
[52]	Energy Management				X						
[7]	QoS		X								
[24]	Power congession			X			X				
[55]	Cost minimization			X							
[16]	Maximizes the power received									X	
[56]	Cost minimization							X			
[58]	Cost minimization								X		
[19]	Cost minimization		X								
[53]	QoS					X					
[68,69]	Life time						X				
[70]	Maximize the network utility						X				
[17]	Optimal number/location of charging points						X				X
[12]	Customer discomfort						X				
[22]	QoS	X	X				X				

Table 15.2 Optimization Solution Types and Related Simulation Tools in Smart Grid Paradigm

References	Solution Types	Tools								
		NS-2	CPLEX	TRNSYS	MATLAB	SIMULINK	MAT-POWER	OPNET	BPEL	JADE
[5,61]	Heuristic	X								
[9,52]	Genetic Based			X						X
[21,52]	Game Theory			X					X	
[10]	PSO			X						
[11,14]	Ant Colony				X	X		X		
[24,25]	Dynamic programming algorithm									
[64]	Markov-based model				X					
[37]	Kalman filter				X					
[15]	NSGA-II			X						
[19]	Bernoulli distribution	X	X		X					
[8]	Branch and Bound	X					X			
[70]	Hierarchical decomposition approach			X						
[20]	Stochastic				X					
[38]	Neural networks									
[51]	Newton-type algorithm				X					

smart grid applications. The authors used NS-2 to simulate the proposed algorithm. Heuristic algorithm has been proposed to solve an optimization formulation in Ref. [61] by using NS-2. Cross layer distributed control algorithm has been used in Ref. [49] to ensure quality of service requirements in wireless senor networks for smart grid applications. The authors used NS-2 to evaluate their proposed scheme. Energy management in a distributed hybrid renewable energy generation system has been addressed in Ref. [52], where authors have used genetic algorithm and game theory. They analyzed the proposed approach by using TRNSYS. In Ref. [10], particle swarm optimization algorithm has been used to solve the optimization formulation for finding the cost optimal building configuration using TRANSYS. Ant colony optimization technique has been used in Ref. [11] to implement maximum power point tracking. The proposed scheme was validated by using MATLAB simulations. In Ref. [24], the authors have suggested dynamic programming algorithm to address the problem of congestion caused by power surpluses produced from domestic solar units on the rooftops. The effectiveness of the algorithm was verified by using Opnet and BPEL simulators. A signal mechanism for the purpose of obtaining a summary of the use of energy in smart buildings has been proposed in Ref. [64] by using Markov based model. The proposed model was simulated using MATLAB. MATLAB has also been used to simulate an energy model for operation optimization and energy saving suggested in Ref. [37] by using Kalman filter approach to improve the forecasting accuracy. In Ref. [15], a multi-criteria tool has been developed to optimize the renovation of buildings. The multi objective optimization formulation has been solved by using a non-dominated sorting genetic algorithm (NSGA-II) through TRNSYS simulator. Bernoulli distribution has been used in Ref. [19] to solve an optimization formulation for sensors relay monitoring mechanism for overhead transmission line in smart grid. The effectiveness of the proposed algorithm was verified using NS-2, CPLEX and MATLAB.

In Ref. [8], branch and bound algorithm has been proposed and simulated using MATPOWER for the identification of outages in power systems with uncertain states and optimal sensor locations. Hierarchical decomposition approach has been suggested in Ref. [70] to solve an optimization formulation by using NS-2 simulator. Stochastic optimization has been used to solve the optimization problem for the detection of power line outages by employing MATLAB simulations [20]. The problem of sensor fault detection and its efficiency analysis in air handling unit has been tackled by using neural networks in Ref. [38]. The authors used TRNSYS simulations to evaluate their proposed algorithm. A real-time optimization approach based on the Newton-type algorithm and least squares method for power system frequency estimation has been proposed in Ref. [51]. The performance of the proposed approach was validated by simulations in MATLAB. The authors argued that the suggested algorithm was robust and accurate under the time varying conditions.

6 CONCLUSIONS AND DISCUSSIONS

Peculiar properties of WSNs and the emerging sociotechnical scenarios are responsible for their application in various fields including but not limited to smart cities, smart grids, environment monitoring, habitat monitoring, greenhouse monitoring, climate monitoring, home automation, industrial automation, water distribution system monitoring, and personal health monitoring. These networks consist of tiny devices provided with low computational power, low energy supply, and low onboard memory space. Due to these inherent resource constraints, it becomes difficult to ensure long lasting operation of these networks. Therefore, it becomes imperative to optimally use the scarce resources available with the sensor nodes. In this chapter, we have analyzed the available literature focused on optimization techniques being used to formulate different practical problems relating to planning, design, deployment, operation, and management of sensor network in the smart grid paradigm. We analyzed the existing literature related to the optimization techniques in smart grid paradigm and showed the trend of the research community with respect to the research emanating from different geographical areas of the world. We also highlighted the relevant work focused on solving different optimization problems specific to various sensor types and the corresponding communication technologies applicable to the smart grid system. The relevant literature has been further investigated to classify different optimization formulations with respect to some commonly used objectives and the corresponding optimization technique. Finally, various constraints have been tabulated. These constraints have been considered while formulating the optimization problems relating to the sensor network applications in smart grid. Therefore, in this chapter the existing literature related to optimization techniques applicable to sensor network in smart grid paradigm has been analyzed. This analysis can provide a foundation for further research in the area and can help to formulate various real-life optimization problems relating to sensor network planning, design, deployment, and operation.

REFERENCES

[1] Yick J, Mukherjee B, Ghosal D. Wireless sensor network survey. Comput Netw 2008;52(12):2292–330.
[2] Liu, Yide. Wireless sensor network applications in smart grid: recent trends and challenges. Int J Distributed Sensor Netw 2012;2012.
[3] Mai T, Sandor D, Wiser R, Schneider T. Renewable electricity futures study. Executive Summary. National Renewable Energy Laboratory (NREL), Golden, CO., Technical Report; 2012.
[4] Ouyang S X., Lin B. An analysis of the driving forces of energy-related carbon dioxide emissions in China's industrial sector. Renewable Sustainable Energy Rev 2015;45:838–49.
[5] Ghalib SA, Gungor VC, Akan OB. A cross-layer design for QoS support in cognitive radio sensor networks for smart grid applications. In: 2012 IEEE international conference on communications (ICC). IEEE; 2012.
[6] Phadke AG. Synchronized phasor measurements in power systems. IEEE Comput Appl Power 1993;6(2):10–5.
[7] Robert W, Munasinghe K, Jamalipour A. A population theory inspired solution to the optimal bandwidth allocation for Smart Grid applications. In: 2014 IEEE wireless communications and networking conference (WCNC). IEEE; 2014.
[8] Zhao Y, Chen J, Goldsmith A, Poor HV. Identification of outages in power systems with uncertain states and optimal sensor locations. IEEE J Selected Topics Signal Process 2014;8(6):1140,1153.
[9] Xian-Chang G, Liao C-S, Chu C-C. Multi-objective power management on smart grid. In: Proceedings of the 2014 IEEE 18th international conference on computer supported cooperative work in design (CSCWD). IEEE; 2014.
[10] Ferrara M, et al. A simulation-based optimization method for cost-optimal analysis of nearly Zero Energy Buildings. Energy Buildings 2014;84:442–57.
[11] Lian JL, Maskell DL. A uniform implementation scheme for evolutionary optimization algorithms and the experimental implementation of an ACO based MPPT for PV systems under partial shading. In: 2014 IEEE symposium on computational intelligence applications in smart grid (CIASG). IEEE; 2014.
[12] Liyan J et al. Multi-scale stochastic optimization for home energy management. In: 2011 4th IEEE international workshop on computational advances in multi-sensor adaptive processing (CAMSAP). IEEE; 2011.
[13] Aguero JR. Applying self-healing schemes to modern power distribution systems. In: 2012 IEEE power and energy society general meeting. IEEE; 2012.

[14] Lin Y-H, Tsai M-S. Development of an improved time–frequency analysis-based nonintrusive load monitor for load demand identification. IEEE Trans Instrum Meas 2014;63(6):1470–83.

[15] Chantrelle FP, et al. Development of a multicriteria tool for optimizing the renovation of buildings. Appl Energy 2011;88(4):1386–94.

[16] Melike E-K, Mouftah HT. DRIFT: differentiated RF power transmission for wireless sensor network deployment in the smart grid. In: 2012 IEEE Globecom Workshops (GC Wkshps). IEEE; 2012.

[17] Melike E-K, Mouftah HT. Mission-aware placement of RF-based power transmitters in wireless sensor networks. 2012 IEEE symposium on computers and communications (ISCC). IEEE; 2012.

[18] Melike E-K, Mouftah HT. Suresense: sustainable wireless rechargeable sensor networks for the smart grid. IEEE Wireless Commun 2012;19(3):30–6.

[19] Wang, Kaixuan, et al. Fault tolerance oriented sensors relay monitoring mechanism for overhead transmission line in smart grid.. IEEE Sensors J 2015;15(3).

[20] Sun X, et al. Novel application of magnetoresistive sensors for high-voltage transmission-line monitoring. IEEE Transac Magnetics 2011;47(10):2608–11.

[21] Rysanek AM, Choudhary R. Optimum building energy retrofits under technical and economic uncertainty. Energy Buildings 2013;57:324–37.

[22] Huang J, et al. Priority-based traffic scheduling and utility optimization for cognitive radio communication infrastructure-based smart grid. IEEE Trans Smart Grid 2013;4(1):78–86.

[23] Al-Anbagi IRFAN, Erol-Kantarci MELIKE, Hussein TM. A reliable IEEE 802.15. 4 model for cyber physical power grid monitoring systems. IEEE Trans Emerg Topics Comput 2013;1(2):258–72.

[24] Lo, Chun H, Nirwan A. Alleviating solar energy congestion in the distribution grid via smart metering communications". IEEE Trans Parallel Distrib Syst 2012;23(9):1607–20.

[25] Khan AA, Mouftah HT. Energy optimization and energy management of home via web services in smart grid. In: 2012 IEEE electrical power and energy conference (EPEC). IEEE; 2012.

[26] Zhigang L, Wang J. Multi-scale data fusion for smart grids. 2012 7th IEEE conference on industrial electronics and applications (ICIEA). IEEE; 2012.

[27] Al-Anbagi SI., Melike E-K, Mouftah HT. Priority-and delay-aware medium access for wireless sensor networks in the smart grid.. IEEE Syst J 2014;8(2):608–18.

[28] Hikma S, et al. Smart Zigbee/IEEE 802.15. 4 MAC for wireless sensor multi-hop mesh networks. In: 2013 IEEE 7th international power engineering and optimization conference (PEOCO). IEEE; 2013.

[29] Melike E-K, Mouftah HT. Wireless sensor networks for cost-efficient residential energy management in the smart grid. IEEE Trans Smart Grid 2011;2(2):314–25.

[30] Akyildiz IF, et al. A survey on sensor networks. IEEE Commun Mag 2002;40(8):102–14.

[31] Ferentinos KP, Theodore AT. Adaptive design optimization of wireless sensor networks using genetic algorithms. Comput Netw 2007;51(4):1031–51.

[32] Irfan AA, Erol-Kantarci M, Mouftah HT. Time slot allocation in WSNs for differentiated smart grid traffic. In: 2013 IEEE electrical power & energy conference (EPEC). IEEE; 2013.

[33] Viani F, et al. Wireless architectures for heterogeneous sensing in smart home applications: concepts and real implementation. Proc IEEE 2013;101(11):2381–96.

[34] Zhu Z, et al. Overview of demand management in smart grid and enabling wireless communication technologies. IEEE Wireless Commun 2012;19(3):48–56.

[35] Fan Z, et al. Smart grid communications: Overview of research challenges, solutions, and standardization activities. IEEE Commun Surveys Tutor 2013;15(1):21–38.

[36] Yan Y, et al. A survey on smart grid communication infrastructures: motivations, requirements and challenges. IEEE Commun Surveys Tutor 2013;15(1):5–20.

[37] Li X, Wen J. Building energy consumption on-line forecasting using physics based system identification. Energy Buildings 2014;82:1–12.

[38] Du Zhimin, et al. Sensor fault detection and its efficiency analysis in air handling unit using the combined neural networks. Energy Buildings 2014;72:157–66.

[39] Melike E-K, Mouftah HT. Tou-aware energy management and wireless sensor networks for reducing peak load in smart grids. In: 2010 IEEE 72nd vehicular technology conference fall (VTC 2010-Fall). IEEE; 2010.

[40] Melike E-K, Mouftah HT. Using wireless sensor networks for energy-aware homes in smart grids. In: 2010 IEEE symposium on computers and communications (ISCC). IEEE; 2010.

[41] Ebrahimi MS, et al. A novel utilization of cluster-tree wireless sensor networks for situation awareness in smart grids. In: 2011 IEEE PES innovative smart grid technologies Asia (ISGT). IEEE; 2011.

[42] Nadeem J, et al. A survey of home energy management systems in future smart grid communications. In: 2013 eighth international conference on broadband and wireless computing, communication and applications (BWCCA). IEEE; 2013.

[43] Niyato D, Ping W. Cooperative transmission for meter data collection in smart grid. IEEE Commun Mag 2012;50(4):90–7.

[44] Kainan Z, Zhu X. Cross-layer network lifetime maximization for hybrid sensor networks. In: 2014 IEEE international conference on smart grid communications (SmartGridComm). IEEE; 2014.

[45] <http://www.neos-guide.org/optimization-tree>.

[46] Rao SS, Rao SS. Engineering optimization: theory and practice. Hoboken, NJ: John Wiley & Sons; 2009.

[47] Marler TR, Arora JS. Survey of multi-objective optimization methods for engineering. Structural Multidisc Optimization 2004;26(6):369–95.

[48] Takama N, Loucks DP. Multi-level optimization for multi-objective problems. Appl Math Model 1981;5(3):173–8.

[49] Shah GA, Vehbi CG, Özgür BA. A cross-layer QoS-aware communication framework in cognitive radio sensor networks for smart grid applications. IEEE Trans Ind Inform 2013;9(3):1477–85.

[50] Jiang X, Pei L, Zheng L. A data reconciliation based framework for integrated sensor and equipment performance monitoring in power plants. Appl Energy 2014;134:270–82.

[51] Sadinezhad I, Vassilios GA. Slow sampling online optimization approach to estimate power system frequency. IEEE Trans Smart Grid 2011;2(2):265–77.

[52] Jun Z, et al. A multi-agent solution to energy management in hybrid renewable energy generation system. Renewable Energy 2011;36(5):1352–63.

[53] Huang J, et al. Priority-based traffic scheduling and utility optimization for cognitive radio communication infrastructure-based smart grid. IEEE Trans Smart Grid 2013;4(1):78–86.

[54] Baobing W, Baras JS. Minimizing aggregation latency under the physical interference model in Wireless Sensor Networks. In: 2012 IEEE third international conference on smart grid communications (SmartGridComm). IEEE; 2012.

[55] Kuznetsova E, et al. An integrated framework of agent-based modelling and robust optimization for microgrid energy management. Appl Energy 2014;129:70–88.

[56] Zhao T, Yang P, Nehorai A. Distributed demand response for plug-in electrical vehicles in the smart grid. In: 2013 IEEE 5th international workshop on Computational advances in multi-sensor adaptive processing (CAMSAP). IEEE, 2013.

[57] Chong W, et al. Dynamic maintenance strategies for multiple transformers with Markov models. In: 2014 IEEE PES innovative smart grid technologies conference (ISGT). IEEE; 2014.

[58] De Santis E, et al. Fault recognition in smart grids by a one-class classification approach. In: 2014 international joint conference on neural networks (IJCNN). IEEE; 2014.

[59] Yong D, et al. A smart energy system: distributed resource management, control and optimization. In: 2011 2nd IEEE PES international conference and exhibition on innovative smart grid technologies (ISGT Europe). IEEE; 2011.

[60] Fahad J, et al. An adaptive optimization model for power conservation in the smart grid. In: 2010 IEEE international conference on systems man and cybernetics (SMC). IEEE; 2010.

[61] Uthman B, Al-Roubaiey A. Mobile radio frequency charger for wireless sensor networks in the smart grid. In: 2014 international wireless communications and mobile computing conference (IWCMC). IEEE; 2014.

[62] Ping Y, et al. Routing renewable energy using electric vehicles in mobile electrical grid. In: 2013 IEEE 10th international conference on mobile ad-hoc and sensor systems (MASS). IEEE; 2013.

[63] Tan YK, Huynh TP, Zizhen W. Smart personal sensor network control for energy saving in DC grid powered LED lighting system. IEEE Trans Smart Grid 2013;4(2):669–76.

[64] Henze GP, et al. An energy signal tool for decision support in building energy systems. Appl Energy 2015;138:51–70.

[65] Virote J, Neves-Silva R. Stochastic models for building energy prediction based on occupant behavior assessment. Energy Buildings 2012;53:183–93.

[66] Romain C, Bianchi P, Jakubowicz J. Distributed convex stochastic optimization under few constraints in large networks. In: 2011 4th IEEE international workshop on computational advances in multi-sensor adaptive processing (CAMSAP). IEEE; 2011.

[67] Fahad J, et al. Engineering optimization models at runtime for dynamically adaptive systems. In: 2010 15th IEEE international conference on engineering of complex computer systems (ICECCS). IEEE; 2010.

[68] Saad LB, Chauvenet C, Tourancheau B. IPv6 (Internet Protocol version 6) heterogeneous networking infrastructure for energy efficient building. Energy 2012;44(1):447–57.

[69] Xin W, et al. Sensor selection schemes in smart grid. In: 2012 IEEE innovative smart grid technologies-Asia (ISGT Asia). IEEE; 2012.

[70] Guo S, Wang C, Yang Y. Joint mobile data gathering and energy provisioning in wireless rechargeable sensor networks. IEEE Trans Mob Comput 2014;13(12). 2836,2852.

[71] Xiaohong H, et al. Smart meter deployment optimization for efficient electrical appliance state monitoring. In: 2012 IEEE third international conference on smart grid communications (SmartGridComm). IEEE; 2012.

[72] Vaidya U, Fardad M. On optimal sensor placement for mitigation of vulnerabilities to cyber attacks in large-scale networks. In: 2013 European control conference (ECC). IEEE; 2013.

[73] George K, Tassiulas L. A delay based optimization scheme for peak load reduction in the smart grid. In: Proceedings of the 3rd international conference on future energy systems: where energy, computing and communication meet. ACM; 2012.

DOCIT: AN INTEGRATED SYSTEM FOR RISK-AVERSE MULTIMODAL JOURNEY ADVISING

16

A. Botea*, M. Berlingerio*, S. Braghin*, E. Bouillet*, F. Calabrese*, B. Chen*, Y. Gkoufas*, R. Nair*, T. Nonner[†], M. Laumanns[†]

IBM Research, Dublin, Ireland; [†]IBM Research, Zurich, Switzerland

1 INTRODUCTION

In many cities, the increasing traffic of private and commercial vehicles is straining the transportation infrastructure, and traffic congestion causes significant losses to the economy. Public transport has a significant potential to reduce traffic and congestion, but only if a substantial modal shift can be achieved. Reasons for the still relatively low adoption of public transport include its perceived unreliability and inconvenience for spontaneous travel.

Unreliability is due to the inherent uncertainty in a public transport system, where vehicles might deviate from their planned schedule due to all kinds of disturbances. Small initial disturbances are amplified by the connections between services, which are crucial for a network's connectivity but can lead to significant passenger delay due to missed connections. Inconvenience is related to passengers needing a considerable lead time to preplan their journey as the level of service and connectivity is not uniform over time. In this context, suitable journey advice, for planning the trip as well as guiding the passenger during the trip, is an important enabling factor for better public transportation service.

Journey planning is the process of advising travelers on how to best use a given transportation system to their journey requests. It is standard practice for transport operators to offer journey planning applications either on their websites or via mobile apps. For a given journey request, these applications typically return one or several itineraries as linear sequences of activities from start to destination, based on efficient shortest path computation engines working on a time-expanded graph model of the service network.

Current journey planning technology assumes a deterministic environment and hence uses a deterministic model. However, due to changing traffic conditions or other disturbances, public transport vehicles are not always on schedule. Any planned itinerary may become suboptimal or even infeasible, for example, due to missed connections. To compensate this shortcoming of preplanned itineraries, modern transport operators have started to offer push services to travelers, informing them of missed connections together with a new, updated itinerary based on the current situation. Although this offers some degree of adaptability, it is obviously only a reactive approach, based on deterministic planning applied on the basis of the current situation.

The question arises whether one can do better by using a stochastic planning approach. Stochastic planning can take contingencies into account during plan computation. A typical public transport network offers multiple alternative services for a traveler to choose at a given location that provide the necessary flexibility to react to disturbances and are the natural basis for a policy-based, uncertainty-aware journey planning [1–3]. A policy is a time-dependent, state-dependent, or history-dependent routing advice at each location. Different classes of policies can be distinguished based on the considered information or the set of actions considered, that is, whether they specify a single service or a set of services in each state. Algorithms to find optimal policies include AO*-based [4] search methods, combinatorial approaches, or dynamic programming, which work well on instances of realistic size, even though the problem is nondeterministic polynomial-time hard (NP-hard) for most policy classes. Resolving the algorithmic challenges for finding policy-based journey plans is a relatively new and active research area.

Building and deploying a system for uncertainty-aware journey planning involves a whole set of new questions regarding data collection and preparation, building appropriate stochastic models, formalization of the notion of journey plans as policies, integration of sophisticated methods for planning under uncertainty, real-time tracking of the state of the system and the travelers, user interaction, and interface design.

In this chapter we discuss these design questions and outline the main considerations and choices we have made to implement the *first* uncertainty-aware multimodal journey planning system in practice. We have faced multiple integration aspects, such as the integration of our system with commercial products, for example, the IBM Intelligent Transportation (IIT)[a] and the IBM Intelligent Operations Center (IOC),[b] the aggregation of heterogeneous transportation data into a unitary knowledge base, and the ability to plug in multiple uncertainty-aware journey planning systems such as Dynamic Intermodal Journey plAnner (DIJA) [1] and the Frequency Planner [2,3], facilitated by the design of a policy format generic enough to capture the different semantics and syntax of different types of policies.

The next section illustrates the differences between standard, deterministic journey plans and plans that take uncertainty into account. We continue with a high-level overview of our system. Then we discuss data integration aspects, followed by the integration of our asset with IBM products. The last section presents concluding remarks and future work ideas.

2 THE POWER OF UNCERTAINTY-AWARE PLANS

The purpose of this section is to illustrate, with a concrete example, differences between standard journey plans and uncertainty-aware journey plans, in terms of their structure and their expressive power. The example and the considerations presented in this section are reused from Ref. [5].

Standard journey plans are sequential, as their structure is a chain of actions applied in the given order. Our uncertainty-aware plans do not have to be restricted to such a simple structure. We allow more than one way of continuing with a journey at a given connection point. More options to follow at a connection point provide a greater flexibility at the time when the journey plan is executed. Depending on the conditions observed at a given connection point during the execution of a plan, one option

[a]http://www.ibm.com/software/products/en/intelligent-transportation/
[b]http://www.ibm.com/software/products/en/intelligent-operations-center/

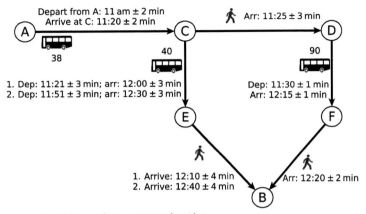

FIGURE 16.1 A Toy Example With Bus Links and Walking Links

can be better than others, and the user can continue the journey accordingly. Such nonlinear plans are often called *policies*.

Contingent plans are one particular type of policies. The structure of a contingent plan is a tree. For simplicity, in this section we use contingent plans as an example of uncertainty-aware multimodal journey plans. We consider just the travel time as the plan quality metric, and discuss differences in terms of worst-case and best-case arrival times.

Consider the example illustrated in Fig. 16.1, where a user wants to travel from A to B. As shown in Table 16.1, there are three ways to complete the journey. Each of these can be seen as a separate sequential plan. In plan 1, it is uncertain whether the connection at stop C will succeed, due to the variations in the arrival and departure times of routes 38 and 40, illustrated in Fig. 16.1.

Table 16.1 Sequential Plans in the Example		
Sequential Plan	**Steps**	**Remarks**
1	11:00 – Bus 38 to C	
	11:21 – Bus 40 to E	Uncertain connection in C
	12 pm – Walk to B	Arrive at around 12:10
2	11:00 – Bus 38 to C	
	11:51 – Bus 40 to E	
	12:30 – Walk to B	Arrive at around 12:40
3	11:00 – Bus 38 to C	
	11:20 – Walk to D	
	11:30 – Bus 90 to F	
	12:15 – Walk to B	Arrive at around 12:20

The best policy is a combination of the sequential plans 1 and 3, as follows. Take bus 38 from A to C. If it is possible to catch bus 40 that departs at 11:21, take it from C to E, and then walk to B. Otherwise, walk to stop D, take bus 90 at 11:30, get off at stop F, and walk to B. The arrival time is 12:10 pm (best case) or 12:20 pm (worst case).

This policy is an example of a contingent plan [6]. A contingent plan can be viewed as a tree of pathways from the origin to the destination. In our work, a contingent plan includes both priorities and probabilities associated with the options available in a state. In the example, the contingent plan contains a branching point at location C. The option of taking bus 40 that departs at around 11:21 am has the highest priority. Walking to stop D is the backup, lower-priority option. The probability p of following the first option can be computed from the distributions of two random variables [1]. One variable models the arrival time of the traveler at stop C, which in our example coincides with the arrival time of bus 38. The other variable is the departure time of bus 40. The probability of taking the backup option is $1 - p$.

As sequential plans are restricted to one unique trajectory each, no sequential plan is capable of matching the performance of the best policy in the running example. For instance, while plan 3 matches the worst-case arrival of the policy, it is weaker than the policy in terms of the best-case arrival time. The sequential plan 2 is weaker in terms of both worst-case and best-case arrival times.

Plan 1 is weaker in terms of the worst-case arrival. But what is the worst-case arrival time of the sequential plan 1 anyways? To answer this question, we note that, in the case of a missed connection, sequential plans have the implicit backup option of waiting for the next trip on the same route. Hence, in the sequential plan 1, in the worst case, the user will wait for the next trip on route 40, which arrives at around 11:51. As this implicit backup option of plan 1 is essentially equivalent to plan 2, we conclude that the worst-case arrival time for plan 1 is about 12:40 pm.

More generally, a sequential plan can be optimal if it is both *safe* (ie, all actions in the sequence will be applicable with probability 1.0) and *fast* (ie, the arrival time is optimal). On the other hand, as shown in the example, there can exist states where one option is good (ie, providing a good arrival time) but uncertain, whereas another option is safe but slower. These are cases where sequential plans lose their ability to implement an optimal policy, and where contingent plans can make a difference, as illustrated in our example.

3 SYSTEM OVERVIEW

Fig. 16.2 illustrates the architecture of our system, which we call Docit. This section overviews the main components, their main functions, and their interactions with each other, with external apps and with commercial products.

In Fig. 16.2, the Snapshot Aggregator creates a knowledge base with all the information available about a multimodal transportation network. The resulting knowledge base is called the *network snapshot*, or simply the snapshot. As discussed later, the snapshot is key input data to important system functions, such as journey planning and journey monitoring. Challenges associated with the data aggregation and the way we address them are discussed in Section 4.

The main functions of the Active Session Manager (ASM) include keeping a record of all travelers currently using the system (the *active sessions*), communicating with travelers, communicating with the planning engine, and providing both current and historical data for the computation of key performance indicators (KPIs). For example, one KPI based on historical data can identify hotspots on the

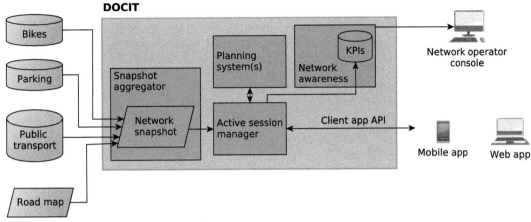

FIGURE 16.2 Overview of the System Architecture

map where travelers miss connections relatively frequently. Hotspots can be presented to the network operators. Experts can take this into account when performing network optimization functions, to better synchronize the two transport lines involved in the missed-connection pattern.

When dynamic network updates are available, the ASM regularly receives a new network snapshot from the Snapshot Aggregator. The IBM IIT product can provide real-time updates on the estimated times of arrival (ETAs) for public transport vehicles, which is a main reason why we have worked on integrating Docit with IBM IIT. In addition, dynamic updates on bike station and car parking data, available in a separate database, are used to estimate waiting times until a bike, a bike parking spot, or a car parking spot becomes available.

Travelers use client apps to interact with the system. Client apps communicate with the ASM through an application programming interface (API). The main functions include submitting journey plan requests, receiving plans, submitting updates on the progress of a trip, submitting requests on the validity status of a plan given the most recent network information available, and receiving notifications on the plan validity status.

Consider that a traveler submits a new journey plan request through a client app, such as a mobile app or a web-based app. The ASM starts an active session for that user. It forwards the request, together with a pointer to the most recent network snapshot available, to the planning engine, and it receives one or several journey plans in response.

After accepting one journey plan, the user can opt for a journey monitoring function, which allows to receive notifications about plan invalidations caused by dynamic events in the transportation network, and to perform replanning. For instance, consider that the traveler is riding a bus toward a connection with a tram line. The connection point is relatively far, in both space and time. An incident on the tram line delays all trams by at least 1 h. The system can notify the traveler as soon as the tram delays caused by the incident are reflected into a freshly computed network snapshot. Without this function, the traveler would find out about the tram service disruption only when arriving at the connection point. That could be too late, however. Clearly, the sooner replanning is performed, the more options are available, enlarging the space of available plans and thus improving the quality of the best available plans.

Brute-force replanning (ie, computing a new plan for every active user every time a new network snapshot is available) is undesirable for two reasons. First off, suggesting a change of plans too frequently, even when the old plan remains perfectly acceptable, can reduce the level of user satisfaction. Second, replanning for *all* users every time might cause a scalability bottleneck.

In addressing these limitations, our approach to replanning works as follows. Every time the ASM receives a new snapshot, it checks whether the current active plans (one plan for each active user) remain valid, given the refreshed knowledge about the network. Replanning is performed only for the plans found to be invalid. The advantage stems from the fact that simulating the steps of a plan is much cheaper than computing a plan. The simulation time is linear in the plan size, whereas searching for a plan often is exponential in the plan size.

The plan validity check involves simulating ahead the rest of the plan, with the new network snapshot, obtained from the Snapshot Aggregator, and the current position of the traveler along the plan, obtained regularly from the mobile app.

The simulation procedure returns a yes/no answer about whether the destination will still be reached in time, given the refreshed data. Recall that, at each connection point in the plan, the initial plan lists one or more options to follow. For instance, one option can be to take a bus (or another public transport vehicle) on a given route. In such a case, the simulation will take the next available trip on the route at hand. Depending on the conditions at that particular state during the trip, zero or more options can be *inapplicable*. For example, missing the last bus of the day on a given route makes that option inapplicable. When all options in a state (connection point) of the plan are inapplicable, we say that we are in a dead-end state. Zero or more options can be *safely ignored* when another option, with a higher priority, can be taken with probability 1. We say that an option that is neither inapplicable nor safely ignored is a *feasible* option.

There are two reasons why a plan can be labeled as invalid with our simulation procedure. The first reason is the presence of a dead-end state that can be reached from the current state through a pathway of feasible options. The second reason is that at least one pathway of feasible options from the current state to a destination leaf node exceeds the arrival time computed initially by more than a threshold value set by the user.

In summary, in a multipathway policy, it is acceptable that one or more pathways fail, as long as the traveler is guaranteed to reach the destination in time along one of the remaining pathways. If no such guarantee is provided, the simulation of the entire policy fails, and the policy is considered invalid.

Docit is focused on risk-averse journey planning, considering the uncertainty present in the network data, and computing plans with a better chance of arriving in time, in case of mishaps such as missed connections. We have integrated two risk-averse journey planning engines, DIJA [1] and the Frequency Planner [2].

As detailed in Section 5, the two planning engines work with significantly different types of policies. From a system integration perspective, the challenge is to design a policy format that is sufficiently general to cover the specific types of policies implemented in various planning systems, such as the two planners integrated into our architecture. A generic policy format offers the freedom to plug in any planning engine that complies with the format. Policy integration aspects are discussed in Section 5.

The ASM provides data to the Network Awareness module, such as the trajectories of active travelers, and statistics about completed sessions. The latter includes, for each session, the journey request, the actual trajectory followed, the arrival time, as well as plan invalidations and replanning rounds experienced in the session at hand. The data is visualized onto the network operator console. We have implemented two ways of visualizing such data: a stand-alone application, called the Dashboard, and visualization functions in the user interface of the IBM IOC product. See more details in Section 6.

For the use of travelers, we have developed both a mobile app and a web-based app. Both apps allow the user to enter a journey plan request, after which the request is submitted to the Docit system. Policies (journey plans) are visualized both on a map and separately, as a graph of locations and journey legs, with details about the timing, the transport modes to follow, and the routes to follow on each public transport link. The mobile app implements additional functions, needed for journey monitoring, such as notifying the server on the actual branch followed by the traveler on a multiple-branch policy, receiving updates of the valid/invalid status of the plan, requesting replanning, and receiving new plans. A detailed discussion of the client apps is beyond the focus of this chapter.

4 AGGREGATING NETWORK DATA

The network snapshot combines data about the public transport, the road map, car parking lots, and bike stations in a shared-bike network.

When aggregating multimodal transportation data, the diversity of the data refers not only to the different types of unimodal subsets but also to the fact that the same kind of data can originate from different sources.

Take, for example, *predictive data*, which is data that covers a time window in the future, such as the remaining part of a day. In multimodal transportation, examples include the following: (1) the ETAs and departure of public transport scheduled vehicles (eg, buses), at various times of the day; (2) the availability of free spots in a car parking lot, at different times of the day (when the predicted availability is zero, also predict the waiting time until a spot will become available); (3) the availability of bikes and free parking spots, at a bike station, at various times of the day [when the predicted availability is zero, include a prediction of the waiting time until the desired resource (bike or parking spot) becomes available]; (4) the predicted travel speed along the segments of a road map.

Predictive data can be obtained from various sources, each one having its own benefits and drawbacks. For example, published schedules (timetables) are one source of public transport ETAs. These have the advantage that they are increasingly available to download, being provided by transportation agencies and city authorities. They cover large time windows, being able to offer long-term predictive data. On the other hand, published schedules are static. They lack stochastic information to model the uncertainty in the ETAs. They also lack accuracy, in those cases when an actual bus trip does not closely respect the published timetable.

Stochastic noise information can be compiled from the second data source we discuss, namely, historical data about actual arrival times. This is potentially more accurate than static timetables, but it is less frequently available as well. The availability depends on the existence of a Global Positioning System (GPS)-reporting infrastructure, as well as the willingness of transport agencies to make the data with the actual journeys available.

Finally, ETAs can also be computed in real time, from GPS data collected from public transport vehicles moving around the city. The advantage is an increased accuracy of the results. The disadvantage is that such a functionality is still relatively uncommon. Furthermore, by their nature, real-time predictions are available only for a short time window into the future. IBM IIT can provide real-time updates on the ETAs and estimated times of departure, as well as historical statistics of these.

This functionality works only for vehicles that have already started their journey, which in turn limits the prediction time window.

Given the strengths and the drawbacks of each data source, combining multiple sources, when they are available, is a natural idea. The resulting snapshot is agnostic to where the data comes from. The structure is the same, and there is no separation line between, say, ETAs computed in real time and static, timetable-based ETAs. When only one subset of the sources is available (eg, just timetable data, without real-time predictions), the system is still able to build a complete snapshot.

For car parking and bike data we have implemented the prediction algorithm reported in Ref. [7]. The planning engines we use in our system make use of the waiting time until a bike or parking spot becomes available, not of the predicted number of bikes or parking spots. Thus, the ability of providing waiting time predictions was a key requirement to be able to integrate the prediction method and the planning engine as part of the same system.

Different from conventional prediction methods, our algorithm provides measures of uncertainty in terms of probability distributions for the input of journey planner. To be specific, whenever the prediction indicates that no parking/bike is available at a given time in the future, we compute the waiting time for the next available parking/bike and its distribution.

In the prediction method our assumption is that the interarrival times follow an exponential distribution with time-varying intensity $\lambda(t)$ at time t, with $t = 0, 1, 2, \dots$. Equivalently, we assume that the parking/bike arrivals follow an inhomogeneous Poisson process. This is because empirical evidence shows that the arrival intensity is a function of time, that is, it is particularly high during busy periods, such as morning or evening rush hours, while it is relatively low in off-peak hours, such as late at night or early morning. We use the following procedure to estimate the $\lambda(t)$ for a given time t, illustrated here for the case of bike prediction.

Days in the historical data are split into two categories, one for working days and one for weekends and public holidays. If desired, a finer separation, including weather data, for instance, can be performed. The following steps are applied separately for each category:

- For each day in the historical data, calculate the duration of all interarrival periods and denote them by $d[1], d[2], \dots, d[k]$. Also count the bikes arrived minus the bikes departed in every period, and denote these by $n[1], n[2], \dots, n[k]$.
- Compute the estimated arrival intensity as the number of increments per unit of time, $\lambda'(t) = n[i(t)]/d[i(t)]$, where $i(t)$ is the index of the interarrival period into which t falls. If no such period exists (ie, t lies before the first or after the last arrival on that particular day), then set $\lambda'(t) = 0$.
- Compute the average of all estimates $\lambda'(t)$ obtained for each day in the training set.

Consequently, the estimated distribution of the waiting time is exponential with mean equal to the average $\lambda'(t)$. This output can be easily combined with the uncertainties from other transportation nodes of the journey planner by computing the convolution of probabilities, to ultimately provide the uncertainty of the entire journey.

4.1 DATA FORMAT

We have adopted a csv-like data format for the network snapshot. Our system queries the data sources available, such as relational databases or files on the disk, and builds a unitary, source-agnostic snapshot as a series of csv files. The snapshot is further loaded into memory, and used in the computation of journey plans, and in the validity checks of existing plans.

Public transport data is based on GTFS,[c] a csv-based format including data such as stops, routes, and trips. Stops are characterized by an ID, a name, and latitude/longitude coordinates. Each route is served by multiple trips, where trips are actual vehicle journeys along that route. Each trip is an ordered sequence of stops, with arrival and departure times associated with each stop. Standard general transit feed (GTFS) supports deterministic arrival and departure times. We have extended the format so that arrival and departure times can optionally include a stochastic noise, modeling uncertainty in the arrival or departure times. The noise is an extra column, of type string, in the csv file of trip timing data. For instance, a value such as "$N(0, 6400)$" represents a Normal distribution with a mean value $\mu = 0$ and the variance 6400.

As GTFS is restricted to public transport, we have extended the format to model road maps, bike station data, and car parking data.

Car parking data and bike station data have a similar structure with each other. A list of car parking lots stores, for each record, as mandatory fields, a name, an ID, lat/lon coordinates, and one or several links between the parking lot and nearby nodes from the road map. The format allows to use optional fields, such as the total capacity. In terms of parking predictive data, our system makes direct use of the waiting time until a spot is available. We represent predicted waiting times for a parking spot as records in a csv file, each record containing the ID of a car parking lot, a discrete, deterministic time of arrival at that parking lot, and a stochastic waiting time until a parking spot becomes available. Bike station data is similar, except that for bikes we have two prediction files, one for predicting parking availability and one for predicting bike availability.

5 THE POLICIES

As mentioned earlier, traditional journey plans are a totally ordered sequence of actions, for example, first take bus 12 and then bus 5A. In contrast, when dealing with stochasticity, it is intrinsic to allow considering multiple options.

The two journey planning engines in use generate different types of policies. In the Frequency Planner's policies, the notion of a "state" boils down to location information. When multiple options are present in a state, the assumption is that the traveler will follow the first one available (eg, take the first bus that arrives on either route 12 or route 14A). In that respect, all options have the same priority. In the Frequency Planner, the history (ie, the actual trajectory from the origin to the current location) is less important, making this type of plans related to Markovian policies.

On the other hand, in DIJA policies, a state depends on more factors, including the time and the history. For example, reaching the same location, at the same time, in two different ways can impact factors such as the amount of walking performed so far. This further impacts how much walking is acceptable in the rest of the journey, to respect a user-specified maximum amount. As the history is important in DIJA policies, the structure of such policies is a tree where the tree nodes (ie, the policy states) are clearly identified with a unique ID, and each edge (ie, a journey leg represented as a statement in the policy JSON format) specifies clearly its starting state and its end state. Options (tree edges) available in a state are prioritized and they can specify a time information (eg, attempt to take bus 10 expected to arrive at 14:45; if missed it, walk to stop 7073).

[c]https://developers.google.com/transit/gtfs/

Both types of policies have their own advantages and disadvantages. Rather than comparing them, our goal is to define a generic policy format that is sufficiently versatile to represent both types of policies. For practical reasons, we want our policies to be easy to read and understand, and to follow an established data format such as JSON.

Therefore, while developing different approaches for stochastic routing, we faced the additional challenge to find a versatile format to represent plans that offer different options. In the frequency-based approach [2,3], a relatively simple way to represent options is sufficient, whereas the more fine-grained DIJA approach [1] requires more timing information. Despite these changing needs, we managed to find a format that is simple but yet flexible enough to cover such a wide range of requirements.

Our policies are a superset of traditional sequential plans. A policy is an unordered list of *statements*. Informally, a statement describes one possible leg in the journey. When several statements apply in a given context, they are precisely the possible "options" to continue the journey. More technically, a statement is a collection of fields that describe a user action (specifically, following a journey leg), a specific context ("state") when the action can be considered, a time interval when the action is valid (optional), group information (optional), and a priority (optional).

> *Actions*: Fields describing the user action include the transport mode, the destination location of the action, and, depending on the mode, other relevant information. For instance, for a bus trip, this includes the route ID, the trip ID (optional), the expected departure time (optional), the expected arrival time (optional), and the headsign (optional).
>
> *State information*: Due to the flexibility of our policy format to cover a range of plan formats, the state information can be as simple as the current location on the map, or as precise as a unique ID of a state in a policy tree, as illustrated earlier. In the latter case, one can use two optional fields in the statement to specify the start state and the end state of the journey leg at hand.
>
> Location-related information includes the unique ID of the location at hand (different from the state ID), as well as additional optional fields, such as the name of the location and its latitude and longitude, for an easier retrieval and display in a client app.
>
> *Groups of statements*: Statements can be grouped together using an optional group ID field. This allows to treat a bunch of nearby stops as one single location, being visited by the union of all buses passing through these stops. Consequently, all statements with the same group ID are executed as if they belong to the same location.
>
> *Timing of statements*: To enable a change of active statements for a location over time, it is possible to add a start and end time as well as an optional start and end date to a statement. This indicates the time interval during which the statement is active. This is required for nonstationary policies such as the DIJA approach [1]. Such activation time intervals work in conjunction with priorities, as explained later in this section.
>
> *Priorities*: We also allow a fine-grained specification of priorities in case the first-come-first-serve strategy should be relaxed. For instance, if two buses arrive at the (almost) same time, then these priorities would specify which bus to board. This is another requirement of the DIJA approach. Groups of statements allow to combine the first-come-first-serve strategy with the strategy based on priorities. The priority can naturally be extended from individual statements to groups of statements. When several statements are grouped together, they all have to have the same priority. The semantics are that groups of statements are tried in the order indicated by their priority, until a group with feasible options is found. Then, inside such a group, the first option that becomes available is taken.

Given that a given location is a bus stop, an example of a statement as simple as possible suggests to board a given bus and get off at another given bus stop. Clearly, using this primitive, we can easily implement traditional sequential plans by translating them into a list of such statements. For instance, the following statement, written in JSON, specifies that at stop 89 we should take bus 12 to stop 32, and from there we should take bus 5A to stop 64:

```
"policy" : [  {"loc_type" : "stop",
  "loc_id" : "89",
  "to_loc_type" : "stop",
  "to_loc_id" : "32",
  "transport_mode" : "public",
  "route_id" : "12"},

  {"loc_type" : "stop",
  "loc_id" : "32",
  "to_loc_type" : "stop",
  "to_loc_id" : "64"
  "transport_mode" : "public",
  "route_id" : "5A"}]
```

Having different location types simplifies the merging of different geographic data sources without merging the ID spaces. For example, a location of type stop would refer to a public transport stop in the corresponding GTFS files. Types of locations currently handled in our format include public transport stops, bike stations, car parking lots, nodes in the roadmap network, and generic locations provided through their lat/lon coordinates. Similarly, the route ID would refer to the route ID from these files, where the transport mode *public* would indicate this relationship.

Now consider the case that this policy additionally contains the following statement, which is a copy of the second statement, but with a different route ID:

```
"loc_type" : "stop",
  "loc_id" : "32",
  "to_loc_type" : "stop",
  "to_loc_id" : "64"
  "transport_mode" : "public",
  "route_id" : "3"
```

As in this example no priorities are assigned to the two statements applicable to stop 32, the semantics are that the user will take the first bus that arrives on any of these routes. Such situations are common in the frequency-based planning approach [2,3]. This is particularly useful when GTFS data provide the frequency of a service, but not specific arrival and departure times. Providing different options via statements can decrease the waiting time, assuming a first-come-first-taken strategy. This allows the realization of tree-like policy structures with multiple alternatives at each stop in a simple way.

Such simple tree-like structures are powerful, but sometimes not expressive enough. This is why we allow optional fields to specify timing information, priorities, history (through unique state IDs), and groups of statements.

DIJA [1] uses a more rigorous timing regime and priorities. Consider a bus route, say route 10, where buses arrive every 30 min. An optimal plan might be as follows: try to take the bus on route 10 arriving at 14:45; if missed, walk to stop 7073, and take another bus from there. In other words, waiting for the bus 10 arriving at 15:15 would be suboptimal. Making use of time intervals and priorities, this plan can be encoded as follows: have a statement B for the bus arriving at 14:45, with a high priority assigned, and with a time interval that, in our example, accounts for the uncertainty in the actual arrival time (eg, 14:40 to 14:50). Have another statement W for the walking leg, with a lower priority.

This mixture of mandatory and optional fields ensures that policies are backward compatible with traditional plans, allowing a simple specification of different alternatives at a stop as in the previous example which is the key semantic requirement, but are yet flexible enough to implement arbitrary state changes.

6 INTEGRATION WITH IBM IIT AND IBM IOC

This section focuses on integrating our system with two commercial products, IIT and IOC.

6.1 IBM INTELLIGENT TRANSPORTATION

IIT is a product for the management and prediction of traffic in cities. It can collect and process data coming from different sources, such as public transport agencies and vehicles, traffic lights, and several types of sensors deployed in a city.

Based on raw data such as GPS data, IIT can dynamically update ETAs for already started trips. Furthermore, actual arrival/departure times can be stored in a database as historical data. These are two of the most relevant features to Docit.

In Docit, better estimates of ETAs can be used to compute plans from scratch, to check the validity of an existing plan, and to replan. As the time window of dynamically computed ETAs is limited, covering vehicle trips started but not completed, this time window overlaps particularly well with the second (plan validity checking) and the third (replanning) tasks. The availability of accurate ETAs, updated in real time, could be key to the effectiveness of such tasks.

Dynamic data are stored in several "third-party" databases with a proprietary schema, that differ from the format required by Docit, presented in Section 4. The Snapshot Aggregator contains a few modules to fetch updates on dynamic data. The Public Transport Updater (PTU) queries IIT databases for public transport ETAs. The Bike Station Updater (BSU) and the Car Park Updater (CPU) obtain updates on bike stations and car parking lots, respectively. Even though the bike and parking databases are technically not part of the IIT product, we discuss the BSU and the CPU together with the PTU, for a simpler structuring of the text. In the network snapshot, dynamic updates are combined with more static parts of the network, such as the list of all stops, which are stored as files on the disk.

This information is used by IIT to enable its features and to create reports for the city operators, and any type of user that is intended to use traffic-related predictions. Alas, IIT stores the data that it collects in several databases according to a proprietary schema that differs greatly from the data format required by Docit (see Section 4 for a discussion of the formats of the data used by our system). Hence, it is required to implement a way to convert the data provided by IIT to our formats. Note that IIT provides information—ETA, traveling speed, etc.—about traveling vehicles only. Hence, the information extracted from IIT is integrated with the one obtained from a canonical GTFS. This way, Docit is able to correctly verify the status of active journeys based on real-time information.

To update such real-time information, we implemented in Docit a component that is in charge of periodically refreshing such information from IIT. Such component, called *NetworkStatusUpdater*, consists of three separate submodules: *PTU*, *BSU*, and *CPU*.

The PTU includes a collection of five SQL queries. Each query involves between 4 and 7 tables in a relational database, to a total of 13 tables. It starts by retrieving all routes and stops available in the data, which are needed to eventually retrieve all public transport trips currently running. These running trips are integrated into the snapshot, combined with static data, such as trips planned to run in the future. Creating the running trip data in our format is a multithread process, as the data are independent from one trip to another, and there can be potentially many trips and stops per trip.

The component computes for each running—or planned to run–vehicle a trip ID. After that, the component proceeds with the creation of the file containing the expected arrival/departure time at each stop for each trip. The latter step is executed in parallel because of the possibly high number of available trips and, in particular, stops per trip. In the current deployment of Docit, this process is executed every 5 min and takes, on average, 35 s to complete.

The BSU queries the bike database periodically—every 10 min in the current implementation – to retrieve the number of bikes and free parking slots available at each station. Given this information, the BSU uses the prediction algorithm presented in Section 4 to generate the prediction of availability and the expected waiting times, for the subsequent 24 h, with a granularity of 5 min. The CPU works similarly.

Computing predictions as described earlier makes use of a series of Generalized Additive Model (GAM) models. The computation of the models themselves is expensive but, fortunately, it can be performed as seldom as once in a few months. Indeed, a new model is necessary when the condition variables associated with a bike station, or car parking lot, are changed. For example, it is required to generate a model for the summer and another one for the winter, as the bike usage patterns differ between a warm and a cold season.

6.2 IBM INTELLIGENT OPERATIONS CENTER

The IBM IOC provides an executive dashboard to city operators. Its functionality spans across multiple aspects of city management, including public safety, transportation, water, social services, and emergency management.

While broad on their own, these functions lack awareness information extracted from the direct experiences of travelers. At the same time, IOC allows to integrate additional functionality into its dashboard relatively easily. The executive dashboard spans agencies and enables drill-down capability into underlying agencies, such as emergency management, public safety, social services, transportation, and water management. Our objective is to integrate such new awareness data, extracted from the experiences of travelers using the transportation network, into a user-friendly and versatile visualization framework, such as IOC. First, we developed a visualization of active journeys registered in the system, updated in real time. Second, we provide a visualization of KPIs based on historical data, such as hotspots of frequently missed connections. Finally, the console can be used to tune system parameters.

Visualizing active journeys is implemented taking advantage of the REST [8] API provided in IOC [9]. Other integration techniques are possible, but we chose to use the REST API because it was the most convenient way to interact with the product. The REST services provide a set of uniform resource identifiers (URIs) that access data in IBM IOC components, such as system properties and KPIs. One can call the services with any HTTP client application, and define an expected response in the form of a JSON object. The JSON format can then be easily parsed by several libraries, which gives the developer greater flexibility in using her/his own environment.

Active journey visualization requires the creation and the configuration of two data sources via the administrative console of IBM IOC. The first one stores, in near real time,[d] the status of the active user sessions, including information such as the current location of the user. The second data source stores the trajectory of the executed part of the journey plan policy. Their content is updated regularly, with data from the ASM, taking advantage of the REST [8] API provided in IOC [9]. Each update is performed with a POST [8] request to the Data Injection Service, which is a service utilized for adding, updating, and canceling records for a data source. The frequency of the updates is a parameter specified in the configuration of Docit. Once the datasets are regularly updated, IOC automatically displays their contents in a map view created for this purpose.

Other functions we created include a system configuration view and the ability to visualize statistics about current and past trips. We extended the actual web interface of IBM IOC with an additional web application providing an interface to internal Docit data. Each different function was wrapped in an independent web clip [10]. In turn, each web clip was integrated in the layout of the web portal that provides IBM IOC. To do that, we created a specialized menu item grouping all the new functionality.

Moreover, we created a specialized map view for Docit. This map connects to the data sources that are periodically updated by our system, providing to the operator a clear and comprehensive view of the currently active user sessions and their trajectories.

7 CONCLUSIONS

A greater adoption of public transportation is considered to be one way of fighting increasing levels of congestion and pollution in urban environments. In real life, transportation networks can be perceived as less reliable than desired, due in part to variability in the arrival and departure times of public transport vehicles.

Existing systems for multimodal journey planning work under deterministic assumptions. We have presented Docit, an uncertainty-aware system for multimodal journey advising. We have described the system main functions, directed to two categories of users, namely, travelers and network operators. Integration aspects that we have faced in this work include data aggregation, integration of diverse uncertainty-aware planning engines, and integration with commercial products. Our system has been integrated the IBM IIT and IOC products, and with databases fed in real time with dynamic transportation network data such as ETAs and estimated times of departure.

As future work, we are interested in adding more transportation modes, such as electric vehicles used in park-and-ride and other relevant scenarios. These can potentially have a substantial impact on fighting carbon emission levels, but they also come with additional research challenges, such as hedging risks related to running out of battery in a trip.

---

[d]Depending on the frequency of the updates received from mobile apps. See an example in the next section.

GLOSSARY

AO* algorithm An artificial intelligence algorithm in which the search space is represented as a AND–OR tree, which means that the tree consists of branches that can be selected alternatively (OR branches) and branches that must be followed together (AND branches).

Contingent plan A journey plan that may contain travel alternatives at each branch

DIJA Dynamic Intermodal Journey Advisor is a stochastic journey planner integrated within Docit

Docit Dynamic Optimization for City Intermodal Transportation is our multimodal journey advisor system

Frequency Planner A stochastic journey planner integrated within Docit

GAM Generalized Additive Model is a generalized linear model in which the linear predictor depends linearly on unknown smooth function of some predictor variables

IIT IBM Intelligent Transportation A solution provided by IBM to manage, control, and analyze city-wide public/private transportation

IOC, Intelligent Operations Center A solution provided by IBM to city/utility/safety operators

Journey plan policy A representation of the travel solutions computed by a journey planning algorithm

Multimodal journey A journey involving more than one mode of transport (ie walking and public transport, or public transport, cycling, and walking)

Network snapshot The information about the public/private transportation network available at a given time. This knowledge base may also contain forecasts about future states, such as predicted arrival and departure times for the rest of the day

Stochastic journey planning A form of journey planning that considers the data on which it is executed as containing a certain level of uncertainty

REFERENCES

[1] Botea A, Nikolova E, Berlingerio M. Multi-modal journey planning in the presence of uncertainty. In: Proceedings of the international conference on automated planning and scheduling, ICAPS-13; 2013.

[2] Nonner T. Polynomial-time approximation schemes for shortest path with alternatives. In: Epstein L, Ferragina P, editors. ESA, Lecture notes in computer science, vol. 7501. Berlin, Heidelberg: Springer; 2012. p. 755–65.

[3] Nonner T, Laumanns M. Shortest path with alternatives for uniform arrival times: algorithms and experiments. In: Proceedings of ATMOS'14; 2014.

[4] Nilsson NJ. Searching problem-solving and game-playing trees for minimal cost solutions. In: IFIP congress (2); 1968. p. 1556–62.

[5] Botea A, Braghin S. Contingent versus deterministic plans in multi-modal journey planning. In: Proceedings of the twenty-fifth international conference on automated planning and scheduling, ICAPS 2015; 2015. p. 268–72.

[6] Peot MA, Smith DE. Conditional nonlinear planning. In: Proceedings of the first international conference on artificial intelligence planning systems. San Francisco, CA: Morgan Kaufmann Publishers Inc; 1992. p. 189–97.

[7] Chen B, Pinelli F, Sinn M, Botea A, Calabrese F. Uncertainty in urban mobility: predicting waiting times for shared bicycles and parking lots. In: 16th international IEEE conference on intelligent transportation systems (ITSC); 2013. p. 53–8.

[8] Fielding RT, Nielsen HF, Berners-Lee T. Internet draft: hypertext transfer protocol – http/1.1. Technical report, W3C; 1999.

[9] IBM. IBM Intelligent Operations Center v1.6 REST APIs; 2013.

[10] Abdel-Hafez H, Balakrishnan S, Caffrey J, Francellino E, Mishra S, Nascimento T, Ravichandran J, Scott C, Vlasov N. IBM Intelligent Operations Center v1.6 programming guide; 2014.

SMART RESTAURANTS: SURVEY ON CUSTOMER DEMAND AND SALES FORECASTING

17

A. Lasek*, N. Cercone*, J. Saunders[†]

*Department of Electrical Engineering and Computer Science, Lassonde School of Engineering, York University, Toronto, ON, Canada; [†]Fuseforward Solutions Group, Vancouver, BC, Canada

1 INTRODUCTION

Smart cities use digital technology in order to improve prosperity, reduce costs, reduce consumption of the resources, and enhance quality and efficiency of urban services. The aim of this chapter is to show an aspect important to cities—yield management with restaurants as an example.

Demand forecasting is one of the important inputs for a successful restaurant yield or revenue management (RM) system. Sales forecasting is crucial for an independent restaurant and for restaurant chains as well.

Nowadays, especially for big restaurant chains, a large number of time-ordered data are collected online and in real time, which results in massive amounts of data. Each time a customer order is placed, the transaction is automatically processed through the point of sale (POS) system and stored in a back-office database. Every POS system has the ability to track sales, cash, and inventory. And it also can track employee productivity, average sales per employee, what menu items are the most popular, and how quickly orders are served from time of input. Other possible reports include the number of customers served on an hourly and daily basis and the number of table turns. Restaurants with many locations can change menu items and prices with the POS system. It makes many restaurant chains collects massive datasets.

As mentioned in Ref. [1] from 1999 information technology eventually became so complicated that it can be integrated tool for decision making management and control of operations for restaurants. Furthermore, restaurant chains eventually grew large enough to support large investments that might be required. According to the authors, IT investments should be evaluated taking into account all aspects of restaurant management: production systems (including demand forecasting), planning (in both the kitchen and the dining room), process controls (including the management of meal times and production kitchen), and enterprise resource planning (enhanced back-office). Nowadays the most valuable are the data that could help to manage the three most important variables for the restaurants: food cost, labor cost, and demand forecast.

The sales transaction data collected by restaurant chains may be analyzed at the store level, the chain level, and the corporate level. For example, Cara Operations Limited—one of Canada's largest

Smart Cities and Homes. http://dx.doi.org/10.1016/B978-0-12-803454-5.00017-1

FIGURE 17.1 Aims of Exploring the Sales Transaction Data at the Store Level

owners of chain restaurants—owns 837 restaurants across the country under 10 brands including Swiss Chalet, Milestones, Kelsey's, etc.

At the level of single store, exploring the large amounts of transaction data allows each restaurant to improve its operations management (eg, labor scheduling), product management (eg, Purchasing Management, product preparation scheduling), and supply chain management, and in consequence reducing restaurant operating costs and increasing quality of serving food (Fig. 17.1), whereas at the corporate level, extraction of relevant information across the restaurants can greatly facilitate corporate strategic planning (Fig. 17.2). Management can assess the impact of promotional activities on sales and brand recognition, assess business trends, conduct price elasticity analysis, and measure brand

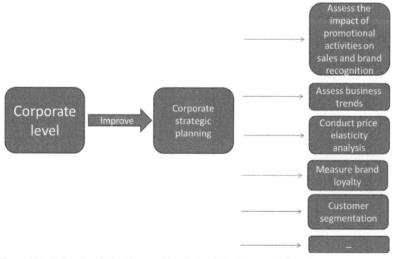

FIGURE 17.2 Aims of Exploring the Sales Transaction Data at the Corporate Level

loyalty [2]. Thus, how to obtain an accurate and timely sales forecast is critical from many different perspectives.

Historically, restaurant managers have used a mental model or a simple rule using recent history to forecast guest counts. As mentioned in Ref. [3] forecasting of restaurant sales has been judgmental based and even in the 1990s this technique was most often used by the majority of the restaurant industry. Judgmental techniques consist of an intuitive forecast based on the manager's experience. But restaurant sales forecasting is a complex task, because it is influenced by a large number of factors, which can be classified as time, weather conditions, economic factors, random cases, etc. This makes judgmental techniques inaccurate. A wide variety of models, varying in the complexity form, have been proposed for the improvement of restaurant forecasting accuracy.

Naïve forecasts are the simplest forecasting models, and provide a baseline, a benchmark against which more sophisticated models can be compared. Naïve method for time series data produces forecasts that are equal, for example, to the last observed value. Or if the time series is seasonal, more appropriate naïve approach may be applied when the prediction is equal to the value from last season (eg, last week, last month, last year, or the appropriate week 1 year ago) [4].

The state of menu-item demand forecasting was established in 1988 and can be found in Ref. [5]. The authors prepared a survey to assess the prediction techniques used by food service directors in a sample of American Dietetic Association Members with Management Responsibilities in Health Care Delivery System. The analysis of the replies showed that Naïve models were used by the majority of respondents. Only less than 25% were using mathematical models to forecast menu-item demand and Moving Average (MA) technique was the most popular mathematical model. But about 75% of the respondents indicated that learning and searching for a forecasting technique that can be used in the food service management is needed [5].

Limited empirical research has been carried out on restaurant sales predicting at the chain level. An investigation of the process of sales forecasting in corporate restaurants was conducted in 2006 and is presented in Ref. [6]. This qualitative study evaluates the sales forecasting process in the commercial restaurants. It consisted of interviewing 12 corporate restaurant forecasting managers responsible for the sales forecasting process for their companies. The results show that commercial restaurants are not enough advanced in forecasting techniques and the study provided recommendations for the companies to improve their sales forecasting processes.

Forecasting is a vast subject, covering a wide range of disciplines, including statistics, computer science, engineering, and economics. Over the years, a basic set of forecasting methods has been developed, but new improvements are still being added. Some of these prediction methods are based on rigorous mathematical and statistical basis, while others are largely heuristic in nature.

Perishability of food makes sales forecasting a unique problem in the restaurant, or generally speaking food service industry, because it relates to food purchased or wasted during production. Forecasting in the food industry is valuable in view of different aspects of the business. Food service managers have to predict sales to plan staff schedules and purchase food and supplies. Most of the meals are prepared right before service. In case of overprediction, when the demand is less than the forecast, there occurs a problem of leftover food and wasted resources, and in consequence it leads to increased food cost. In addition, overforecasting increases labor costs, because the extra handling of food requires additional use of employees. On the other hand, underforecasting leads to depletion of food before customer demand will be satisfied, thereby reducing food revenue and market share. Moreover, underforecasting causes insufficient number of staff and, consequently, customer dissatisfaction with service [7]. Thus accurate forecast is a definite goal of food managers, who aim to achieve a successful business.

Yet there does not exist any review of forecasting methods for the restaurant industry or any reports that describe the performance of various forecasting methods in restaurant RM applications. The aim of this chapter is to survey and classify restaurant sales forecasting techniques published over the past 20 years. Some of the content of this review was published as part of a conference paper [8].

The rest of this chapter is organized as follows. Section 2 contains basic information on yield management for restaurants and an overview of the role of forecasting for RM. Based on literature review we specified seven categories of restaurant forecasting techniques:

1. multiple regression;
2. Poisson regression;
3. exponential smoothing and Holt–Winters model;
4. autoregressive (AR), MA, and Box–Jenkins models;
5. neural networks;
6. Bayesian network;
7. hybrid methods.

They are arranged in roughly chronological order and discussed in Section 3. Section 3 is divided into eight subsections, one subsection for each category of restaurant forecasting techniques, and the eighth subsection describes application of Association Rule Mining for the restaurant industry. Every subsection provides a brief verbal and mathematical description of each technique and gives a literature review of a representative selection of publications in the given category. Section 4 presents a summary of all described methods and the discussion of advantages and disadvantages of each of the methods. Section 5 includes summarizing of our research and some remarks.

2 REVENUE MANAGEMENT

The first definition of RM was given by Smith et al. in 1992 as "selling the *right* inventory item to *right* customer at *right* time at *right* price" [9]. The determination of "right" entails both achieving the most revenue possible for the company and delivering the greatest utility to the customer [10]. Without this balance, the strategy of RM will in the long term discourage those customers who feel that the restaurant acts dishonestly.

A pioneer in RM was airline industry [11]. Other examples of industries in which RM is implemented nowadays are hotel industry, car rental companies, tour operators, cruise ship lines, transport companies, advertising (radio, TV broadcasters, and the most important nowadays: online advertising, eg, Google), energy transmission company, production, financial services and clothing retailers, restaurants, and others (Fig. 17.3).

Now RM in the airline industry is highly developed professional practice based on a scientific basis. The core methodology of RM developed for use in the airline industry (and related industries, eg, hotels) over the past 25 years is briefly described in, for example, Ref. [12]. There are also given some insights into use of simulation in the RM. This airline success leads to the rapidly growing interest in using RM techniques in other industries. With every new application in the industry, there occur

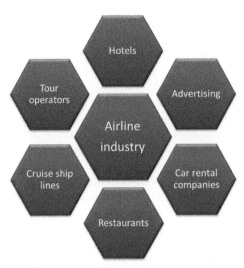

FIGURE 17.3 Examples of Industries With Revenue Management

new challenges in modeling, forecasting, and optimization as well, so research in this area continues to grow. For example, in some of these industries customers have accepted dynamic pricing, while in restaurants definitely not (daily special is the one exception). Although the details of RM problems may significantly change from one sector to another, the emphasis is always placed on making better demand decisions—rather not manually, with guesswork and intuition, but with scientific models and technologies.

Restaurant sector is sufficiently similar to the hotels' and airlines' ones in the area of RM. A short comparison of restaurant and hotel RM can be found, for example, in Ref. [13]. However, the restaurants also have unique features that pose particular challenges, requiring operators to creatively develop appropriate RM strategies. Among the unique features of restaurants there are relative flexibility of capacities and flexible meal duration, and these are important topics to consider when implementing RM in restaurant. Unlike airlines and hotels, restaurants have a little more flexible capacity, for example, a restaurant may have outside patio for additional seating during days with a good weather. What is more, the total available number of seats per day in the restaurant is not set, since customers' duration of meals are unpredictable [14]. However, restaurant can manage table turn rate.

In practice, RM means determining prices according to forecasted demand so that price-sensitive customers who are willing to purchase at off-peak times can do it at lower prices, while customers who want to buy at peak times (price-insensitive customers) will be able to do it [15]. For restaurant industry RM can be defined as selling the *right* seat to the *right* customer at the *right* price and for the *right* duration [16]. The goal for restaurant RM is to maximize revenue by manipulating price and meal duration. The price is quite obvious target for manipulation and many operators already offer price-related promotions to expand or shift peak period (eg, early bird specials, special menu promotions). More complex manipulation of price include setting price for a particular part of the day, day-of-week pricing, and price premiums or discounts for different types of party sizes, tables, and customers. Management of meal duration is a bit more complicated. Reduction of meal duration can be achieved by modifying the service process, changing the number of employees, or altering menu [16].

The RM strategy is the most effective when it is applied to operations that have the following attributes: relatively fixed capacity, demand that is time variable and predictable, perishable inventory, and the relevant costs and pricing structure [10]. All these attributes can be found in the restaurant industry.

The definition of capacity depends on the industry. Capacity is usually based on physical characteristics (eg, hotel capacity is usually measured by rooms, and airlines capacity is measured by seats), but in some cases it is also measured by time (eg, consulting companies), or working hours (eg, golf courses) [15]. Capacity of restaurants can be measured in seats, kitchen size, menu items, and number of employees. Kitchen capacity, the menu design, and members of staff capabilities are just as important as the number of seats in the restaurant. Capacity is usually determined in the short term, although many industries have a certain flexibility to be able to either reconfigure their capacity or add additional short-term opportunities. For example, airlines can reconfigure the seats or change the size of the plane, which is assigned to the route [15]. The number of places in the restaurant is generally fixed in the short term, although usually there is a possibility to add some number of tables or seats depending on reconfiguring the dining room, using outdoor seats at patio during nice weather, or adding patio heaters or shade to extend the range of conditions over which patios can be utilized. Adjusting the efficiency of a kitchen is usually more expensive, although output cuisine can often be improved by changing the menu or menu complexity or by increasing the level of employment. Despite all these possibilities for changes, restrictions in the kitchen and in the dining room can make the changes fruitless. Thus, restaurants' capacity is basically constant [10].

The restaurant demand consists of people who make reservations and guests who walk in and all guests in total are a set from which managers can choose the most profitable mix of customers. One particular factor for restaurant operators is that they must take into account the length of time a party stays once it is seated. If the restaurant managers can accurately predict the duration of the meal, they can make better reservation decisions and give a better estimate of the waiting time for walk-in guests. Reservations are precious, because they give the company the possibility to sell and control their inventory early on. Moreover, companies that take reservations have the ability to accept or reject the reservation request, and they may use this possibility depending on the periods of high or low demand. Both of these types of demand can be managed, but they require different strategies. To forecast the demand and make a RM, the restaurant operator has to analyze the rate of bookings and walk-ins, guests' desired times for dining, and probable meal duration. Tracking patterns of guests' arrivals requires an effective reservation system [10].

Companies that use the RM have the perishable capacity that cannot be stored for later use, for example, a vacant seat in the airplane or unfilled cabin on a cruise cannot be recovered. For this reason, many industries consider offering discounts or promotions to be able to fill their unused capacity [15].

Restaurant's inventory can be thought of as its supply of raw food or prepared meals. Instead, restaurant inventory should be considered as time, as the period during which a seat or a table is available. If the seat or the table is not occupied for a period of time, that part of the inventory perishes. Instead of counting table turns or income for a given part of the day, restaurateurs should measure revenue per available seat hour, commonly referred to as RevPASH. This is a measure defined by Klimes in Ref. [16], which captures the time factor involved in the restaurant seating. This is a consequent extension of the measurement commonly practiced in other industries that use RM, which focuses on the revenue per available inventory unit. For example, hotels measure revenue per available room-night (RevPAR) and airlines measure revenue per available seat-mile (RPSM).

As was mentioned before, industries that use RM, including restaurants, have appropriate costs and pricing structure. The combination of relatively high fixed costs and low variable costs gives them even more motivation for the fulfillment of their unused capacity, for example, restaurants must generate sufficient revenue to cover variable costs and offset at least some of the high fixed costs. And the relatively low variable costs give these industries some pricing flexibility and give them the opportunity to cut prices in periods of low demand [15]. However, the cost structure is quite different for restaurants compared to hotels and airlines. Variable cost (food and labor) in a restaurant is about 50%, which is much lower than in the mentioned industries.

Customer demand has two components: its timing and the duration of customer experience [15]. Customer demand will be different depending on the time of year, day of week, and time of day. For example, in a restaurant, it may be higher on weekends, or at specific times during periods of lunch or dinner. Time can be sold directly or indirectly. When companies sell time *explicitly* (eg, by the day as in the case of hotels), they are better able to control their capacity, because they know how long customers will be using the capacity, while companies that sell time indirectly (eg, restaurants) usually sell service experiences. An attempt to control customer duration is a very delicate issue due to the potential impact on customer's experience and satisfaction. For example, in case of restaurant it has to reckon with the length of time a party stays once it is seated.

A RM system requires predictions of quantities such as customer demand, price sensitivity, and cancellation probabilities, and its performance depends vitally on the quality of these forecasts. But the most critical element in a strategy for restaurant RM is good prediction of future demand. Restaurant managers have always struggled with the question of how many guests will show up this day. Customer demand varies by the time of year, month, week, day, and the day part. Restaurant demand may be higher on weekends (especially on Fridays and Saturdays), during holidays, during summer months, or at particular periods such as lunch or dinner time. Restaurant operators want to be able to forecast time-related demand so that they can make effective pricing and table-allocation decisions [10].

From a practical point of view it is useful to predict demand (guest count) at different time intervals. They are listed in Table 17.1. Note that demand estimates made during the day can be very accurate but often local labor laws prohibit sending staff home without working a minimum number of hours, and it can be impractical to call in staff on short notice. Also, if the restaurant is overstaffed, it is expensive for the restaurant, and the staff is unhappy because they have to share the tips.

Many RM systems in practice rely primarily on historical sales data to build forecasts. While this leads to highly efficient systems for data collection and automatic forecasting, relying solely on historical data has its weaknesses. For example, in media RM predictions for rating of new programs must be done, despite the fact that their ratings have often little relationship with the ratings observed for previous programs. Similarly, when the restaurant chain opens a store in a new city, there is often little or no historical data on which to base predictions. Moreover, even if the product remains constant, significant

Table 17.1 Time Intervals in Which Restaurant Demand Should Be Predicted

No.	Time Interval	Range of the Interval
1	Month	Next 12 months
2	Week	Next 4–6 weeks
3	Day	Next 7–10 days
4	Day part	Next 1–3 days

Table 17.2 Variables That Can Be Used as Predictors

No.	External Variable	Range or an Example of the Variable
1	Time	Month, week, day of the week, hour
2	Weather	Temperature, rainfall level, snowfall level, hour of sunshine
3	Holidays	Public holidays, school holidays
4	Promotions	Promotion/regular price
5	Events	Sport games, local concerts, conferences, other events
6	Historical data	Historical demand data, trend
7	Macroeconomic Indicators (useful for monthly or annual prediction)	CPI, unemployment rate, population
8	Competitive issues	Competitive promotions
9	Web	Social media comments, social media rating stars
10	Location type	Street/shopping mall
11	Demographics of location (useful for prediction by time of a day)	The average age of customers

CPI, consumer price index.

changes in the economy, competing technologies, or industry structure may cause the historical data not enough useful in predicting the future.

Thus, sales forecasting is the answer to the question how high will be sale under certain circumstances. The circumstances include the nature of sellers, buyers, and the market (eg, competitors). Sales prediction is very complex, precisely due to the impact of internal and external environment. However, a reliable sales forecasting can improve the quality of business strategy. Thus, important factors are historical sales data, promotions, economic variables, location type, or demographics of location. All variables that are useful in predicting demand and can be crucial in improving the accuracy of forecasts are listed in Table 17.2. A multicriteria decision-making method used to rank alternative restaurant locations is presented in Ref. [17]. In Ref. [18] important attributes for restaurant customers are presented, which can help in determination and prediction of customers' intentions to return.

Ref. [19] provides empirical evidence on the impact of online consumer reviews in the restaurant industry. The author investigated this issue with reviews from Yelp.com and restaurant data from the Washington State Department of Revenue. Yelp.com is the dominant source of consumer opinions in the restaurant industry. The researcher presented a few proposals concerning the impact of consumers' opinion about the restaurant industry. First, a one-star increase in Yelp rating leads to an increase of 5–9% in revenue for independent restaurants, depending on the specification. This effect applies only to independent restaurants; ratings do not affect chain restaurants. Because chains already have relatively little uncertainty about quality, their demand does not respond to consumer reviews. Second, restaurants with chain affiliation have declined in market share as Yelp penetration has increased, which suggests that online consumer reviews substitute for more traditional forms of reputation. The reaction of consumers is larger when ratings contain more information. However, consumers also respond more strongly to information that is more visible, which suggests that the way information is given matters.

Ref. [20] brought an overview of work related to the Restaurant Revenue Management (RRM) that was made until 2010. The author notes that, despite the fact that history of the topic of increasing restaurant profitability dates back about 50 years, there are a surprising number of important and unanswered questions. These questions are not only those that might interest scientists but also those that may be relevant to the performance of individual restaurant, and the restaurant industry as a whole. And a part of these questions is related to the area of restaurant forecasting.

The RM application in restaurant industry is described in Ref. [21]. This is a case where the scientists worked with a local restaurant to test their ideas, make recommendations for improvement, and watch the results. This article describes how RM strategy was developed for the 100-seat casual restaurant in Ithaca, New York. Five tasks are listed for managers who want to develop an RRM program: (1) establish the baseline of performance, (2) understand the drivers of that performance, (3) develop a strategy for RM, (4) implement this strategy, and (5) monitor results of the strategy. An example from hotel industry was mentioned, where hotel RM system produced daily forecasts for the next 2–3 months and all forecasts were classified as *hot*, *warm*, or *cold*. Business periods of high demand were considered to be hot, low-demand periods were considered to be cold, and other times were considered to be worm. And hotel managers developed different strategies for each of the expected demand levels. Analogously the authors of the paper found that similar approach could be used in the restaurant industry. The restaurant required one set of policies for slow times and a completely different set for the busy times. During the hot periods managers should try to reduce the duration of meals and possibly raise the prices in the menu, for example, by eliminating discounts. During cold times, managers need to focus on increasing the number of clients and possibly increase average checks.

Maximizing productivity is extremely important for large and popular restaurants to increase their profits and remain competitive. The challenge of their floor managers is to decide when and where to seat each arriving guest. So, a tool that could help floor managers make better decisions would be very valuable to the restaurant. Authors of article in Ref. [22] claim that restaurants can increase their revenues by optimizing their nesting decisions, for example, when to save tables waiting for larger parties, even when there are smaller parties currently in the line. They developed mathematical programming models for RRM and showed that by saving tables in anticipation for larger parties, management can increase revenue, even when there are smaller parties waiting in the line. They created two classes of optimization models to maximize restaurant revenue, while controlling the average waiting time and the perception of fairness that may violate the *first-come-first-serve* rule. In the first class of models, the total programming, stochastic programming, and approximate dynamic programming methods were used in order to dynamically decide when to seat an incoming party in the restaurant that does not accept reservations. In the second class of presented models, researchers used the stochastic gradient algorithm to decide when, if at all, to accept a reservation. As shown by results using simulated data and those using real data from a restaurant, both of these models lead to significant improvements in revenues compared to using the *first-come–first-serve* policy.

In Ref. [23] authors provide an introduction to research in restaurant pricing. Restaurant operators set prices in a cost-based, demand-based, or competition-based way. As a part of a demand-based pricing, a restaurant aims to determine the right price for a menu item by understanding how price affects demand. Usually, there are three approaches that are used to estimate the impact of price on demand: estimation of customer willingness to pay, menu engineering, and price elasticity analysis.

Most companies have customer segments that are price sensitive and others that are not. Some clients may be willing to change the time of their use of service capacity in exchange for a lower price.

On the other hand, some customers are less price sensitive and they are willing to pay a premium for the desired place at the desired time. Managers need to be able to identify different segments, and then adjust the price within a specified time to meet the needs for a given segment [15].

In 2013 the authors Kimes and Beard gave probably the most recent summary of the researches about RM in restaurant industry in a paper [24]. This article presented a framework for understanding the various aspects of RRM and to identify areas that need improvements and future research. Among other things, studies on how best to design the restaurant reservation system that take into account demand, meal duration, customer value, and table assignment would be very beneficial from scientists' and managers' point of view.

3 LITERATURE REVIEW
3.1 MULTIPLE REGRESSION

Multiple regression is a simple, yet powerful technique used for predicting the unknown value of a dependent variable X_t from the known value of two or more explanatory variables (predictors) $V_1, ..., V_k$. Multiple Regression uses least squares to predict the future and allow forecasters to experiment the effects of a different combination of the independent variables on the prediction. The equation for multiple regression is as follows:

$$X_t = \propto_0 + \propto_1 V_{1t} + \cdots + \propto_k V_{kt} + \varepsilon_t$$

where ε_t is the error. Coefficients $\propto_0, ..., \propto_k$ can be estimated using least squares to minimize sum of errors [25].

For example, multiple regression models can be used in econometrics, where regression equation(s) model a causal relationship between the dependent variable (eg, restaurant sales) and external variables such as disposable income, the consumer price index (CPI), and unemployment rate. One of the advantages of econometric models created for predicting restaurant sales is that the researchers can logically formulate a cause-and-effect relationship between the exogenous variables and future sales or demand. Econometric models have however some drawbacks. Geurts and coworkers [26] noticed that the future values of the independent variables themselves have to be predicted, which can cause data in an econometric model to be inaccurate and the model to be weak in its ability to forecast. Also the relationship found between the dependent and independent variables may be pretended or their causal relationship can change over time, causing the need for constant update, or even a complete redesign of the model.

An example of using multiple regression is presented in Ref. [27]. The purpose of this study was to identify the most appropriate method of forecasting meal counts for an institutional food service facility. The forecasting methods included naïve models, MAs, exponential smoothing methods, Holt's and Winters' methods, and linear and multiple regressions. The result of this study showed that multiple regression was the most accurate forecasting method.

Also in Ref. [28] multiple regression model was used to demonstrate its potential for predicting future sales in the restaurant industry and its subsegments. Authors considered in this study the macroeconomic predictors such as percentile change in the CPI, in food away from home, in population, and in unemployment. They collected data from 1970 to 2011 from a variety of sources, including the

National Restaurant Association (NRA), the United States Department of Agriculture (USDA), the Bureau of Labor Statistics, and the US Census Bureau. The model, trained and tested on aggregated data from the past 41 years, appears to have reasonable utility in terms of forecasting accuracy.

In Ref. [29] the author used several regressions and Box–Jenkins models to forecast weekly sales at a small campus restaurant. The result of testing indicates that a multiple regression model with two predictors, a dummy variable and sales lagged 1 week, was the best forecasting model considered.

Regression model was also used in a specific situation described in Ref. [30], where the restaurant was open and close during different times of the week or year.

3.2 POISSON REGRESSION

Restaurant guest count is an example of variable that takes on discrete values. When the dependent variable consists of count data, Poisson regression can be used. This method is one from a family of techniques known as the generalized linear model (GLM). The foundation for Poisson regression is the Poisson distribution error structure and the natural logarithm link function:

$$\ln X = \propto_0 + \propto_1 V_1 + \cdots + \propto_k V_k$$

where X is the predicted guest count, $V_1, \ldots, V_k$ are the specific values on the predictors, ln refers to the natural logarithm, $\propto_0$ is the intercept, and $\propto_i$ is the regression coefficient for the predictor V_i.

The method is used, for example, in Refs. [4,31]. In Ref. [32] authors noticed that Poisson Regression can be used to predict the number of customers being served at a restaurant during a certain time period.

3.3 BOX–JENKINS MODELS (ARIMA)

Time series models are different from Multiple and Poisson Regression models in that they do not contain cause–effect relationship. They use mathematical equation(s) to find time patterns in series of historical data. These equations are then used to project into the future the historical time patterns in the data. There are three types of time series patterns: trend, seasonal, and cyclic. A trend pattern exists when there is a long-term increase or decrease in the series. The trend can be linear, exponential, or different one and can change direction during time. Seasonality exists when data is influenced by seasonal factors, such as a day of the week, a month, and one-quarter of the year. A seasonal pattern exists of a fixed known period. And a cyclic pattern occurs when data rise and fall, but this does not happen within the fixed time and the duration of these fluctuations is usually at least 2 years [33].

The AR model specifies that the output variable depends linearly on its own previous values. The notation AR(p) refers to an AR model of order p. The AR(p) model for time series X_t is defined as follows:

$$X_t = c + \sum_{i=1}^{p} \varphi_i X_{t-i} + \varepsilon_t$$

where $\varphi_1, \ldots, \varphi_p$ are the parameters of the model, c is a constant, and ε_t is white noise. ε_t are typically assumed to be independent and identically distributed (IID) random variables sampled from a normal distribution with zero mean: $\varepsilon_t \sim N(0, \sigma^2)$, where σ^2 is the variance [10].

Another common approach in time series analysis is a MA model. The notation MA(q) indicates the MA model of order q:

$$X_t = \mu + \varepsilon_t + \theta_1 \varepsilon_{t-1} + \cdots + \theta_q \varepsilon_{t-q}$$

where μ is the mean of the series, $\theta_1, \ldots, \theta_q$ are the parameters of the model, and $\varepsilon_{t-1}, \ldots, \varepsilon_{t-q}$ are white noise error terms [10].

In other words, a MA model is a linear regression of the current value of the series of the data against current and previous, unobserved white noise error terms, or random shocks. These random shocks at each point are assumed to be mutually independent and to come from the same, usually a normal distribution.

MA method is very simple, based on the idea that the most recent observations serve as better predictors for the future demand than do older data. Therefore, instead of having the forecast as the average of all data, a window with an average of only q previous observations is used.

MA reacts faster to the underlying shifts in the demand if q is small, but small span results in a forecast more sensitive to the noise in the data.

If the date shows up or down trend, the MA is systematically under projections or above forecast. To handle such cases, improvements such as a double or triple MA have been developed, but for this kind of data exponential smoothing methods are usually preferred, described in the next section.

AR and MA models were used to make a prediction for many different time series data. One of the examples is presented in Ref. [34], which is the first research looking into the casino buffet restaurants. Authors examined in this study eight simple forecasting models. The results suggest that the most accurate model with the smallest Mean Absolute Percentage Error (MAPE) and root mean square percentage error (RMSPE) was a double MA.

Another tool created for understanding and predicting future values in time series data is model ARMA(p; q), which is a combination of an AR part with order p and a MA part with order q. The general autoregressive moving average (ARMA) model was described in 1951 in the thesis of Whittle [35]. Given a time series of data X_t, the ARMA model is given by the following formula:

$$X_t = c + \varepsilon_t + \sum_{i=1}^{p} \varphi_i X_{t-i} + \sum_{i=1}^{q} \theta_i \varepsilon_{t-i}$$

where the terms in the equation have the same meaning as earlier.

An autoregressive integrated moving average (ARIMA) model is a generalization of an ARMA model. ARIMA models (Box–Jenkins models) are applied in some cases where data show evidence of nonstationarity (stationary process is a stochastic process whose joint probability distribution does not change over time and consequently parameters, eg, the mean and variance, do not change over time) [36].

Most of real-time series data turns out to be nonstationary. In such cases, the stationary time series models may not fit the data well and can produce poor prognosis. Techniques for dealing with nonstationary data try to make such data stationary by applying suitable transformations, so that stationary time series models can be used to analyze the transformed data. The resulting stationary predictions are then converted back to their original nonstationary form. One of such techniques is the differentiation of successive points in the time series. An *autoregressive integrated moving-average process*, ARIMA(p, d, q), is one whose dth differenced series is an ARMA(p, q) process.

Interesting case of big data mining project for one of the world's largest multibrand fast-food restaurant chains with more than 30,000 stores worldwide is illustrated in Ref. [2]. Time series data mining is discussed at both the store level and the corporate level. To analyze and forecast large number of data researchers used Box–Jenkins seasonal ARIMA models. Also an automatic outlier detection and adjustment procedure was used for both model estimation and prediction.

A system designed to generate statistical predictions on menu-item demand in hospitals with intervals of 1–28 days prior to patient meal service is described in Ref. [37]. Authors used 18 weeks of supper data for analysis of menu-item preferences and to evaluate the performance of the forecasting system. There were three interdependent levels in the system: (1) Forecasting patient census, (2) predicting diet category census, and (3) forecasting menu-item demand. To assess the effectiveness of mathematical forecasting system and manual techniques a cost function was used. The costs of menu-item prediction errors resulting from the use of exponential smoothing method and Box–Jenkins model were about 40% less than the costs associated with manual system.

A paper considering the unique seasonal pattern in university dining environments is given in Ref. [7]. This study determines the degree of improvement in accuracy of each tested forecasting model in situation when the data are seasonally adjusted. Researchers compare the seasonally adjusted data and raw data and verify if seasonally adjusted data improves the accuracy. The data for this study is collected at a dining facility at a Southern university during two consecutive spring semesters. The customer count data of the 2000 spring semester was used as a base for forecasting guest counts in the 2001 spring semester. The data includes guest counts for dinner meals from Monday to Saturday, since the dining facility is closed on Sundays. Researchers selected six different forecasting methods including naïve model, MA method, Simple Exponential Smoothing, Holt's Method, Winters' Method, and Linear Regression. More sophisticated forecasting techniques, such as Box–Jenkins or neural networks, were not tested here. The accuracy of these models was assessed by Mean Squared Error (MSE), Mean Percentage Error (MPE), and MAPE. The results show that Winters' method outperforms the other five methods when raw data is used. It turned out that seasonally adjusted data is much more effective in forecasting customer counts and significantly improve accuracy in most of used methods. All the other five mathematical forecasting methods outperform the naïve model when using seasonally adjusted data. And the MA method is the most accurate method of forecasting when seasonally adjusted data is used.

Overall, the main result of this study indicates that the use of seasonally adjusted data is critical for better forecasting accuracy in case of the university dining operations, where seasonal pattern certainly occurs. Thus the researchers strongly recommend employing the MA model with seasonally adjusted data to predict the number of customers in this kind of places.

Note, however, that the prediction of abnormal, extremely high or low demands is not considered in this study. In real situation, forecasting may be more complicated due to unexpected variables such as special events or unusual weather. Thus, the authors recommend for food service managers to employ techniques such as MA with the judgments from their own experience to get better forecasting results under their unique environment.

3.4 EXPONENTIAL SMOOTHING AND HOLT–WINTERS MODELS

Exponential smoothing, proposed in the late 1950s, is another technique that can be applied to time series data to make forecasts. Whereas in the simple MA the past observations are weighted equally, exponential smoothing uses exponentially decreasing weights over time. The more recent the observation, the higher is the associated weight. For the sequence of observations $\{x_t\}$ beginning at time $t = 0$, the simplest form of the exponential smoothing algorithm is given by the following formula:

$$s_0 = x_0$$

$$s_t = \propto x_t + (1 - \propto)s_{t-1}, \, t > 0$$

where $\{s_t\}$ is the estimation of what the next value of x will be, $\propto$ is the smoothing factor, and $0 < \propto < 1$.

Triple exponential smoothing (suggested in 1960 by Holt's student, Peter Winters) takes into account seasonal changes and trends. As we mentioned in the previous section, seasonality is a pattern in time series data that repeats itself every L period. There are two types of seasonality: *multiplicative* and *additive* in nature.

For time series data $\{x_t\}$, beginning at time $t = 0$ with a cycle of seasonal change of length L, triple exponential smoothing is given by the following formulas:

$$F_{t+m} = (s_t + mb_t)c_{t-L+1+(m-1)\bmod L}$$

$$s_0 = x_0$$

$$s_t = \propto \frac{x_t}{c_{t-L}} + (1 - \propto)(s_{t-1} + b_{t-1})$$

$$b_t = \beta(s_t - s_{t-1}) + (1 - \beta)b_{t-1}$$

$$c_t = \gamma\frac{x_t}{s_t} + (1 - \gamma)c_{t-L}$$

where F_{t+m} is an estimate of the value of x at time $t + m$ ($m > 0$), $\propto$ is the data smoothing factor, β is the trend smoothing factor, γ is the seasonal change smoothing factor, $0 < \propto, \beta, \gamma < 1$, $\{s_t\}$ represents the smoothed value of the constant part (level) for time t, $\{b_t\}$ is the estimation of the linear trend for period t, and $\{c_t\}$ represents the sequence of seasonal factors.

Exponential smoothing was one of the most common and simple methods for food and beverage sales forecasting (eg, Refs. [38,39]).

The results of the study [3] show that for the actual sales in the restaurant, which is independently owned and located in a medium-sized university town, Box–Jenkins and exponential smoothing models performed as well as or better than an econometric model. Authors as an initial database used monthly observations of sales from 6 years from this restaurant. And they tested model on the next 7 months of data. Since time series models are usually more economical in terms of time and skill levels of the users, the results of this study are important for forecasting in the restaurant industry. Results of this paper suggest that for a restaurant manager with limited resources, an exponential smoothing model could usually give a very satisfactory forecast of sales. Forecast accuracy can be improved by using a Box–Jenkins model; however, this model is more complex.

Another study that compares different forecasting models to predict the meal participation at university residential dining facilities is presented in Ref. [40]. Authors used data collected from two dining rooms over 15 weeks to test naïve forecasting techniques, three different versions of MA, and simple exponential smoothing. The analysis of these prognostic models using Mean Absolute Deviations (MADs), MSEs, and MAPEs indicated that all the simple mathematical forecasting techniques provided better forecasts than naïve methods and MA methods gave the best results.

Another assessment of which forecasting model would most accurately predict meal demand is presented in Ref. [41]. The uniqueness of this study lies in the fact that the data do not relate to ordinary restaurant, but to congregate lunch programs located throughout the United States, designed for serving older adults. But lunch meal demand forecasting in this program has much in common with forecasting demand for restaurants. Food service managers in these programs are responsible for operational expenditure. Funds for that are made available to local meal providers as reimbursements based on meals served, in contrast to meals produced; thus the ability to accurately predict demand meal can become critical in an effort to maintain fiscal responsibility. Accurate forecasts allow food service managers to purchase adequate amounts of food and supplies, produce the right amount of meals, and properly plan a schedule for staff. Overproduction can lead to food waste and excessive costs, and on the other hand underproduction may create shortages that will affect customer satisfaction. Forecasting techniques, including naïve, three versions of MA, and simple exponential smoothing, are applied to data collected within 4 months from seven locations in large urban agglomeration. The author used MADs and MSEs to analyze all these methods. The results of the evaluation indicated that for all locations simple mathematical forecasting techniques provided better meal demand prediction than did the naïve method. In four of the seven sites, exponential smoothing outperformed the other methods, while in other places, MA models provided the best forecasts.

The extension of this paper can be found in Ref. [42]. This time data from eight different congregate meal sites in the Southern California area were used over the 6-month period. All sites utilized a nonselect 6-week cycle menu. As in the previous study the simple mathematical forecasting techniques provided a better prediction of meal demand than did the naïve method in all cases. The results of this study indicate that exponential smoothing outperformed the other prediction methods in all locations. The second aim of this paper was to answer the following research question: Could the accuracy of the forecast be improved when the historical customer count information will be obtained from the same day of the menu cycle instead of from the same day of the week? The results of this paper indicate that even though the lunch sites studied used a cycle menu, the effectiveness of the prediction method was not significantly improved when the same day of the menu cycle was used as the source of the historical data, as opposed to the same day of the week.

The aim of study [43] was to develop, test, and evaluate mathematical forecasting using entree demand data collected from a university dining hall food service. The used time series models are applicable to all types of food and beverage operations, because the used data contained typical gastronomy production variable: menu changes. Researchers employed naïve methods (used as the base case), simple MA, and simple exponential smoothing. They modeled both the total number of guests during the meal period and entrees on the menu (entree combinations). Once the guest count data were used, the forecast was then multiplied by a preferences statistic, which was calculated for each entree to forecast the percent of this entree that would be chosen when the entree combination was offered. Results of this study are consistent with other researches related to food service forecasting. The time series models outperformed the Naïve Model. The simple MA model gave smaller but more numerous forecast errors, whereas simple exponential smoothing did the reverse.

3.5 ARTIFICIAL NEURAL NETWORKS

All the forecasting methods we have discussed in previous subsections have the same strategy: make a functional assumption for the relationship between the observed data and various factors and then estimate the parameters of this function using historical data. In contrast, neural network methods and other machine learning algorithms use interactions in a network architecture to automatically estimate the underlying unknown function that best describes the demand process. Artificial Neural Networks (ANNs) are a class of models inspired by biological neural networks and studies of the human nerve system, in particular the brain. The methods are based on approaches that mimic the way the human brain learns from experience and are used to estimate or approximate functions that can depend on a large number of inputs and are generally unknown. In theory, with the suitably constructed architecture and with the properly carried out training procedure, neural networks are able to approximate any non-linear function after a sufficient degree of *learning*.

ANNs are systems of connected *neurons*, where the connections have numeric weights that can be tuned based on experience from historical data, which causes that neural networks are adaptive to inputs and capable of learning.

ANNs are used for sales forecasting due to the promising results in the areas of control and pattern recognition. In Fig. 17.4 there is an example of neural network architecture for guest count prediction.

In *feedforward* networks (or *perceptrons*)—the most important class of neural networks used for forecasting – the nodes of the network are arranged in succeeding layers, and the arcs are directed from one layer to the next, left to right as shown in Fig. 17.4. Training data is "fed" to the first (the input layer), and the forecast is "read" from the last (the output layer). Usually, in demand forecasting applications, each node in the input layer corresponds to the explanatory, and each node in the output layer corresponds to a single explanatory variable, for example, guest count. There are a number of hidden layers in between.

A set of training data is used to calibrate the weights and the value of the threshold functions. When these parameters are defined, the network can be used for forecasting. Thus, the three main steps are defining the network, training, and forecasting. There exist many procedures to automatically prune or

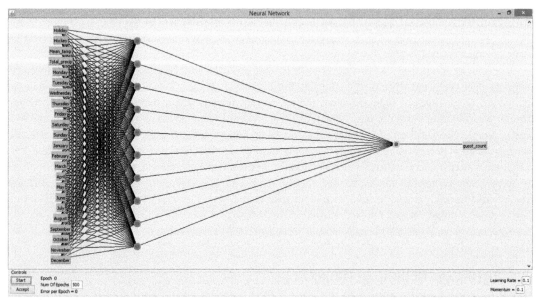

FIGURE 17.4 Example of Neural Network Architecture for Guest Count Prediction

grow the network topology on the basis of the observed data and network's predictive performance, and many ways of improving training of neural networks, particularly in terms of speeding up the learning time or avoiding overfitting.

Ref. [44] compares ANNs and traditional methods including Winters' exponential smoothing, Box–Jenkins ARIMA model, and multivariate regression. The results indicate that generally ANNs are more successful compared to the more traditional statistical methods. Analysis of experiments shows that the neural network model is able to capture the trend and seasonal patterns, as well as the interactions between them. Despite many positive features of ANNs, constructing a good network for a given project is a quite difficult task. It consists of choosing an appropriate architecture (the number of hidden layers, the number of nodes in each layer, the connections between nodes), selecting the transfer functions of the middle and output nodes, designing a training algorithm, selecting initial weights, and defining the stopping rule.

In the study in Ref. [45] authors combined an ANN and a genetic algorithm to design and develop a sales forecasting model. They collected historical sales data from a small restaurant in Taipei City and used them as the output for the forecasted results while associated factors including seasonal impact, impact of holidays, number of local activities, number of sales promotions, advertising budget, and advertising volume were chosen as input data. All the training and test data were preprocessed in such a way that the data were mapped between $[1, -1]$. And at the end an inverse transformation was performed on the results of prediction in order to restore the actual value of the forecasted sales. First, this approach applies the ANN to select the relevant parameters of the current sales condition as the input data. Then it uses a genetic algorithm to optimize the default weights and thresholds of the ANN. Researchers used empirical analysis to examine the effectiveness of the model. The results indicate that this is a scientifically practical and effective sales forecasting method that can achieve rapid and accurate prediction.

Fuzzy neural network with initial weights generated by genetic algorithm can be found, for example, in Ref. [46]. This study first proposes a genetic algorithm initiated fuzzy neural network that is able to learn the fuzzy IF–THEN rules for promotion provided by marketing specialists. This network can both learn fuzzy IF–THEN rules and incorporate fuzzy weights. The main reason for proposing this method is that the effect of promotion on sales is always very vague or fuzzy. The results from the fuzzy neural network are further integrated with the forecast from ANN using the time series data and the length of promotion from another ANN. In addition, this study also aims to develop an intelligent sales forecasting system based on the aforementioned concepts. Forecasting future demand is central to the operation of the convenience store companies, and the reliable prediction of sales can be a great benefit in improving the quality of a business strategy as well as decreasing cost, which means increased profit. Managers need sales forecasts as essential input to many decision activities.

3.6 BAYESIAN NETWORK MODEL

Ref. [47] proposes a service demand forecasting method that uses a customer classification model to consider various customer behaviors. A decision support system based on this method was introduced in restaurant stores. Authors automatically generated categories of customers and items based on purchase patterns identified in data from 8 million purchases at a Japanese restaurant chain. The data was collected from 5 years from 48 stores. Researchers produced a Bayesian network model including the customer and item categories, conditions of purchases, and the properties and demographic information of customers. Based on that network structure, they could systematically identify useful knowledge and predict customers' behavior. Details of this demand forecasting technique are given in Ref. [48].

3.7 HYBRID MODELS

In the literature a hybrid approach to sales forecasting for restaurants is also proposed. The motivation of the hybrid model comes from the following prospects. It is often difficult in practice to determine whether one specific method is more effective in prediction then others. Thus, it is difficult for researchers to select the appropriate forecasting model for their unique situations. Usually, different approaches are tested and the one with the most accurate result is chosen. However, the final selected model is not always the best to use in the future. The problem of model selection can be facilitated by using combined methods [49].

Furthermore, it may not even be necessary to determine which method gives the best forecast: a linear combination of the predictions from two different models, with a properly chosen set of weights, may be constantly superior to any one of the component methods. This idea was proposed in the article in Ref. [50], and then investigated by other forecasting researchers. The intuition behind this result is that if the errors produced by two methods of prediction are negatively correlated, then combining them will reduce the overall forecast error.

Researchers use hybrid model in many different areas of forecasting. As an example of a hybrid system with excellent performance the model of daily product sales in a supermarket proposed in Ref. [51] can be shown. Authors combined ARIMA models and neural networks to a sequential hybrid forecasting system, where output from an ARIMA-type model is used as input for a neural network.

In Ref. [52] a research is presented not directly from the restaurant industry, but from a similar area of convenience stores, which provide multiple services, including daily fresh foods. Forecasting future demand is crucial for the functioning of such convenience store industry and reliable sales forecasts can be very beneficial in improving the quality of business strategy, as well as increasing profits by reducing costs. Sales forecasts are needed as an important contribution in various fields such as marketing, production, and sales. Here freshness and fast speed of turnover are important factors. Each store needs to have fresh food sales predicted accurately in order to maintain the high quality of sold products. In this study, an improved hybrid sales forecasting model of fresh foods, called Enhanced Cluster and Forecast Model, has been successfully developed. The model combines self-organizing map neural networks and radial basis function neural network. The model is evaluated for a 6-month sales dataset of daily fresh foods at a chain of 75 convenience stores distributed throughout Taiwan. According to the authors' knowledge traditional convenience stores managers do not use a model approach for forecasting, or some of them use a fuzzy neural model. The proposed hybrid model not only is more efficient but also makes the model easier to build with greater accuracy. The proposed sales forecasting model is suitable for forecasting system in the real world.

For restaurant industry a hybrid methodology that takes advantage of the unique strength of ARIMA and ANN models in linear and nonlinear modeling and combines these two methods is proposed in Ref. [49]. In the real world time series data are rarely pure linear or nonlinear. They often contain both linear and nonlinear patterns. In this case, neither ARIMA nor ANNs may be suitable for modeling and forecasting time series. Hence, by combining ARIMA with ANN models, researchers can model data more accurately. Experimental results on sets with real data indicate that the combined model can be an effective way to improve accuracy of predictions achieved by each of the models used separately.

3.8 ASSOCIATION RULES (MARKET BASKET ANALYSIS)

In this subsection we want to mention an additional method that can help in restaurant forecasting. In Ref. [53] the problem of mining association rules and related time intervals is studied, where an association rule holds in either all or some of the intervals. As an example association rules in a given database of restaurant transactions are considered.

In Ref. [54] authors apply a simplified version of market basket analysis (MBA) rules to explore menu-item assortments, which are defined as the sets of most frequently ordered menu-item pairs of an entre and side dishes.

In some cases, MBA does not provide useful information if data item is the name of goods. In Ref. [55] authors propose a new MBA method that integrates words segmentation technology and association rule mining technology. Characteristics of items can be generated automatically before mining association rules by using word segmentation technology. This method has been applied to a restaurant equipped with electronic ordering system to give recommendations to customers, where the experiments were done. The experiment results show that the method is efficient and valid.

4 DISCUSSION OF METHODS AND DATA MINING ALGORITHMS

The summary of all approaches is presented in Table 17.3.

Table 17.3 Summary of Sales/Demand Forecasting Methods

Method	Description	Examples of Papers
Multiple Regression	Multiple Regression uses least squares to predict the unknown value of a dependent variable from the known value of two or more explanatory variables (predictors)	[26–30,34]
Poisson Regression	Poisson regression uses the Poisson distribution error structure and the natural logarithm link function	[32]
Box–Jenkins model (AR, MA, ARIMA)	The AR model specifies that the output variable depends linearly on its own previous values. The simple MA weights the past observations equally	[2,7,29,37,43]
Exponential smoothing and Holt–Winters models	Exponential smoothing uses exponentially decreasing weights over time	[3,27,38–41,43]
ANNs	ANNs use interactions in a network-processing architecture to automatically identify the underlying function that best describes the demand process	[45,46]
Bayesian Network Model	Bayesian Network can represent the probabilistic relationships between the variables	[47,48]
Hybrid Model	Hybrid models combine two different methods in one	[49,51]
Association Rules	Association Rules algorithms find frequent patterns in the data	[53–55]

ANN, artificial neural network; *AR*, autoregressive; *ARIMA*, autoregressive integrated moving average; *MA*, moving average.

It is difficult for forecasters to choose the right technique for their unique situations. Model selection is one of the most delicate tasks in the predictive analysis. Intuition, judgment, experience, and repeated testing are needed to find a model that generalizes well and has good forecasting power. Typically, a number of different models are tried and tested, and the one with the most accurate result is selected.

In our opinion techniques that take into account external factors mentioned in Table 17.1 are the best. Not only the choice of method but also preparing the relevant input data affects the high efficiency of the model.

Thus one of the tasks in model selection is deciding which variables should be included as an input feature. In general, it is undesirable to include too many variables. Correlations between independent variables can lead to an incorrect coefficient estimate. The question is the following: Which subset of the possible explanatory variables is the best to use? To answer this question we can use one of the feature selection algorithms. A simpler methodology, often used in practice, is to start with an initial subset, and then try to add one variable at a time, each time verifying whether it increases predictive power. Similarly, we can start with a full set and remove one variable at a time, every time testing for loss of predictive power.

A common problem with fitting the model to training data is *overfitting*. Overfitting may be limited by considering models that are "reasonable" from a subjective, business point of view, instead of blindly trying to find the best fit model based on historical data. In the case of neural networks, the problem is more subtle and difficult to detect. Since there is no clear functional form between independent variable and explanatory variables, and since neural networks can approximate any function, there is a high danger that we might overtrain and adapt the network to noise.

Understanding the different techniques—their advantages and limitations, and the relationships between them—is important when choosing the appropriate method in a particular application and for development of new methods, when none of the existing models seems right. The purpose of this chapter, and particularly this section, is to help readers in this task.

In Table 17.4 there is a brief description of the advantages and disadvantages of methods of demand and sales prediction.

Most of the prediction algorithms in practice of RM are variations of standard methods, and many of them are not particularly complicated or mathematically sophisticated. Also, many researchers use multiple algorithms that allow users to select one or several methods or, alternatively, the system can combine forecasts from different methods in one hybrid model. In terms of forecasting methods, the emphasis in the RM systems is placed on speed, simplicity and reliability. Many forecasts have to be made and time spent on their production is limited. Finally, in a situation of practical forecasting, tasks related to data—such as data collection, preprocessing, and purification—require no less effort than choosing the technique of forecasting.

Prediction is usually performed overnight in a batch process and then fed to the optimization modules, so the time interval for completing all operations takes a few hours.

To provide a base for research in model development and implementation assessment of the state of practice in forecasting production demand in food service operations is needed. In a study [56] there are results from a short survey conducted to document the forecasting techniques utilized by food service directors in 1990. The study found that only about 16% of the food service operators used mathematical models for forecasting demand and the most frequently used mathematical model was the MA technique. But judgment based on the past records was the most frequently used forecasting method and typically production demand was determined 1 week in advance. This study implied that

Table 17.4 Advantages and Disadvantages of Sales/Demand Forecasting Methods

Method	Input Data	Output	Advantages	Disadvantages
Multiple Regression	Exogenous variables such as disposable income, the consumer price index, unemployment rate, personal consumption expenditures, housing starts	For example, restaurant sales/customer demand	+ The decision maker can logically formulate the model based on a cause-and-effect relationship between the causal variables and future sales	– Multiple regression analysis can fail in clarifying the relationships between the predictor variables and the response variable when the predictors are correlated with each other – The relationship found between the dependent and independent variables may be spurious or can change over time, making it necessary to constantly update or totally redesign the model
ARIMA (Box–Jenkins models)	Historical time series demand/sales data	Long-term or short-term predictions of future demand/sales	+ Do not need any external data	– The input series for ARIMA needs to be stationary, that is, it should have a constant mean, variance, and autocorrelation through time – These methods require improvements if the data are influenced by heterogeneity (eg, promotion)
Exponential Smoothing model, Holt–Winters models	Historical time series demand/sales data	Long-term or short-term predictions of future demand/sales	+ Exponential smoothing generates reliable forecasts quickly, which is a great advantage for applications in industry + Do not need any external data	– Method is influenced by outliers (sales/demand that are unusually high or low)
Bayesian Network Model	Particular set of variables	The probability of the variable, for example, high sale	+ All the parameters in Bayesian networks have an understandable interpretation	

(*Continued*)

Table 17.4 Advantages and Disadvantages of Sales/Demand Forecasting Methods (*cont.*)

Method	Input Data	Output	Advantages	Disadvantages
Neural Networks	For example, associated factors including seasonal impact, impact of holidays, number of local activities, number of sales promotions, advertising budget, and advertising volume can be used as input data. All the training and test data used in this study are required to be preprocessed. The input and output data used for training and the input data used for testing have to be preprocessed so that the data were mapped between $[1, -1]$	Sales amount can be chosen as the output data for the Forecasted results. An inverse transformation should be conducted on the results of the simulated forecast to restore the actual value of the forecasted sales condition	+ Have high tolerance of noisy data + Ability to classify patterns on which they have not been trained + Can be used when there is little knowledge of the relationships between attributes and classes + They are well suited for continuous-valued inputs and outputs, unlike most decision tree algorithms + Are parallel; parallelization techniques can be used to speed up the computation process + Can model complex, possibly nonlinear relationships without any prior assumptions about the underlying data generating process + Overcome misspecification, biased outliers, assumption of linearity, and reestimation	− Neural networks involve long training times and are therefore more suitable for Applications where this is feasible − They require a number of parameters that are typically best determined empirically, such as the network topology or structure. Constructing a good network for a particular application is not a trivial task. It involves choosing an appropriate architecture (the number of hidden layers, the number of nodes in each layer, and the connections among nodes), selecting the transfer functions of the middle and output nodes, designing a training algorithm, choosing initial weights, and specifying the stopping rule − Neural networks have been criticized for their poor interpretability
Association Rule Mining (Market Basket Analysis)	Transactional database (TDB) or Relational database (RDB). Given a minimum support (min_{sup}) and a minimum confidence (min_{conf})	All association rules that satisfy both min_{sup} and min_{conf} from a dataset D	+ Association rules that satisfy both min_{sup} and min_{conf} can help with discover factors that influence high/low demand	

ARIMA, autoregressive integrated moving average.

in the 1980s and 1990s food service operators were limited to forecasting methods that are simple and fast.

The situation has changed over the years and different state of things can be found in the paper [57] published in 2008. In this paper are presented the results and a discussion of the interviews with 12 corporate restaurant forecasting managers responsible for the sales forecasting process in their companies. Firms were tested using a qualitative research method design and long interviews to gather information on methods, techniques, and technological systems used in the sales process and forecasting. The main findings included a wide variety of hardware and software used among the participants, as well as variety of methods and techniques used by them to forecast sales. The results also show that managers have experienced different levels of satisfaction with the systems, approach, and techniques that were in use.

As far as the techniques used to accomplish the sales forecast were concerned, there was a wide variety of method listed by managers. Of the quantitative techniques used to develop sales forecasts, five companies used regressions analysis, decomposition, and straight-line projections, while three companies used exponential smoothing, MAs, and trend-line analysis. Two companies used the life-cycle analysis and one company used an expert system. None of the restaurants utilized the Box–Jenkins technique, simulations, or neural networks to develop the forecast. Of the qualitative techniques used to develop sales forecasts, five companies used a jury of executive opinion, while three companies used the sales force composite. None of the companies used customer expectations to develop the sales forecast.

5 CONCLUDING REMARKS

Demand prediction plays a crucial role in planning operations for restaurant's management. Having a reliable estimation for a menu item's future demand is the basis for other analysis. Various forecasting techniques have been developed, each one with its particular advantages and disadvantages compared to other approaches.

The evolution of the respective forecasting methods over past 20 years has been revealed in this chapter. A review and categorization of consumer restaurant demand techniques is presented in the chapter. Techniques from a wide range of methodologies and models given in the literature are classified here into seven categories: (1) multiple regression; (2) Poisson regression; (3) exponential smoothing and Holt–Winters model; (4) AR, MA, and Box–Jenkins models; (5) neural networks; (6) Bayesian network; and (7) hybrid methods. The methodology for each category has been described and the advantages and disadvantages have been discussed. This chapter conducts a comprehensive literature review and selects a set of papers on restaurant sales forecasting.

It is almost universally agreed in the forecasting literature that no single method is best in every situation.

ACKNOWLEDGMENT

This work was supported by a Collaborative Research and Development (CRD) grant from the Natural Sciences and Engineering Research Council of Canada (NSERC), grant number: 461882-2013.

REFERENCES

[1] Daryl A, Dyer C. A framework for restaurant information technology. Cornell Hotel Restaurant Adm Q 1999;40(3):74–84.

[2] Liu L-M, Bhattacharyya S, Sclove SL, Chen R, Lattyak WJ. Data mining on time series: an illustration using fast-food restaurant franchise data. Comput Stat Data Anal 2001;37:455–76.

[3] Cranage DA, Andrew WP. A comparison of time series and econometric models for forecasting restaurant sales. Int J Hospitality Manage 1992;11(2):129–42.

[4] Wulu JT Jr, Singh KP, Famoye F, Thomas TN, McGwin G. Regression analysis of count data. J Indian Soc Agric Stat 2002;55(2):220–31.

[5] Miller JL, Shanklin CW. Forecasting menu-item demand in foodservice operations. J Am Diet Assoc 1988;88(4):443–9.

[6] Green YNJ, Weaver PA. A sales forecasting benchmarking model. Int J Hospitality Tourism Adm 2006;6(4): 3–32.

[7] Ryu K, Shawn Jang S-C, Sanchez A. Forecasting methods and seasonal adjustment for a university foodservice operation. J Foodservice Business Res 2004;6(2):17–34.

[8] Lasek A, Cercone N, Saunders J. Restaurant sales and customer demand forecasting: literature survey and categorization of methods. In: EAI international conference on big data and analytics for smart cities, Toronto; October 13–14, 2015.

[9] Smith BC, Leimkuhler JF, Darrow RM. Yield management at American Airlines. Interfaces 1992;22(1):8–31.

[10] Kimes SE, Chase RB, Choi S, Lee PY, Ngonzi EN. Restaurant revenue management applying yield management to the restaurant industry. Cornell Hospitality Q 1998;39(3):32–9.

[11] Talluri KT, van Ryzin GJ. The theory and practice of revenue management. New York: Springer Science + Business Media, Inc; 2005.

[12] Talluri KT, van Ryzin GJ, Karaesmen IZ, Vulcano GJ. Revenue management: models and methods. In: Mason SJ, Hill RR, Mönch L, Rose O, Jefferson T, Fowler JW, editors. Proceedings of the 2008 winter simulation conference; 2008.

[13] Kimes SE. Restaurant revenue management: could it work? J Revenue Pricing Manage 2005;4:95–7.

[14] Heo CY, Lee S, Mattila A, Hu C. Restaurant revenue management: do perceived capacity scarcity and price differences matter? Int J Hospitality Manage 2013;35:316–26.

[15] Kimes SE, Wirtz J. Revenue management: advanced strategies and tools to enhance firm profitability. Found Trends Marketing 2015;8(1):1–68.

[16] Klimes SE. Implementing restaurant revenue management: a five-step. Cornell Hospitality Q 1999;40(3): 16–21.

[17] Tzeng G-H, Teng M-H, Chen J-J, Opricovic O. Multicriteria selection for a restaurant location in Taipei. Int J Hospitality Manage 2002;21(2):171–87.

[18] Qu H. Determinant factors and choice intention for Chinese restaurant dining. A multivariate approach. J Restaurant Foodservice Marketing 1997;2(2):35–49.

[19] Luca M. Reviews, reputation, and revenue: the case of Yelp.Com. Harvard Business School NOM Unit working paper no. 12-016; September 16, 2011.

[20] Thompson GM. Restaurant profitability management. The evolution of restaurant revenue management. Cornell Hospitality Q 2010;51(3):308–22.

[21] Kimes SE, Barrash DI, Alexander JE. Developing a restaurant revenue management strategy. Cornell Hotel Restaurant Adm Q 1999;40:18–29.

[22] Bertsimas D, Shioda R. Restaurant revenue management. Operations Res 2003;51:472–86.

[23] Kimes SE, Phillips R, Summa L. Pricing in restaurants. The Oxford handbook of pricing management. Oxford: Oxford University Press; 2012. 106–120.

[24] Kimes SE, Beard J. The future of restaurant revenue management. J Revenue Pricing Manage 2013;12(5): 464–9.

[25] Hastie T, Tibshirani R, Friedman J. The elements of statistical learning data mining, inference, and prediction. In: Springer series in statistics. Berlin, Heidelberg: Springer; 2nd ed. 2008.

[26] Michael D, Geurts J, Kelly P. Forecasting retail sales using alternative models. Int J Forecasting 1986;2(3): 261–72.

[27] Ryu K, Sanchez A. The evaluation of forecasting methods at an institutional foodservice dining facility. J Hospitality Financial Manage 2003;11(1):27–45.

[28] Reynolds D, Rahman I, Balinbin W. Econometric modeling of the U.S. restaurant industry. Int J Hospitality Manage 2013;34:317–23.

[29] Forst FG. Forecasting restaurant sales using multiple regression and Box–Jenkins analysis. J Appl Business Res 1992;8(2):15–9.

[30] Morgan MS, Chintagunta PK. Forecasting restaurant sales using self-selectivity models. J Retailing Consumer Serv 1997;4(2):117–28.

[31] Coxe S, West SG, Aiken LS. The analysis of count data: a gentle introduction to Poisson regression and its alternatives. J Pers Assess 2009;91(2):121–36.

[32] Sellers KF, Shmueli G. Predicting censored count data with COM-Poisson regression. Working paper. Hyderabad: Indian School of Business; 2010.

[33] Hyndman RJ, Athanasopoulos G. Forecasting: principles and practice. Otexts.com; 2014.

[34] Hu C, Chen M, Chen McCain S-L. Forecasting in short-term planning and management for a casino buffet restaurant. J Travel Tourism Marketing 2004;16:79–98.

[35] Whittle P. Hypothesis testing in time series analysis, Vol. 4 of Statistics Upsala, Almqvist & Wiksells; 1951.

[36] George B, Gwilym J. Time series analysis: forecasting and control. San Francisco: Holden-Day; 1970.

[37] Messersmith AM, Moore AN, Hoover LW. A multi-echelon menu item forecasting system for hospitals. J Am Diet Assoc 1978;72(5):509–15.

[38] Miller JJ, McCahon CS, Miller JL. Foodservice forecasting using simple mathematical models. J Hospitality Tourism Res 1991;15(1):43–58.

[39] Miller JJ, McCahon CS, Miller JL. Foodservice forecasting: differences in selection of simple mathematical models based on short-term and long-term data sets. J Hospitality Tourism Res 1993;16(2):93–102.

[40] Blecher L, Yeh RJ. Forecasting meal participation in university residential dining facilities. J Foodservice Business Res 2008;11(4):352–62.

[41] Blecher L. Using forecasting techniques to predict meal demand in Title IIIc congregate lunch programs. J Am Diet Assoc 2004;104(8):1281–3.

[42] Blecher L, Yeh RJ. An analysis of forecasting methods using "same day of the week" versus "same day of the menu cycle" to predict participation in congregate lunch programs. Foodservice Res Int 2004;14(3):201–10.

[43] Miller JL, McCahon CS, Bloss BK. Food production forecasting with simple time series models. J Hospitality Tourism Res 1991;14:9–21.

[44] Alon I, Qi M, Sadowski RJ. Forecasting aggregate retail sales: a comparison of artificial neural networks and traditional methods. J Retailing Consumer Serv 2001;8:147–56.

[45] Chen P-C, Lo C-Y, Chang H-T, LoCho Y. A study of applying artificial neural network and genetic algorithm in sales forecasting model. J Convergence Inf Technol 2011;6(9):352–62.

[46] Kuo RJ. A sales forecasting system based on fuzzy neural network with initial weights generated by genetic algorithm. Eur J Operational Res 2001;129(3):496–517.

[47] Koshiba H, Takenaka T, Motomura Y. A service demand forecasting method using a customer classification model. In: CIRP IPS2 conference; 2012.

[48] Ishigaki T, Takenaka T, Motomura Y. Customer-item category based knowledge discovery support system and its application to department store service. In: IEEE Asia-Pacific services computing conference; 2010.

[49] Zhang GP. Time series forecasting using a hybrid ARIMA and neural network model. Neurocomputing 2003;50:159–75.

[50] Bates JM, Granger CWJ. The combination of forecasts. Operational Res Q 1969;20:451–68.

[51] Aburto L, Weber R. A sequential hybrid forecasting system for demand prediction machine learning and data mining in pattern recognition. In: Lecture notes in computer science, vol. 4571. Berlin, Heidelberg: Springer; 2007. p. 518–32.

[52] Lee W-I, Shih B-Y, Chen C-Y. A hybrid artificial intelligence sales-forecasting system in the convenience store industry. Hum Factors Ergon Manufacturing Serv Ind 2012;22(3):188–96.

[53] Li Y, Ning P, Wang XS, Jajodia S. Discovering calendar-based temporal association rules. Data Knowl Eng 2003;44(2):193–218.

[54] Ting P-H, Pan S, Chou S-S. Finding ideal menu items assortments: an empirical application of market basket analysis. Cornell Hospitality Q 2010;51(4):492–501.

[55] Xie W-X, Qi H-N, Huang M-L. Market basket analysis based on text segmentation and association rule mining. In: First international conference on networking and distributed computing; 2010.

[56] Miller JL, Repko CJ. Survey of foodservice production forecasting. J Am Diet Assoc 1990;90(8):1067–71.

[57] Green YNJ, Weaver PA. Approaches, techniques, and information technology systems in the restaurants and foodservice industry: a qualitative study in sales forecasting. Int J Hospitality Tourism Adm 2008;9(2):164–91.

Author Index

Subject Index

CPI Antony Rowe
Chippenham, UK
2018-01-03 21:27